CONSTRUCTION PATHWAYS

CONSTRUCTION PATHWAYS

BUILDING SK[...]

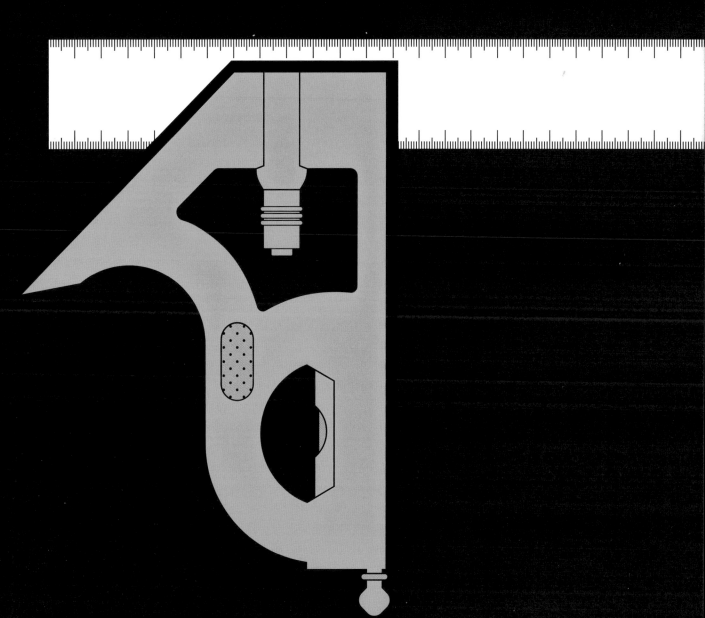

Construction Pathways
1st Edition
Shayne Fagan

Content manager: Sandy Jayadev
Content developer: Talia Lewis
Senior project editor: Nathan Katz
Cover designer: Olga Lavecchia
Text designer: Justin Lim
Permissions/Photo researcher: Helen Mammides
Editor: Pete Cruttenden
Proofreader: James Anderson
Indexer: Max McMaster
Art direction: Linda Davidson
Typeset by KnowledgeWorks Global Ltd.

Any URLs contained in this publication were checked for currency during the production process. Note, however, that the publisher cannot vouch for the ongoing currency of URLs.

This first edition published in 2022

© 2022 Cengage Learning Australia Pty Limited

Copyright Notice
This Work is copyright. No part of this Work may be reproduced, stored in a retrieval system, or transmitted in any form or by any means without prior written permission of the Publisher. Except as permitted under the *Copyright Act 1968,* for example any fair dealing for the purposes of private study, research, criticism or review, subject to certain limitations. These limitations include: Restricting the copying to a maximum of one chapter or 10% of this book, whichever is greater; providing an appropriate notice and warning with the copies of the Work disseminated; taking all reasonable steps to limit access to these copies to people authorised to receive these copies; ensuring you hold the appropriate Licences issued by the Copyright Agency Limited ("CAL"), supply a remuneration notice to CAL and pay any required fees. For details of CAL licences and remuneration notices please contact CAL at Level 11, 66 Goulburn Street, Sydney NSW 2000, Tel: (02) 9394 7600, Fax: (02) 9394 7601
Email: info@copyright.com.au
Website: www.copyright.com.au

For product information and technology assistance,
    in Australia call **1300 790 853**;
    in New Zealand call **0800 449 725**

For permission to use material from this text or product, please email
aust.permissions@cengage.com

National Library of Australia Cataloguing-in-Publication Data
ISBN: 9780170454025
A catalogue record for this book is available from the National Library of Australia.

Cengage Learning Australia
Level 7, 80 Dorcas Street
South Melbourne, Victoria Australia 3205

Cengage Learning New Zealand
Unit 4B Rosedale Office Park
331 Rosedale Road, Albany, North Shore 0632, NZ

For learning solutions, visit **cengage.com.au**

Printed in China by 1010 Printing International Limited.
1 2 3 4 5 6 7 25 24 23 22 21

BRIEF CONTENTS

CHAPTER 1	Work effectively and sustainably in the construction industry	1
CHAPTER 2	Plan and organise work	29
CHAPTER 3	Carry out measurements and calculations	51
CHAPTER 4	Prepare to work safely in the construction industry	79
CHAPTER 5	Apply WHS requirements, policies and procedures in the construction industry	131
CHAPTER 6	Undertake basic estimation and costing	157
CHAPTER 7	Undertake a basic construction project	181
CHAPTER 8	Handle and prepare bricklaying and blocklaying materials	205
CHAPTER 9	Use bricklaying and blocklaying tools and equipment	231
CHAPTER 10	Handle carpentry and construction materials	257
CHAPTER 11	Use carpentry tools and equipment	317
CHAPTER 12	Undertake basic installation of wall tiles	365

CONTENTS

Guide to the text	ix
Guide to the online resources	xii
Foreword	xiii
Preface	xiv
Acknowledgements	xv
List of figures	xvi
List of tables	xxii

CHAPTER 1   Work effectively and sustainably in the construction industry   1
Overview   1
Identify industry structure, occupations, job roles and work conditions   2
Take responsibility for own workload   7
Work effectively in a team   9
Identify and follow environmental and resource efficiency requirements   13
Summary   16
References and further reading   16
Get it right   17
Worksheets   19

CHAPTER 2   Plan and organise work   29
Overview   29
Determine and plan basic work task activities   30
Organise performance of basic work tasks   37
Summary   41
References and further reading   41
Get it right   43
Worksheets   45

CHAPTER 3   Carry out measurements and calculations   51
Overview   51
Obtain measurements   52
Obtain and record linear measurements accurate to 1 mm   56
Perform basic calculations   58
Summary   67
References and further reading   67
Get it right   69
Worksheets   71

CHAPTER 4   Prepare to work safely in the construction industry   79
Overview   79
Health and safety legislative requirements of construction work   80
Construction hazards and risk-control measures   84
Health and safety communication and reporting processes   93

	Incident and emergency response procedures	100
	Summary	108
	References and further reading	108
	Get it right	109
	Worksheets	111

CHAPTER 5 **Apply WHS requirements, policies and procedures in the construction industry** — 131
- Overview — 131
- Identify and assess risks — 132
- Identify hazardous materials and other hazards on worksites — 135
- Plan and prepare for safe work practices — 136
- Apply safe work practices — 141
- Follow emergency procedures — 144
- Summary — 145
- References and further reading — 145
- Get it right — 147
- Worksheets — 149

CHAPTER 6 **Undertake basic estimation and costing** — **157**
- Overview — 157
- Gather information — 158
- Estimate materials, time and labour — 164
- Calculate costs — 165
- Document details and verify where necessary — 170
- Summary — 172
- References and further reading — 172
- Get it right — 173
- Worksheets — 175

CHAPTER 7 **Undertake a basic construction project** — **181**
- Overview — 181
- Review and prepare to undertake basic construction project — 182
- Manufacture components for basic construction project — 186
- Assemble project components — 189
- Clean up — 191
- Summary — 193
- References and further reading — 193
- Get it right — 195
- Worksheets — 197

CHAPTER 8 **Handle and prepare bricklaying and blocklaying materials** — **205**
- Overview — 205
- Prepare for work — 206
- Handle, sort and stack materials — 212
- Clean up — 218
- Summary — 219
- References and further reading — 219
- Get it right — 221
- Worksheets — 223

CHAPTER 9 **Use bricklaying and blocklaying tools and equipment** — **231**
- Overview — 231
- Plan and prepare for work — 232
- Identifying and selecting hand tools — 232

Identifying and selecting power tools	244
Clean up	247
Summary	248
References and further reading	248
Get it right	249
Worksheets	251

CHAPTER 10 Handle carpentry and construction materials — **257**

Overview	257
Plan and prepare	258
Selecting appropriate materials	259
Construction materials and components	259
Transporting material	286
Clean up	288
Summary	290
References and further reading	290
Get it right	291
Worksheets	293

CHAPTER 11 Use carpentry tools and equipment — **317**

Overview	317
Planning and preparing to use tools and equipment	318
Identifying and selecting hand, power and pneumatic tools	319
Identifying, selecting and using plant and equipment	339
Using tools safely	349
Cleaning up after using tools and equipment	351
Summary	353
References and further reading	353
Get it right	355
Worksheets	357

CHAPTER 12 Undertake basic installation of wall tiles — **365**

Overview	365
Prepare for basic installation of wall tiles	366
Set out basic wall tiling	373
Preparation and installation	375
Grout tiles	379
Clean up	380
Summary	382
References and further reading	382
Get it right	383
Worksheets	385

Glossary	397
Index	400

Guide to the text

As you read this text you will find a number of features in every chapter to enhance your study of **Construction Pathways** and help you understand how the theory is applied in the real world.

CHAPTER-OPENING FEATURES

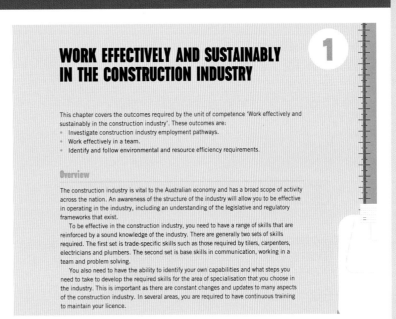

Identify the key chapter topics and competencies with the **Learning outcomes** at the start of each chapter.

Establish a broad outline of the chapter with the **Overview**.

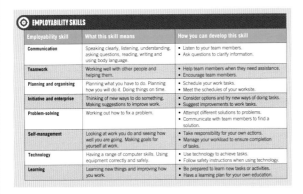

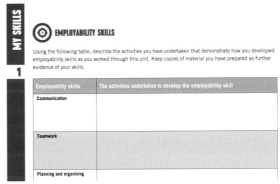

Review the key **Employability skills** on which you will be assessed at the start of every chapter. Consider the suggested activities for developing your skills for the relevant chapter.

At the end of the chapter, record the activities you have undertaken to develop these skills in the **Employability skills activity log**.

FEATURES WITHIN CHAPTERS

LEARNING TASK 1.3 TRADE APTITUDE ANALYSIS

Visit https://www.aapathways.com.au and search for 'Practice Aptitude Quizzes'. From here, you can identify an apprenticeship in a trade area that you may be interested in. Download and complete the practice quiz. Once you have checked your results, discuss with your teacher the areas that you need to develop further

Learning tasks encourage you to practically apply the knowledge and skills that you have just read about.

EXAMPLE 3.2

CALCULATING THE VOLUME OF SPOIL

Refer to Figure 3.23 for these calculations.
Step 1 Calculate the length of the edge beam.

Length − width of footing = centre line length of footing

A → 7.800 − 0.300 = 7.500
B ↓ 3.600 − 0.300 = 3.300
C ← 7.800 − 0.300 = 7.500
D ↑ 3.600 − 0.300 = 3.300
Total lineal metres = 21.600

Example boxes provide step-by-step instructions on how to perform specific tasks/processes.

During the cutting process, water and small pieces of masonry may be thrown up into the operator's face; therefore, it is essential that wrap-around safety goggles or glasses be worn to prevent possible permanent eye injury.

Safety boxes highlight key workplace health and safety tips at a glance.

COMPLETE WORKSHEET 1

Worksheet icons indicate when it is appropriate to stop reading and complete a worksheet at the end of the chapter.

END-OF-CHAPTER FEATURES

At the end of each chapter you will find several tools to help you to review, practise and extend your knowledge of the key learning objectives.

SUMMARY

In Chapter 5 you have learnt about the application of work health and safety requirements, policies and procedures. In addition, the chapter has:
- explored different aspects of identifying and assessing risks through reporting procedures and following safe work procedures
- identified hazardous materials and other hazards on site
- discussed the different types of work practices on a construction site and how they are applied
- identified how to follow basic emergency procedures.

REFERENCES AND FURTHER READING

NSW Government, *Work Health and Safety Act 2011 No 10*.
Safe Work Australia, *Emergency Plan Template*, https://www.safeworkaustralia.gov.au/doc/emergency-plan-template.
Safe Work Australia, *Guide for Major Hazard Facilities – Information, training and instruction for workers and other people at the facility*, https://www.safeworkaustralia.gov.au/doc/guide-major-hazard-facilities-information-training-and-instruction-workers-and-other-persons.
Safe Work Australia, *Model Code of Practice: Construction work*, https://www.safeworkaustralia.gov.au/search/site?search=Code+of+Practice+Construction+Work.
Safe Work Australia, *Model Code of Practice: How to manage work health and safety risks*, https://www.safeworkaustralia.gov.au/search/site?search=Code+of+Practice%3A+How+to+manage+work+health+and+safety+risks.
Safe Work Australia, *Model Code of Practice: Managing risk of plant in the workplace*, https://www.safeworkaustralia.gov.au/doc/model-code-practice-managing-risks-plant-workplace.
Safe Work Australia, *Model Code of Practice: Managing the work environment and facilities*, August 2019, https://www.safeworkaustralia.gov.au/doc/model-code-practice-managing-work-environment-and-facilities.

The **Chapter summary** highlights the important concepts covered in each chapter and link back to the key competencies.

The **References and further reading** sections provide you with a list of each chapter's references, as well as links to important text and web-based resources.

END-OF-CHAPTER FEATURES

Review all of the Australian Standards referenced throughout the chapter with the list of **Relevant Australian Standards**.

Relevant Australian Standards
AS 1418 Cranes including hoists and winches
AS 2294.1 Earth-moving machinery – Protective structures – General
AS 2550 Cranes – Safe use
AS 2958.1 Earth-moving machinery – Safety, Part 1: Wheeled machines – Brakes
AS 3745 Planning for emergencies in facilities
AS 4687 Temporary fencing and hoardings
AS 4991 Lifting devices
AS/NZS 1873 Power-actuated (PA) hand-held fastening tools
AS/NZS 1891.1 Industrial fall-arrest systems and devices – Harnesses and ancillary equipment
AS/NZS 1891.4 Industrial fall-arrest systems and devices – Selection, use and maintenance
AS/NZS 60745 Hand-held motor operated electric tools – Safety – General requirements

Worksheets help assess your understanding of the theory and concepts in each chapter.

WORKSHEET 1

To be completed by teachers
- Student competent ☐
- Student not yet competent ☐

Student name: _____
Enrolment year: _____
Class code: _____
Competency name/Number: _____

Task
Read through the sections from the start of the chapter up to 'Plan and prepare for safe work practices', then complete the following questions.

1 Identify two different hazards and provide an example of each.

2 What is the purpose of a hazard report form?

3 What two possible areas are compromised if proper work practices are not followed?

Identify the incorrect skill or technique demonstrated in the **Get it Right** feature and explain the correct technique that should be used instead.

GET IT RIGHT

The photo below shows a person using the wet saw unsafely. Identify the unsafe practices and provide reasoning for your answer.

Guide to the online resources

FOR THE INSTRUCTOR

Cengage is pleased to provide you with a selection of resources that will help you to prepare your lectures and assessments when you choose this textbook for your course. Contact your Cengage learning consultant for more information.

INSTRUCTOR RESOURCE PACK

Premium resources that provide additional instructor support are available for this text, including a Word-based Testbank, PowerPoints and all artwork from the text.

These resources save you time and are a convenient way to add more depth to your classes, covering additional content and with an exclusive selection of engaging features aligned with the text.

The Instructor Resource Pack is included for institutional adoptions of this text when certain conditions are met.

The pack is available to purchase for course-level adoptions of the text or as a standalone resource.

Contact your Cengage learning consultant for more information.

SOLUTIONS MANUAL

The Solutions Manual includes solutions to end-of-chapter worksheets and answers to in-text activities.

MAPPING GRID

The downloadable competency mapping grid demonstrates how the text aligns to the Certificate II in Construction Pathways.

END-OF-CHAPTER WORKSHEETS AND GET IT RIGHTS

Download writeable PDFs of all chapter Worksheets and Get it Right features.

END-OF-CHAPTER EMPLOYABILITY SKILLS ACTIVITY LOG

Download a writeable PDF template of the Employability Skills Activity Log shown at the end of every chapter.

FOREWORD

In Australia, the construction industry provides employment in a range of vocations and services. The industry is one of the biggest employers of tradespeople in the country, and provides important services to consumers, business and other industries.

In order for building and construction enterprises to keep pace with global change and sustainable practices, the vocational and training (VET) sector must respond quickly and efficiently to meet industry needs. This text is designed to meet the requirements of the CPC – Construction, Plumbing and Services Training Package by providing information and activities that reflect the ever-changing skill set required to undertake safe and effective activities. The knowledge and skills derived from this text will provide learners with the tools for future learning and will prepare new and existing workers for a long and rewarding career in the industry.

PREFACE

Construction Pathways has been structured on the CPC – Construction, Plumbing and Services Training Package requirements.

The chapters have been placed in a sequence to develop the fundamental components for the learner. The intent of this publication is to provide an appropriate resource for students who are seeking to start a career in the construction industry via either school- or TAFE-based programs.

The chapters have been placed into two broad groups.

The first six chapters cover units of competence designed to develop the fundamental skills and knowledge required to work in the industry:

1. Work effectively and sustainably in the construction industry
2. Plan and organise work
3. Carry out measurements and calculations
4. Prepare to work safely in the construction industry
5. Apply WHS requirements, policies and procedures in the construction industry
6. Undertake basic estimation and costing.

Note that while Chapter 1 is not a part of the units required to complete the Certificate II qualification, the concepts and information covered in the chapter reinforce and support the material in Chapter 5.

The second group of chapters focuses on the basic skills and knowledge required in specific trade areas:

7. Undertake a basic construction project
8. Handle and prepare bricklaying and block laying materials
9. Use bricklaying and block laying tools and equipment
10. Handle carpentry and construction materials
11. Use carpentry tools and equipment
12. Undertake basic installation of wall tiles.

To maximise skills development, Chapter 7 should be taught either concurrently with this group of chapters, or towards the end of the program.

Each chapter also includes learning task boxes and end-of-chapter worksheets to reinforce and support student learning. Key terms and concepts are also highlighted throughout the text; their definitions have been collated in a glossary at the end of the book.

Once the learner has worked their way through the material covered in this text, they should have attained sufficient skill and knowledge to commence working on a construction site in a safe and effective manner.

ACKNOWLEDGEMENTS

Cengage Learning would like to thank the following reviewers for their incisive and helpful feedback:
- Nav Jagdeo – Orange International College
- Michael Landers – TAFE NSW Western Sydney
- Vincent Digges – TAFE NSW
- Shane Wright – Box Hill Institute of TAFE.

Shayne Fagan would like to thank Michael Landers and the Stonemasonry Teaching section at Miller TAFE College NSW and the teaching staff from the Wall and Floor Tiling Teaching section at Macquarie Fields TAFE College for their feedback and input into this edition.

Cengage would like to extend special thanks to Richard Moran and Edward Hawkins for their numerous contributed photos.

Every effort has been made to trace and acknowledge copyright. However, if any infringement has occurred, the publishers tender their apologies and invite the copyright holders to contact them.

LIST OF FIGURES

1.1	Construction industry structure	3
1.2	Typical type of residential construction	3
1.3	Typical example of commercial construction	3
1.4	Typical example of industrial construction	4
1.5	Typical example of civil infrastructure construction	4
1.6	Common trades on a residential construction project	5
1.7	Common sources of friction and disharmony	10
1.8	Three skill capability areas to be effective in the workplace	11
2.1	Typical work order with sufficient information to undertake the required preparation and completion of the activity	32
2.2	Sample Gantt chart used to map out sequencing and duration of works	33
2.3	Typical installation guide provided for installers from manufacturers	36
2.4	Project planning cycle of review	39
3.1	Online estimating programs	54
3.2	Typical scale rule	56
3.3	Standard four-fold rule	56
3.4	Typical retractable metal blade tape measure	56
3.5	Open-reel and closed-case long tape measures	57
3.6	A trundle wheel	57
3.7	A laser distance-measuring device	57
3.8	Parallax error when reading a tape measure incorrectly	58
3.9	Dimensions required to calculate the area of a rectangle	59
3.10	Dimensions required to calculate the area of an L-shaped area	59
3.11	Dimensions required to calculate volume	59
3.12	Comparison of volume and mass of different materials	59
3.13	A number line demonstrating addition	60
3.14	A number line demonstrating subtraction	60
3.15	A diagrammatical demonstration of division	60
3.16	An example of division applied to a wall frame	61
3.17	An example of multiplying	61
3.18	Right-angle triangle	62
3.19	Right-angle triangle with expanded view of Pythagoras' theorem	63
3.20	Right-angle triangle identifying elements for trigonometry calculations	63
3.21	Centre line length of trench	64
3.22	Line to indicate length to external edge of footing	64
3.23	Diagram of plan with footing elements labelled for easy identification	64
3.24	Cross-sectional view of footing system	64
3.25	Floor plan of typical concrete floor slab	65
3.26	A section through the concrete floor slab and edge beam	65
3.27	Floor plan to calculate number of bricks	65
3.28	Floor plan to calculate floor tiles	66
3.29	Floor plan of concrete floor slab with section through slab and edge beam	66
4.1	Typical code of practice – *How to safely remove asbestos* (front cover)	82
4.2	White Card sample as issued in South Australia	83
4.3	Risk matrix diagram	85
4.4	The hierarchy of control measures	86
4.5	The risk-management process	88
4.6	Safety helmet	89
4.7	Fabric sun brim accessory for a safety cap and bucket hat	89
4.8	Clear wide-vision goggles	90
4.9	Clear-framed spectacles	90
4.10	Full-face welding mask	90
4.11	Face shield	90
4.12	Hearing protection	91
4.13	Mini dust mask, for nuisance dust only	91
4.14	P1/P2 disposable mask	91
4.15	Half-face respirator with P2 class dust filters fitted	91
4.16	Gloves	92
4.17	Barrier cream	92
4.18	Foot protection	92
4.19	Example of a picture sign	93
4.20	Example of a word-only message	93
4.21	Example of a combined picture and word sign	93
4.22	Digging prohibited	95
4.23	No pedestrian access	95
4.24	Eye protection must be worn	95
4.25	Hearing protection must be worn	95
4.26	Fire hazard	95
4.27	Toxic hazard	95
4.28	Electric shock hazard	95
4.29	Danger signs	96
4.30	First aid	96
4.31	Emergency (safety) eye wash	96
4.32	Fire alarm call point	96

#	Title	Page
4.33	Electrical safety signs and tags	96
4.34	Sample accident report form	98
4.35	Evacuation diagram	102
4.36	Type B first aid kit	103
4.37	Stored contents of the kit	103
4.38	The elements necessary for a fire	103
4.39	Portable fire extinguisher guide	105
4.40	Using a fire extinguisher – PASS	105
4.41	A typical fire blanket packet	106
4.42	A fire blanket in use	106
4.43	Hose reel	106
5.1	Typical hazard report form	134
5.2	Sign indicating an explosive power tool is in use	135
5.3	Manufacturer's instruction for fitting earplugs correctly	137
5.4	Temporary fencing and signage around deep excavation site	138
5.5	Standard checklist assessing the fencing requirements of a construction site	139
5.6	Typical safety data sheet supplied by a manufacturer	140
5.7	Site housekeeping inspection checklist	143
6.1	Typical floor plan	159
6.2	Typical drawing of elevations	160
6.3	Sample of National Classification system	161
6.4	Example of typical window schedule	162
6.5	Paint SDS demonstrating manufacturer's specifications for coverage	163
6.6	Online search results for individual item costs	164
6.7	Online search results for items that are categorised for building and construction	164
6.8	Common format used to calculate material required for a project	165
6.9	Floor plan for Garage Project	165
6.10	Calculation of concrete required for Garage Project	165
6.11	Labour constant of pre-mixed concrete for Garage Project	166
6.12	Calculation of labour requirements for Garage Project	166
6.13	Floor plans and diagrams for Garage Project	167
6.14	Total labour constants for Garage Project	168
6.15	Total labour and material calculations for Garage Project	169
6.16	Overhead and profit calculations for Garage Project	171
7.1	Safety fence	184
7.2	Typical construction PPE	185
7.3	Domestic dwelling construction using a slab on the ground	187
7.4	Domestic dwelling construction using bearers and joists	187
7.5	Plans for pergola	189
7.6	Industry certification for building components such as windows	189
7.7	Plan view of setting out with string lines and profiles	190
7.8	Position of pegs when setting out structure	190
7.9	Section view of post secured to footing	191
8.1	Brick types (from top left): clinkers, callows, sandstocks, commons and face	207
8.2	Brick classifications (from top left): solid unit, cored unit, hollow unit, horizontally cored unit and special purpose unit	208
8.3	Calcium silicate bricks	208
8.4	Range of brick colours	208
8.5	Range of brick textures available	208
8.6	Concrete blocks stacked ready for transporting	208
8.7	Retaining wall with concrete-filled hollow cement blocks	209
8.8	Wall constructed of breeze blocks	209
8.9	ACC blocks and large panels	209
8.10	Typical stone wall configuration	210
8.11	Natural stone wall with different coloured stone	210
8.12	Faux stone wall construction	210
8.13	Typical course chart	211
8.14	Common hand tools used to cut bricks and masonry	212
8.15	A scutch hammer being used to shape a brick	212
8.16	Diamond-dust blade wet-cutting brick saw	212
8.17	Front-end loader – bobcat	212
8.18	Typical brick barrow and brick carrier	213
8.19	Blocks on a pallet and wrapped in plastic	213
8.20	Mortar guide based on intended use of mortar	214
8.21	General purpose and off-white 20 kg bags of cement	215
8.22	Tonne-sized bags of sand	216
8.23	20 kg bags of hydrated lime	217
8.24	Range of available coloured oxides	217
8.25	Jointing finishes	217
9.1	Lump hammers with timber handle and fibreglass handle	233
9.2	Typical brickies hammer	233
9.3	Single-ended scutch hammer with replaceable comb	233
9.4	Brick bolsters	233
9.5	Medium-sized cold chisels	234
9.6	Scutch chisels	234
9.7	Plug chisels	234
9.8	Bricks being placed into position guided by the line of the string	235
9.9	Nylon string line	235
9.10	Block set	235
9.11	Block set holding a string line in position	235
9.12	Metal pins and string	236

9.13	Chalk line	236
9.14	A profile pole being used to guide the position of the bricks	236
9.15	Gauge rod set up for standard brickwork	237
9.16	Standard brickies trowel	237
9.17	Trowel being used to create a furrow in the mortar	237
9.18	Typical bucket trowel	238
9.19	Tuck pointers	238
9.20	Joint raker with wheels	238
9.21	Alternative method to achieve a square finish to the mortar	238
9.22	Brick jointer	238
9.23	Brick jointer being used to finish off mortar between brickwork	239
9.24	Heavy-duty plastic bucket with metal handle	239
9.25	Plastic drum used to store water	239
9.26	Long-handled round-point shovel and short-handled square-mouth shovel	239
9.27	Bag used by brickies to carry hand tools	240
9.28	Mortar board	240
9.29	Raised mortar board stand	240
9.30	Single metal plank used by brickies	241
9.31	Scaffolding around a house	241
9.32	Lifting tongs used to carry multiple bricks or large blocks	241
9.33	Typical wheelbarrow used to transport mortar	241
9.34	Larry used to mix mortar	242
9.35	Floor scraper used to clean floor slabs before and after the construction of a brick wall	242
9.36	Brush used to clean down brickwork	242
9.37	Typical plumb bob	243
9.38	Using a level to check for plumb and straight	243
9.39	Using a level to check for level and straight	243
9.40	Water level	243
9.41	Laser level with rotary laser mechanism	244
9.42	Static laser level	244
9.43	Plumb laser and laser receiver	244
9.44	Typical mortar mixer	245
9.45	Brick saw	245
9.46	Brick elevator transporting bricks to a higher location	246
9.47	Saw cuts	246
10.1	A contemporary residential building, using a variety of materials	259
10.2	Section through a log	260
10.3	Diagram of a cube of hardwood (magnification × 250). The pits in the cell walls have been omitted.	260
10.4	Tallowwood tree and timber	260
10.5	Diagram of a cube of softwood (magnification × 250). The pits in the cell walls have been omitted.	261
10.6	Radiata pine trees and timber	261
10.7	Rot stop wood preservative and end seal	262
10.8	Cross-cutting timber with a hand saw	263
10.9	Safe use of a power saw to rip timber	263
10.10	Wearing leather gloves to prevent splinters	263
10.11	Lifting correctly to prevent back injury	264
10.12	Carrying timber safely and comfortably	264
10.13	Hiab lifting a sling of timber	264
10.14	Bowing and bending caused by poorly aligned gluts	265
10.15	A sling of framing timber stored out of the weather	265
10.16	Typical plywood composition	266
10.17	hyJOIST®	266
10.18	Correct method for cutting of sheet material	267
10.19	Mechanical handling of sheets stacked on a pallet	267
10.20	Manual handling of sheet material	268
10.21	Material used for coarse aggregate in concrete	269
10.22	Storage of aggregates	270
10.23	Deformed reinforcement bars	270
10.24	Trench mesh reinforcement	271
10.25	Sheet reinforcement	271
10.26	Concrete cancer	272
10.27	Tools and/or equipment used for paint application	273
10.28	Cleaning and proper storage of brushes	273
10.29	An example of a fire-rated sealant	274
10.30	Some of the many steel products	275
10.31	Various tools used to cut steel: (a) hacksaw, (b) straight-blade tin snips, (c) bolt cutters	276
10.32	Typical heat loss of an uninsulated brick veneer cottage in temperate regions of Australia	277
10.33	Setting plasterboard sheets with base coat	278
10.34	Tools used for plasterboard work: (a) flat steel trowel, (b) keyhole saw, (c) broad knife, (d) hand sander, (e) corner tool, (f) small tool, (g) utility knife	279
10.35	Always handle plasterboard at the edges to avoid sheet breakage.	279
10.36	Friable asbestos coating around pipework	280
10.37	Friable asbestos coating around structural steel that has been coated on site	280
10.38	Non-friable asbestos roofing	281
10.39	Non-friable asbestos sheeting	281
10.40	Asbestos being wetted down	281
10.41	Handling asbestos using appropriate PPE	281
10.42	Asbestos material being wrapped in plastic by licensed personnel with appropriate PPE	281

Fig.	Description	Page
10.43	Asbestos loaded in bin with plastic lining ready to be wrapped	282
10.44	Images of different patterned finishes on glass	282
10.45	A range of different finishes to glass bricks	283
10.46	Glass window	283
10.47	Glass door	283
10.48	Frameless glass shower screen	283
10.49	Glass splashback for a kitchen	283
10.50	Good and bad outcomes of glass-cutting	284
10.51	Plastic pump-up vacuum cup, 200 mm	284
10.52	Double pad vacuum lifter, lifting capacity 60 kg	284
10.53	Glass being lifted using a crane and specialised equipment	284
10.54	Specialised wrist guard PPE for use when glass is being handled	285
10.55	Liquid nails for timber to brickwork	285
10.56	Stud adhesive used for Gyprock™	285
10.57	Adhesive for gluing tiles to concrete floors	286
10.58	Crane lifting trusses	286
10.59	All-terrain forklift trucks	287
10.60	Rubber-tyred pallet trolley with hydraulic lift	287
10.61	Rubber-tyred metal-framed platform trolley	287
10.62	Correct manual lifting technique	288
10.63	Polypropylene banding being used to strap timber	288
11.1	Combination square	319
11.2	Roofing square	319
11.3	Quick/pocket square	319
11.4	Try square	319
11.5	Timber mallet; the mallet head may be made of brush box and the handle of spotted gum	320
11.6	Claw hammers	320
11.7	Plasterboard hammers	320
11.8	Warrington hammer	320
11.9	Sledgehammer	320
11.10	Nail punches and floor punches	321
11.11	A selection of chisels used to cut into timber	321
11.12	(a) Jointer or try plane, (b) jack plane, (c) smoother plane, (d) block plane	321
11.13	Bench grinder with guide	322
11.14	Oil stones	322
11.15	Diamond stones	322
11.16	(a) Pad saw, (b) tenon saw, (c) panel saw, (d) crosscut saw, (e) rip saw	322
11.17	Round or hexagonal steel shank with specialist ends	322
11.18	Pincers	322
11.19	Four-fold rule	323
11.20	Marking gauge	323
11.21	Utility knife	323
11.22	Typical timber plank	323
11.23	Typical aluminium plank	323
11.24	Typical timber saw stool	324
11.25	(a) G clamp, (b) quick-action clamp, (c) sash clamp, (d) spring clamp, (e) F clamp	324
11.26	Drop-forged single open-end spanner	324
11.27	Chrome vanadium double open-end spanner	324
11.28	Podger for scaffolding and formwork centre adjustments – may be used for levering	324
11.29	Double-end ring spanner	325
11.30	Ring and open-end combination	325
11.31	Half-moon ring spanner	325
11.32	Square-drive ratchet handle and socket	325
11.33	Adjustable shifting spanner	325
11.34	(a) Insulated combination pliers, (b) insulated diagonal cutters, (c) needle-nose pliers and (d) external straight circlip pliers	325
11.35	Tin snips with a straight edge blade for general cutting	326
11.36	Jewellers' snips for curved work	326
11.37	Red, yellow and green aviation snips	326
11.38	Hand pop rivet gun	326
11.39	Bolt cutters	326
11.40	Wood float – available in various lengths and widths	326
11.41	Steel float – available in a variety of shapes and sizes	327
11.42	Concrete screed	327
11.43	Bull float and extension handle	327
11.44	Concrete edging tools	327
11.45	Typical section through a reinforced water hose	327
11.46	Putty knife	328
11.47	A broad knife and a filling knife	328
11.48	Hacking knife	328
11.49	Shave hook	328
11.50	Standard-type brush	328
11.51	Roller, roller frame and metal or plastic tray	329
11.52	Abrasive papers	329
11.53	Hand tools for breaking, cutting and grubbing (a) Crowbar, (b) Fork, (c) Mattock, (d) Pick, (e) Spade, (f) Long-handled round-mouth shovel, (g) Spud bar	330
11.54	(a) short-handled square-mouth shovel, (b) long-handled square-mouth shovel, (c) short-handled round-mouth shovel, (d) long-handled round-mouth shovel	330
11.55	(a) auger post-hole digger, (b) post-hole shovel, (c) trenching shovel	330
11.56	Reading the bubble of a spirit level	331
11.57	An RCD-protected portable power board	331
11.58	Electrical tags	332
11.59	Circular saw	332
11.60	Drop saw	333

11.61	Compound mitre saw set to cut a compound mitre cut	333
11.62	Electrical planers and blades: (a) 155 mm, (b) 82 mm	334
11.63	Cutting action of a jig saw	334
11.64	Jig saw cutting a piece of ply	334
11.65	Cordless and powered sabre saws	335
11.66	(a) Plunge router, (b) standard router, (c) trimmer router	335
11.67	Pistol-grip drill with side handle drilling steel	336
11.68	Tungsten carbide-tipped drill bit (masonry bit)	336
11.69	Battery-powered impact driver	336
11.70	Chuck operated with a key	336
11.71	Keyless chuck type	337
11.72	Main components of a belt sander	337
11.73	Sander actions	337
11.74	Main components of an orbital sander	337
11.75	Nails being loaded into nailer	338
11.76	Compressed air nailer	338
11.77	Gas nailer	338
11.78	A stripped-down indirect-acting EPT	339
11.79	A direct-acting EPT	339
11.80	Explosive-powered tool danger warning sign	339
11.81	Typical builder's temporary power pole and board	340
11.82	Portable site generator	340
11.83	Trailer-mounted diesel generator	341
11.84	Portable site compressor	341
11.85	Trailer-mounted site compressor	342
11.86	Electric demolition hammer with moil point	342
11.87	Air breaker used for heavy work	343
11.88	Tilting drum mixer	343
11.89	Mobile horizontal drum mixer	343
11.90	Inclined drum mixer with a capacity of up to 8.0 m^3	343
11.91	Mobile pan mixer with a capacity of 0.2 m^3 to 1.5 m^3	344
11.92	Section through the vibrating head of the poker vibrator	344
11.93	Portable immersion vibrator (petrol-driven)	344
11.94	Twin screedboard vibrator	345
11.95	Standard builder's barrow	345
11.96	Ball barrow	345
11.97	Two-wheeled barrow	346
11.98	Various types of industrial vacuum cleaners	346
11.99	Mobile cherry picker work platform	347
11.100	Lightweight mobile scaffolding	347
11.101	Single ladder	348
11.102	Extension ladder	348
11.103	Step ladder	348
11.104	Platform-type step ladder	348
11.105	Dual-purpose step/extension	349
11.106	Trestles – used with a plank to create a working platform	349
11.107	Labelling faulty equipment	350
11.108	(a) Stiff straw or millet broom, (b) stiff yard broom of straw or polypropylene, (c) broad soft-bristle floor broom of animal hair or polypropylene	351
12.1	5 kg and 20 kg bags of tile adhesive	368
12.2	Ceramic tiles	368
12.3	Mosaic tiles	368
12.4	Fully vitrified tiles	369
12.5	Porcelain tiles	369
12.6	Terracotta tiles	369
12.7	Slate tiles	369
12.8	Granite tiles	369
12.9	Chalk line	370
12.10	Spirit level used to set out level line	370
12.11	Spirit level used to set out a plumb line	371
12.12	Tape measure used to check tiles when setting out	371
12.13	Notched trowel used to apply adhesive	371
12.14	Bucket used to temporarily store adhesive	372
12.15	Aluminium straight edge used to support tiles	372
12.16	Spacers used to keep tiles in the correct position	372
12.17	Tiles being cut with manual tile cutter	372
12.18	Tiles being cut with a tile-cutting machine	372
12.19	Paint scraper	373
12.20	Dustpan and broom	373
12.21	Sponge being used to clean down tiles	373
12.22	Diamond pattern	374
12.23	Pinwheel pattern	374
12.24	Herringbone pattern	374
12.25	Stretcher bond pattern	374
12.26	Tiles installed plumb to the orange reference point, although adjacent walls may not be plumb	374
12.27	Tiles being laid out with a gauge rod	375
12.28	Wall junction with multiple decorative tiles with different tile lengths	375
12.29	Typical waterproofing applied to both floors and walls of a wet area to be tiled	375
12.30	Plywood sheeting lining a wall frame	376
12.31	Fibre-cement sheeting fixed to a wall to take wall tiles	376
12.32	Compressed fibre cement sheeting	376
12.33	Plasterboard products can be used in both wet and dry areas of a building	377
12.34	Example of a manufacturer's fixing requirements	377
12.35	Tiles and equipment cluttering work area, leading to a poor-quality outcome	378
12.36	Finding the high point to start levelling from	378
12.37	Setting a straight edge into position to support tiles	378

12.38	Straight edge in position ready to support tile installation	378
12.39	Applying primer to the substrate	379
12.40	Applying adhesive to the substrate	379
12.41	Limiting the amount of adhesive to prevent from drying before tiles are fixed	379
12.42	Placing tiles in the correct position as planned	379
12.43	Checking alignment of tiles as they are progressively installed	380
12.44	Applying grout to the wall tiles	380
12.45	Spreading and working the grout in between the tiles	380
12.46	Removing excess grout from tiles once joints have been filled with grout	381
12.47	Sponging off excess grout left on tiles	381
12.48	Final wipedown leaves tiles clear of any grout film	381
12.49	Extract from a manufacturer's SDS for an adhesive product	381

LIST OF TABLES

1.1	Examples of organisational requirements	6
1.2	An example of the roles in a work team for a construction project	7
1.3	Range of industry areas that have training packages	12
1.4	Generic skills needed in the construction industry	12
1.5	Examples of environmentally hazardous materials	14
3.1	Requirements of measurement techniques and specific methods of calculation	53
3.2	Formulas commonly used in the construction industry for area and perimeter calculations	60
3.3	Formulas commonly used in the construction industry for volume calculations	60
3.4	Converting millimetres to metres	61
3.5	Units of measure	62
4.1	Five reasons for the development of WHS Laws	80
4.2	Current state and territory WHS Acts and Regulations	81
4.3	The seven categories of signs	94
4.4	Contents of a first aid kit B	102
5.1	Examples of common hazards	133
5.2	Table indicating the Australian Standards related to a particular piece of equipment	141
5.3	Typical list of contacts in an emergency plan	144
7.1	Strategies to comply with environmental requirements	184
7.2	Common tools and equipment	186
8.1	Categories of bricks	207
8.2	Calculating mortar material quantities	214
10.1	Natural durability classification of heartwood of some commonly used timbers	262
10.2	Ignition performance of tested materials	267
10.3	Types and uses of steel	275
10.4	Types and uses of insulating materials	277
10.5	Types and uses of plasterboard	278
10.6	Environmental Protection Agency state guidelines	280
11.1	Diameter and relative spacings	345
11.2	Power tool safety checklist	350
12.1	Recommended trowel notch size and coverage of adhesive	368
12.2	Manufacturer's specifications	377

WORK EFFECTIVELY AND SUSTAINABLY IN THE CONSTRUCTION INDUSTRY

This chapter covers the outcomes required by the unit of competence 'Work effectively and sustainably in the construction industry'. These outcomes are:
- Investigate construction industry employment pathways.
- Work effectively in a team.
- Identify and follow environmental and resource efficiency requirements.

Overview

The construction industry is vital to the Australian economy and has a broad scope of activity across the nation. An awareness of the structure of the industry will allow you to be effective in operating in the industry, including an understanding of the legislative and regulatory frameworks that exist.

To be effective in the construction industry, you need to have a range of skills that are reinforced by a sound knowledge of the industry. There are generally two sets of skills required. The first set is trade-specific skills such as those required by tilers, carpenters, electricians and plumbers. The second set is base skills in communication, working in a team and problem solving.

You also need to have the ability to identify your own capabilities and what steps you need to take to develop the required skills for the area of specialisation that you choose in the industry. This is important as there are constant changes and updates to many aspects of the construction industry. In several areas, you are required to have continuous training to maintain your licence.

Employers value people who fit into their workplace, solve day-to-day problems, manage their time and are keen to continue learning. These types of skills are known as employability skills. The employability skills you develop while working through this unit will be assessed at the same time as the skills and knowledge.

It is important to show evidence that you have developed these skills. Your trainer or assessor will discuss with you how to record your employability skills. The following table provides some examples of things you might do to develop employability skills in this unit.

EMPLOYABILITY SKILLS

Employability skill	What this skill means	How you can develop this skill
Communication	Speaking clearly, listening, understanding, asking questions, reading, writing and using body language.	• Listen to your team members. • Ask questions to clarify information.
Teamwork	Working well with other people and helping them.	• Help team members when they need assistance. • Encourage team members.
Planning and organising	Planning what you have to do. Planning how you will do it. Doing things on time.	• Schedule your work tasks. • Meet the schedules of your worksite.
Initiative and enterprise	Thinking of new ways to do something. Making suggestions to improve work.	• Consider options and try new ways of doing tasks. • Suggest improvements to work tasks.
Problem-solving	Working out how to fix a problem.	• Attempt different solutions to problems. • Communicate with team members to find a solution.
Self-management	Looking at work you do and seeing how well you are going. Making goals for yourself at work.	• Take responsibility for your own actions. • Manage your workload to ensure completion of tasks.
Technology	Having a range of computer skills. Using equipment correctly and safely.	• Use technology to achieve tasks. • Follow safety instructions when using technology.
Learning	Learning new things and improving how you work.	• Be prepared to learn new tasks or activities. • Have a learning plan for your own education.

Identify industry structure, occupations, job roles and work conditions

The construction industry is one of the largest industry sectors in Australia, and comprises a complex array of occupations, roles and workplace conditions.

Scope and nature of the construction industry

Australia's construction industry provides a range of outputs that are vital to society. The industry ensures that there is appropriate shelter for people, along with services such as water, electricity and gas so everybody can live in a safe and healthy built environment. The construction industry also provides the infrastructure that supports other industries to operate.

The construction industry is highly diverse and a major component of the economy. According to the Australian Bureau of Statistics, the industry contributes more than 7.7% of the gross domestic product and employs more than 1 117 000 people.

The construction industry also generates employment in industry areas that provide services to the construction industry. For example, the manufacturers of excavators for use in the industry will also offer maintenance and repair services for these machines.

The scope of work in the construction industry is large, and can be split into three main areas: residential construction, non-residential construction and infrastructure construction (see Figure 1.1). Each of these areas will engage tradespeople such as plumbers, electricians, carpenters and tilers.

Residential construction is common in the spread of suburbs. The type of construction used in these structures typically comprises concrete slabs, lightweight framing in either timber or steel, cladding in brick, timber and composite sheeting, and fit-outs using linings, cupboards, ceramic tiles and basic furnishings (see Figure 1.2). The jobs usually involve specialised skills, but there will be many occasions where you will need to multi-task across several areas.

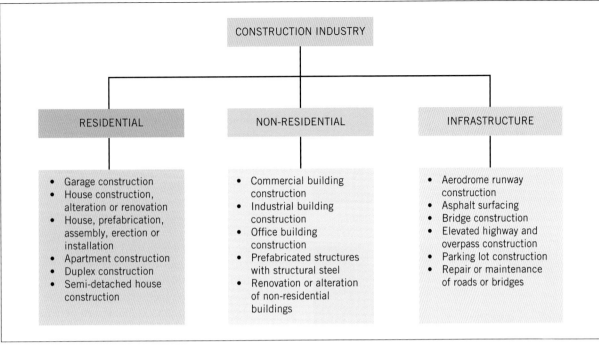

FIGURE 1.1 Construction industry structure

FIGURE 1.2 Typical type of residential construction

FIGURE 1.3 Typical example of commercial construction

Non-residential construction can be further broken down into two areas: the commercial sector and the industrial sector:
1. The commercial sector covers shopping centres, hotels, schools, office blocks, hospitals and multi-story construction. The type of construction used in these structures includes reinforced-concrete-framed construction, pre-cast concrete panels, pre- and post-tensioned concrete, and steel-framed construction. The jobs involved in this type of construction are specialised in relation to concrete and large structural members (see Figure 1.3).
2. The industrial sector mainly covers the construction of factories, warehouses, storage areas and small industrial complexes. The type of construction used includes pre-cast tilt-up panels and portal-steel-framed construction. The jobs involved in this type of construction are specialised in relation to concrete and large structural members (see Figure 1.4).

Infrastructure in the civil sector involves the construction of large projects such as airports, bridges, roads, tunnels and dams. The type of construction used includes earth works and large concrete pours. The jobs involved in this type of construction involve operating heavy machinery, surveying, and formwork and falsework (see Figure 1.5).

It is necessary to have a broad understanding of the different areas in the construction industry. This will give you an opportunity to distinguish between the size and scale of the type of works that need to be performed. It will also give you an understanding of the different aspects of a trade when applied in these different areas.

FIGURE 1.4 Typical example of industrial construction

FIGURE 1.5 Typical example of civil infrastructure construction

LEARNING TASK 1.1

JOB ROLES IN THE CONSTRUCTION INDUSTRY

Visit the Australian Apprenticeships Pathways site (https://www.aapathways.com.au/career-research/job-pathways) and click on the following links:
- Construction, Plumbing and Services CPC
- Resources and Infrastructure RII.

For each of the sectors, identify two different job roles for the three industry areas: residential, non-residential and infrastructure. Provide a brief description of each job role and identify the types of tasks performed in each job role.

Construction job roles, occupations and trade callings

Several trades operate across all three areas of the industry, though individual tradespeople tend to stay within a single sector as they develop their expertise. Figure 1.6 demonstrates the breadth of trades employed within the residential construction industry.

These main trades will also rely on the other job roles to be effective and productive; that is, there are additional job roles that sit under each trade area. For example:
- **Trade:** bricklayer, blocklayer
- **Supporting job roles:** construction assistant, builder's labourer, trades assistant, trades assistant (brick and blocklaying).

Supporting job roles are not always on the construction site. There are job roles for the off-site manufacture of components for a construction project, such as wall frames, kitchen cabinets and windows. These components will then need to be transported, and even though the transport operator is not directly working on the structure of the building, their job role and purpose has a direct impact on the overall construction industry.

Further, there is a whole workforce of administration roles that support all operations in the construction industry. These roles are involved from the design phase to the completion of a construction project, and include architects, construction managers, project managers, compliance inspectors and safety officers.

Employment impacts of technology, work processes and environmental issues

There have been many technological advances in the construction industry, from tools and equipment to materials, software, work processes and building methods, and all of these have had environmental implications. Such improvements have added to the productivity of people who work in the construction industry.

Tools and equipment

The evolution of the nail gun from hammer and nail is one of the obvious demonstrations of technology in the construction industry. However, the introduction of the nail gun has changed from the first nail guns – many nail guns are now gas-driven without needing to be connected to an air compressor.

Technological developments have also led to benefits in work health and safety, such as the use of vacuum units attached to power tools that generate large amounts of harmful dust or the introduction of the wet cutting of masonry units. The adoption of controlled measures improves outcomes for all workers in minimising their exposure to harmful materials.

Materials

There have been several improvements in relation to the materials used in the construction industry. The removal of asbestos from common construction materials has had a direct impact on many people's lives. We have also seen improvements with the quality of adhesives now used, which have reduced the time spent on the construction site.

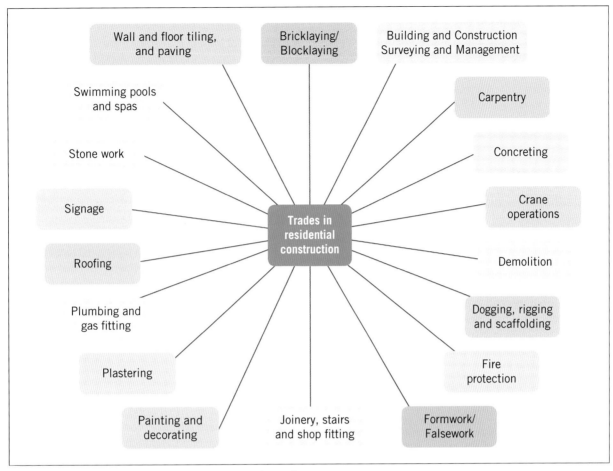

FIGURE 1.6 Common trades on a residential construction project

Software

The introduction of computers and the internet has also led to improvements in the construction industry. Many aspects of managing a construction site are now being serviced through specifically designed software to improve the accuracy of the work and distribution of information. Plans for a site now can be sent through in an electronic format and building companies will work from these to price up the project and determine the materials required.

Work processes

Work process improvements have impacted the construction industry in several ways. One of the big shifts has been the off-site manufacturing of building components. An example of this is wall frames and trusses that are manufactured in a factory setting and then transported to the job site. Such work was traditionally done on the site. The increase of off-site manufacturing has reduced the time spent by workers on site, and the off-site manufacturing of building components has created jobs for people who may not be fully qualified as tradespeople.

Building methods

Building methods have changed over time as well. Some examples of this are the introduction of plasterboard, which has replaced fibrous plaster; the introduction of sheet material to build kitchen cabinets instead of being framed up by sticks of timber; and the introduction of pre-cast concrete tilt up panels, which are typically erected on site using a crane to form the walls of a structure. All of these examples have had an impact on the way that work on a construction site occurs, and has also required a different approach to project planning and coordination.

Environmental issues

Environmental issues in relation to the construction industry have evolved considerably over time. Many of the construction materials that have been used in the past, particularly asbestos, are now prohibited from use, while new products have been developed for specific purposes. A major change on construction sites has been to manage the wastage of materials and, where possible, to recycle them. Plasterboard is particularly well suited for recycling (see Chapter 10).

Understanding the trends in organisation and industry practices will allow you to develop skills that will be in demand in the future, increasing your value as a skilled employee. You can research industry trends in specific bulletins, industry newsletters and industry websites, and by talking with others at your worksite. You can also be proactive about the development of new technologies and processes for your industry by looking at the processes you are currently undertaking and thinking about how they might be improved.

Impact of employment conditions, organisational requirements, responsibilities and duties

All workplaces employ people through the use of contracts. These employment contracts outline the responsibilities and duties of employees and the responsibilities and duties of the organisation, including policies, procedures and legislated conditions such as pay rates and leave provisions.

The construction industry has four main categories of employment conditions:
1. daily hire employee
2. full-time weekly hire employee
3. part-time weekly hire employee
4. casual employee.

Each of these categories fall under the Building and Construction General On-site Award 2010. The Australian Building and Construction Commission (ABCC) plays a fundamental role in ensuring that the construction industry provides a fair, efficient and productive environment for all people who operate in the construction industry.

There are several related awards for off-site job roles, including:
- Black Coal Mining Industry Award 2010
- Cement, Lime and Quarrying Award 2020
- Electrical, Electronic and Communications Contracting Award 2020
- Joinery and Building Trades Award 2020
- Manufacturing and Associated Industries and Occupations Award 2020
- Mining Industry Award 2020
- Plumbing and Fire Sprinklers Award 2020
- Premixed Concrete Award 2020.

One set of conditions relates to the establishment of an apprenticeship. Each state in Australia has its own legislation that prescribes the conditions of an apprenticeship and traineeship. It is essential that you are aware of the conditions for an apprenticeship for your own state and the authority that administers the legislation.

All employers and organisations will have specific employer and employee roles and responsibilities, along with work conditions. Many of the conditions are common as they are drawn from the related award, but there are also work health and safety obligations, codes of practice, organisational requirements and licensing requirements.

Organisational requirements that may be related to your employment are listed in Table 1.1.

There are different legislative arrangements in place for the range of different job roles within the construction industry. Many states have different licensing requirements and conditions for areas of high risk in the construction industry, which require that such work is carried out by an individual who can demonstrate a level of competence. Licensing not only

TABLE 1.1 Examples of organisational requirements

Organisational requirement	What this means
Access and equity principles	Policies and approaches that ensure employment and work conditions are responsive to the diverse needs of all workers. Equity means making things equal and not having different rules for different people.
Anti-discrimination, equal employment opportunity and other policies	All employment and work conditions must treat people equally and ensure everyone has the same rights and responsibilities regardless of their gender, race, age, cultural background, ethnic origin, disability, political beliefs or religious beliefs.
Business, performance, quality assurance plans, and systems and processes	All policies and procedures are written according to best-practice principles and all work must be carried out to the best possible standard.
Ethical standards of the company	All work must be carried out in an ethical manner with a high degree of professionalism and integrity, by respecting laws and acting honestly and truthfully.
Company goals and objectives	All work must be done in line with the goals, aims and objectives of the company. For example, you will be expected to act honestly and with integrity, work to a high standard and provide quality customer service.
Organisation policy, guidelines and requirements	An organisation's policies and procedures are written to cover all the information listed above. A worker's role and responsibility is to follow the organisation's policies and procedures at all times.

helps to maintain a standard of how a task is performed but also assists in ensuring that the final product will meet all building codes and regulations.

Safe work methods and practices, WHS and the use of PPE

Each state has legislation that relates to the work health and safety of employees. The purpose of the legislation is to create and maintain a safe working environment. There are as many aspects to the legislation as there are many different types of workplaces. The legislation supports a proactive role in driving a culture and workplace practices to make safety the first priority. There are several standard approaches that are highlighted in the legislation that will guide the practices to ensure a workplace is as safe as possible. One of the most common aspects is the use of personal protective equipment (PPE). Specific information about WHS and PPE is discussed in later chapters.

COMPLETE WORKSHEET 1

Take responsibility for own workload

No matter which industry or workplace you are employed in, you need to be able to accept responsibility for the tasks that you are required to perform. Taking ownership of a task and related activity is a quality that is respected by work colleagues and is highly valued by employers.

Taking responsibility for a task is broader than just performing the task itself but includes all associated activities, such as preparation of the task by planning and organising all the elements required for the task, as well as considering WHS, environmental concerns and impacts on other workers and trades.

LEARNING TASK 1.2
BUILDING AND CONSTRUCTION GENERAL ON-SITE AWARD

Visit https://fairwork.gov.au. Click on 'Awards & agreements', navigate to lists of awards and agreements, and complete the following table:

What is the amount of tool allowance for a carpenter per week?	
What is the definition of an apprentice?	
What percentage of the standard rate is a third-year apprentice entitled to?	
What is the minimum weekly wage for a CW/ECW 3?	
What are three different allowances found under Item 22: Other allowances?	

For each job role in the award, provide a brief description of tasks that are performed.

Plan work activities and deadlines with group members and other affected workers

Work teams can consist of a range of different roles that are involved with a construction project. An example is given in Table 1.2. When establishing work schedules and setting deadlines it is important that the appropriate team members are consulted. Note that the range of job roles that may be involved is not limited to those listed in the table.

TABLE 1.2 An example of the roles in a work team for a construction project

Role	Responsibility
Architect	Engaged by the client to design and control the building process on the client's behalf. Other responsibilities include lodging the development application with designs or models, supervising the drafting of plans and specifications, preparing tender documents to engage a builder, and engaging consultants for the structural, mechanical and other specialist areas.
Builder	The main contractor selected by a tender process to carry out construction as per plans and specifications. The builder will also subcontract labour to deal with specific parts of the construction.
Building Inspector	For domestic construction, the inspector visits the site on behalf of the council to check structural work and pass completed stages of a job.
Client	Finances the project. This may be the developer or people involved in a joint venture who require a building for a specific purpose.
Construction manager	Responsible for all building contracts the company has won or is tendering for. Usually working from the main office, the construction manager forms tender documents and controls the building activities for the company.
Draftsperson	Engaged by the architect to prepare plans and details before and during construction.

Role	Responsibility
General foreperson	Responsible for delegation of duties to workers on the site and also responsible for site safety. Other responsibilities are coordination of subcontractors and progress of work.
Land surveyor	Responsible for the site setting out and control of the vertical alignment of a building.
Principal certifying authority	For domestic construction the PCA carries out similar duties to the local government building inspector; however, it is directly engaged by the client to conduct inspections.
Project manager	Usually employed on very large sites to control the running of the site and to make sure the project runs to schedule and to budget.
Quality assurance officer	Engaged to control the quality of construction on the site and ensure that the work meets the standards laid down in the specifications and tender documents.
Quantity surveyor	Responsible for preparation of a bill of quantities for the architect to enable the building to be priced by an estimator. Also required to prepare quantities for variations to the contract and assessing progress claims.
Site manager	Runs the day-to-day activities and liaises directly with the foreperson and the project manager.

Having sound planning processes and communicating the outcome of such a process allows all staff to contribute in the execution of the activities identified for the project. This coordination and communication gives the project a higher chance of meeting the deadlines without creating a rushed or panicked work environment.

The skills needed to plan and prioritise your workloads and set reasonable deadlines include:
- literacy skills such as reading and writing
- numeracy skills such as measuring and calculating
- communication skills including active listening and speaking
- organisational skills such as planning and prioritising.

Learning and practising these skills is critical to your success in any industry. A practice that will assist you to plan and schedule effectively is to keep records of how long a task takes and the resources required to complete the task. Such records will become a good reference point for any future planning activities that you may be involved with.

Identify variations and difficulties affecting own performance and report these issues

To be able to identify variations that affect your performance or quality requirements, you need to have a sound understanding of your levels of performance and quality standards. To have a degree of self-awareness of your performance levels does require some time to develop. However, the quality standards need to be established very early on in your work, as the quality of the work that you perform is why you are employed and what the customer is paying for.

If there are any variations to the performance levels, you need to call these out at the commencement of the task. It is not always the individual's ability that may cause this situation. The information that has been provided may not provide clarity about what is required. There have been many instances where the parts of the documentation supplied for the work contradict each other. It is not uncommon for a range of documents to provide information in relation to the same item. These documents may include:
- diagrams, signs, sketches, memos, safety data sheets (SDSs) and safe work procedures
- instructions issued by authorised organisational or external personnel
- manufacturer specifications and instructions
- work specifications, requirements and plans
- regulatory and legislative requirements, Australian Standards and building codes
- work bulletins and schedules
- verbal, written or graphical instructions.

As a project progresses, customer variations may occur. This may be addition to or reduction of the design, material or other aspects of the project. If such changes do occur then the information needs to be clearly communicated to all the trade areas that may be affected, although through this process it is easy for miscommunication to occur.

In addition to this, quality may suffer due to several factors, such as condition of the work area, condition of the material and condition of equipment that is to be used. The quality standards of a project or task are usually defined in the specification.

Once a variation has occurred, it is essential that it is reported. You also need to ensure that it is reported to the appropriate person on the site. In the first instance you should report directly to your line manager, who is usually the supervisor of your work team. Your supervisor will ensure that the correct manager is advised, depending on the nature of the variation. On many sites, if there are any safety related issues they will be reported to the safety officer on the site.

Any reports should follow the organisation's reporting processes. Effective reporting will require skills such as:
- verbal communication skills so that information can be clearly given and received in a conversation
- written communication skills such as being able to write a report

- technology skills including the ability to use a computer or mobile technology
- numeracy skills such as being able to take measurements and make calculations.

Request additional support to achieve or improve work outcomes or quality

Occasionally you need to ask for help or support to complete a work task. You need to be honest with yourself to recognise if you need assistance and be specific about the type of help you need. By knowing your abilities and the task expectations, you can judge if assistance is required. It is important that you are not too embarrassed or nervous to ask for help. Requesting appropriate assistance will demonstrate to your supervisor and teammates that you are a responsible worker and that the quality outcome of the task is important to you and your work ethic.

Some tasks you could need assistance with include:
- manual-handling tasks
- new tasks
- unusual tasks
- tasks requiring skills you don't yet have
- tasks that require equipment you don't have
- tasks where new or different materials are to be used.

Sometimes it is hard to know when to request assistance. Monitoring your activities makes it easier to know if you need assistance. At times you will need to make a judgement call, and it is best to err on the side of caution and raise your concerns before attempting to undertake the task.

Once you have identified that you need assistance, you need to ensure when you are communicating your needs that you clearly articulate your concerns and offer suggestions as to what is required. In most instances, this will be a verbal discussion with your line manager.

COMPLETE WORKSHEET 2

Work effectively in a team

Teams complete the majority of work done in the construction industry. There is a range of different teams that you will be involved with during the construction of a project. There are broader teams that consist of different trades working together as a higher-level team. You will also be required to work in small teams of co-workers who you interact with on a daily basis to complete a given task.

Being able to work in the different teams on a building site is one of the key factors in being a preferred employee or a successful subcontractor for builders and construction companies.

Identify site goals and team contributions to a construction activity

Teams are a group of people working together to common goals. Teams are often referred to as crews or gangs on a construction site. A goal is usually a body of work that needs to be completed by a certain time. There are large goals that will need several people to work together to achieve the desired outcome by a given time.

The task and time frame are established from the overall schedule that has been set at the commencement of the project. As the tasks are broken down into smaller chunks, individual teams led by a supervisor will establish an approach to achieve the work.

It is at this stage that you need to be involved and aware of what the expectations are. Typically, at the commencement of a piece of work or day there is a brief toolbox talk to discuss what is required from the day's efforts. During toolbox talks ensure that you clearly understand what is required of you, and if you are unsure then ask questions of your supervisor or co-workers. Remember that if you are unsure, there is a good chance that there is someone else in the team who is also unclear about what is required.

Identify individual contributions to tasks and confirm them with team members

Once your team has clarity about the task and the time frame, you need to make sure that you know the particular role that you will be undertaking to assist the team to complete the task. The allocation of specific tasks is usually allocated based on an individual's ability and experience in performing the task and achieving the quality standard required.

It is not uncommon for a senior tradesperson to do much of the work with a less experienced worker providing support for the particular task. Examples of this may be holding the end of a tape measure and writing down information while your co-worker takes the measurements. In many situations, set-up to perform the task will be a big part of the work that is conducted on a building site, and once the work has commenced you may be in a supporting role to keep the supply of material available for the more experienced worker.

As you are exposed to more tasks and situations you will be called on to take a bigger role in undertaking a task.

Provide assistance and encouragement to other team members

One of the biggest advantages of using work teams is the easy access to help and the encouragement of other team members.

Giving encouragement and assistance is an essential part of the team experience. Assistance can be giving help and advice when it is needed. Some forms of advice include:

- verbal advice, such as telling the person how to do something
- demonstration, such as showing another person how to do something.

Sometimes a team member may want to talk through options or issues without wanting you to give them any advice. Listen to them and offer advice if it is requested or required.

Encouragement should be given where appropriate. Encourage co-workers to ask questions, provide feedback or voice their opinions over issues. Workers should also be encouraged to actively join in team activities, suggest improvements to work activities and be involved in work processes. And just as you encourage other team members, they should also encourage and assist you.

Encouragement to team members can be expressed verbally or with body language, such as a nod, smile or wave. On some worksites, the best form of encouragement you can give is to refrain from making adverse comments when things go wrong.

Using assistance and encouragement helps build the team relationships. It also builds the confidence and skills of the individual, which not only helps the individual but also the team and the organisation.

The main feature is that you and other members of the team are clear about what is required at an individual level. There will be times when activities have not been covered or have been missed, but they will usually be dealt with on a needs basis.

Initiate team improvements

Being an effective team requires the team members to have a collaborative approach to improving the way they operate. This approach needs to be supported by the managers of the different teams. While there are many different aspects that a team could focus on for improvement, the process for this improvement must be supported by an environment where all team members are valued and their perspectives or observations are welcomed and considered.

Encouragement and support for each team member will allow for interaction among individuals in the team. The ability to be able to communicate effectively is important for a team to improve.

Ultimately, only team members can make the changes to practices that need to be adopted. This is because tasks are performed by individuals in the team environment and through interaction with each other.

Causes of disharmony and other barriers to achievement

There are times when there can be friction between individuals in a team. There may be a whole range of reasons for this friction (see Figure 1.7). Having an

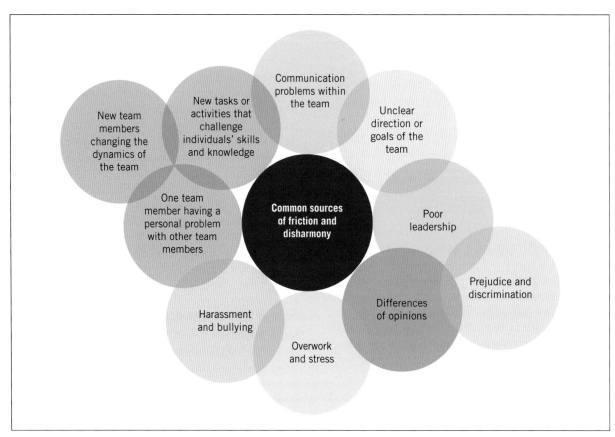

FIGURE 1.7 Common sources of friction and disharmony

understanding of the issue will provide an opportunity for the individuals to resolve the situation.

Any workplace conflicts need to be dealt with as quickly as possible. If the issue is not addressed, the situation may become worse and the effectiveness of the team will suffer.

The first step in resolving any conflict is to acknowledge there is an issue and set about to identify what it may be. This may involve a discussion with the whole team or just the individuals who are in conflict. An action plan then needs to be developed to resolve the cause.

If the conflict cannot be addressed within the team in an appropriate manner, the issue may have to be addressed by a person external to the team. There are specialists who work in this area, who can assist in identifying the issue and help to build the action plan and monitor progress. For such a process to be effective, there is a need for all of the participants to be committed to improving the performance of the team and to address any issues that are affecting the team.

COMPLETE WORKSHEET 3

Identify own development needs

Identifying your own current and future skill development needs is an important step in progressing your career. Your skills development needs may be workplace-specific, such as learning new skills, or they may be general developmental skills, such as improving your language and literacy skills.

Knowing what your current skill levels are can help you to identify situations for further development. Planning your skill development is important to ensure a balanced approach to learning new skills. By planning your learning, you are also able to take advantage of your personal learning styles and preferred learning methods.

Identify your own existing and required skills

The construction industry is very diverse in the range of skills required. The Core Skills for Work Development Framework identifies three skill capability areas that are required in the modern workplace (see Figure 1.8). These are:
- core skills for work
- language, literacy and numeracy (LLN) skills
- technical or discipline-specific skills.

There is broad range of skills and knowledge under each of the three areas. It is possible to develop all three areas at the same time, but LLN skills will equip you to improve your chances of successful development in the other two areas.

Core skills for work

The purpose of core skills for work is to clearly identify a set of non-technical skills, knowledge and

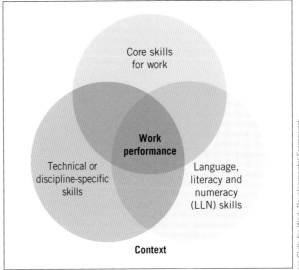

FIGURE 1.8 Three skill capability areas to be effective in the workplace

understandings that underpin successful participation in work. This area has 10 core skills split in three clusters:
- Cluster 1 – Navigate the world of work:
  - Manage career and work life
  - Work with roles, rights and protocols.
- Cluster 2 – Interact with others:
  - Communicate for work
  - Connect and work with others
  - Recognise and utilise diverse perspectives.
- Cluster 3 – Get the work done:
  - Plan and organise
  - Make decisions
  - Identify and solve problems
  - Create and innovate
  - Work in a digital world.

Language, literacy and numeracy skills

Under the framework, there are five core LLN skills:
- learning
- reading
- writing
- oral communication
- numeracy.

Technical or discipline-specific skills

Training packages are used as the basis for most of the programs delivered in the VET system, including Australian Apprenticeships, training courses offered by registered training organisations, VET in Schools programs, recognition of existing skills, and occupational licensing.

The qualifications that are developed fall under a range of different industry areas. Table 1.3 lists some of these areas.

The aims of training packages are to:
- help the VET system achieve a better match between skills demand and supply

TABLE 1.3 Range of industry areas that have training packages

Aerospace	Electricity supply generation
Aged services	Electricity supply transmission distribution and rail
Agriculture and production horticulture	Electrotechnology
Ambulance and paramedic	Forest management and harvesting
Animal care and management	Furnishing
Business services	Manufacturing and engineering
Children's education and care	Maritime
Civil infrastructure	Metalliferous mining
Client services	Naval shipbuilding
Coal mining	Process manufacturing, recreational vehicle and laboratory
Community sector and development	Property services
Complementary health	Technicians support services
Construction, plumbing and services	Textiles, clothing and footwear
Disability support	Timber and wood processing
Drilling	Timber building solutions

- encourage flexible and relevant workforce development and learning
- provide for the national recognition of the vocational outcomes of learning
- guide and support individuals in their choice of training and career (Department of Education, Skills and Employment).

The training packages are developed in consultation with industry advisory bodies to ensure the relevance of the content to the qualifications offered.

Common generic skills needed in the construction industry are listed in Table 1.4.

TABLE 1.4 Generic skills needed in the construction industry

Generic skill	Example
Communication skills	Verbal communication, reading, writing and numeracy
Interpersonal skills	The ability to work in a team
Interpretation skills	Understanding plans, specifications, maps and diagrams
Organisational skills	Planning work and coordinating team members
Task-specific skills	Plastering, painting, building, concreting etc.
Technological skills	Using mobile phones, two-way radios and the internet

It is important that you engage in a process of determining your current capabilities in the different areas. As an apprentice you will have a range of different support mechanisms to assist you in this process. Typically, an employer will provide you with on-the-job training and experience; this will be supported by a registered training organisation (RTO), which will provide support and development opportunities in all three aspects of the Core Skills for Work Development Framework.

Consult with appropriate personnel to determine own learning needs

Learning needs can vary from site to site due to the nature of the work. Identifying your own learning needs means:

- identifying your current knowledge and skills
- identifying future requirements such as what you might be doing in five years
- determining what you need to learn to achieve the required goals
- establishing how to get the skills you need.

If you are employed as an apprentice or trainee, such conversations should be occurring on a regular basis. These discussions could occur with a range of different people, depending on the size of the organisation – it may be your direct supervisor or the manager of training for the organisation.

These people can also help you to determine the most appropriate way to develop your skills and knowledge. They may suggest the following actions:

- Complete an assessment of current skills
- Undertake on-the-job training or mentoring
- Organise for job rotation, task sharing or refresher training
- Enrol in formal vocational education and training
- Enrol in non-accredited training and short courses
- Undertake individual guided study.

In many organisations and workplaces, a large amount of training occurs in the workplace. Learning in many organisations has been identified as 70:20:10. This model represents the proportion of learning that takes place in different modes: 70% in the workplace, 20% through interactions with other people and 10% in formal educational settings (Based on Training Industry, 2014).

Determine opportunities to learn and develop skills and knowledge

Due to the rapid pace of innovation and changes to Australian industries, it is important to keep your skills and knowledge relevant and up to date with the latest developments. In many states there is a requirement for ongoing professional development to demonstrate that you are keeping abreast of developments to maintain particular professional licences.

There are several ways to keep up to date. Many people undertake formal training that is available through an RTO, industry bodies and industry websites. In many states, if you wish to undertake specific tasks on a construction site you will need to hold a licence. These types of licensing arrangements are best achieved by undertaking specialised training for that area of skill and knowledge.

If you have a clear direction of what you wish to pursue, you should research what opportunities are available to help you to achieve your goals. For example, in some circumstances manufacturers will provide training or publish information that can support you in developing an understanding of a new product.

Many people choose to undertake formal training so there is a record of the training and so they have a certificate that confirms the successful completion of the training. In Australia the National Training Framework has a recognised structure of training, and in many states the training may be funded by the government.

COMPLETE WORKSHEET 4

LEARNING TASK 1.3 TRADE APTITUDE ANALYSIS

Visit https://www.aapathways.com.au and search for 'Practice Aptitude Quizzes'. From here, you can identify an apprenticeship in a trade area that you may be interested in. Download and complete the practice quiz. Once you have checked your results, discuss with your teacher the areas that you need to develop further in order to provide you with the capabilities to complete the particular trade course.

Identify and follow environmental and resource efficiency requirements

Resources at worksites are the materials and products that are used in the construction project. Resource use and efficiency can have an impact on the profitability and safety of worksites. Environmental hazards related to resources must be monitored and documented correctly.

Improvements in resource efficiency can be made due to advances in technology, systems or procedures. Environmental legislation and regulations are applicable to some resources used regularly on construction worksites, and work plans and resource management plans will include these requirements.

Identify environmental and resource efficiency requirements that apply to your role

Every construction worksite is different and uses different resources depending upon the nature of the work being undertaken. During the site induction process, you should be advised of the resource management systems and procedures that are in place. You will also need to have a good understanding of the materials that will be used on that particular project.

Resources that are used on worksites need to be monitored for safety and efficiency. Peak efficiency means the equipment is operating at its best level and is in optimum condition. Manufacturer's manuals and operational manuals detail the optimum performance conditions, and will include product maintenance schedules.

Ensuring the equipment you are using is working at peak efficiency can be achieved through:
- air testing pipes
- using efficient fittings
- following tool maintenance routines to ensure efficient usage
- using transportation systems such as load sharing and planning deliveries.

Some methods to ensure the efficient use and monitoring of material resources might include:
- the strategic use of materials to reduce offcuts and wastage (measure twice, cut once)
- using procedures and materials that reduce or eliminate resource use such as energy-efficient tools, recycling and minimising water use.

A reason for monitoring resource efficiency is to improve environmental and safety outcomes. In addition, the adoption of such practices can improve the profitability of a project. Other reasons include:
- minimising environmental risks such as stormwater pollution
- implementing waste-minimisation systems through the recycling of materials
- improving environmental performance
- achieving efficient production on site
- achieving efficient consumption of resources on the worksite.

When you monitor how resources are used, you should also document the related work processes. The workplace procedures you follow to document the processes are specific to each organisation. These procedures may include:
- recording resources used on site
- inspecting resources, materials and products on arrival from suppliers
- recording details on how the resources are used such as what the resource is, why it is required, who ordered it or used it and when the resource was used.

Your supervisor will typically detail your organisation's requirements for resource and environmental monitoring.

Follow requirements to identify and report environmental hazards

Environmental hazards relating to resources used on the worksite must be controlled by workplace compliance measures. These measures may include workplace policies, procedures, specifications, plans, codes of practice and governmental legislation and regulations. The exact information you need to comply with will be outlined in your worksite policies.

Environmental hazards include any substance, material, product or resource that may be dangerous to any aspect of the environment such as people, animals, water or plants. Table 1.5 provides some examples of environmentally hazardous materials.

TABLE 1.5 Examples of environmentally hazardous materials

Substance	Example
Corrosive	Acids and alkalis
Explosive	Some chemicals
Flammable	Liquid, gas or solid fuels and timbers
Infectious	Diseases such as anthrax
Toxic	Poisons

Your workplace procedures, job safety analysis (JSA) sheets and safety data sheets (SDSs) will outline the exact dangers and the required control measures that must be taken.

Identifying any hazards not covered by the procedures, JSAs or SDSs is the responsibility of all people on the worksite. Once you have identified a hazard you need to report the hazard to the correct people. In the first instance, if there is no emergency then you should contact your direct supervisor. Your supervisor will then contact the appropriate authorities, which may include the:
- safety officer
- site manager
- project manager
- fire brigade
- police
- ambulance
- WorkCover, WorkSafe or similar agency
- Environmental Protection Authority or other government agency.

Reporting the hazard allows for the creation of new procedures or JSAs. These allow others on your worksite to be safe from the new hazard and help to prevent the hazard recurring if a similar incident happens in the future.

Follow requirements to identify and report resource efficiency issues

The majority of medium-to-large worksites have formally documented plans to ensure the continuous improvement of environmental practices. These plans outline the environmental requirements that must be met and how the worksite resources should be managed efficiently.

Some of these plans will include:
- stormwater protection plans such as run-off management or water collection on some sites
- regeneration policies including site clean-up procedures to ensure the required level of regeneration or reparation is done to the site
- waste management such as recycling or reusing materials, and using landfill as a last resort
- dangerous goods management such as the correct storage, transport and disposal of toxic, chemical, dangerous and hazardous goods
- amenities management including correct procedures for grey-water management and sewage management.

Following these plans is vital to ensure you protect the environment and comply with the resource management standards of your organisation.

Make suggestions for improvements to workplace practices

Workplace improvements are generally made by the people undertaking the tasks or activities. This is because the workers are more likely to recognise where an improvement can be made and what the best method for improving a task or activity might be.

Possible improvements you might discover include:
- new materials, products and work practices that could be used at the worksite
- new safety devices or PPE that could increase workplace safety
- energy-conservation techniques or alternatives to the current energy source that are more efficient, such as solar energy
- water efficiencies including using grey water where appropriate
- greenhouse gas emission controls such as turning off equipment when not in use or keeping equipment well maintained so that it is more efficient and less polluting
- transport advances such as car-pooling, scheduling deliveries to minimise fuel use, and maintaining vehicles
- reducing or recycling waste materials.

When you encounter a situation, process or work practice that you believe could be improved upon, you need to discuss the possible improvement with others in your workplace. The best place to have these discussions is in a team meeting. You could also discuss the improvements with supervisors, site managers, team members and others.

Your worksite may require you to make a formal report, request for improvement report or other documented means for suggesting an improvement.

Report breaches of workplace environmental practices

Every worksite needs to comply with environmental regulations and legislation. There are three levels of legislation, which are created by:
- federal government
- state and territory governments
- local government.

Other forms of legislation or best practice codes that may affect your work in the construction industry could include:
- your organisation's environmental or recycling policies and procedures
- industry codes such as the Building Code of Australia (BCA)
- international codes such as the environmental agreements outlined in the Kyoto Protocol.

To identify which environmental legislation and regulations you need to comply with, check your organisational policies and procedures. These policies and procedures outline the exact requirements you need to follow. If you breach environmental legislation, regulations or Acts, you could be fined or charged with an offence. Your organisation and supervisor can also be fined. All workers need to be aware of the environmental legislation applicable to their worksite.

If you identify any breach of environmental requirements, you must report it straight away. You should also report any aspects that may lead to a breach to prevent that breach from happening. This includes any near misses that may occur.

Identifying a breach or near-breach of the regulations or legislation is essential to fix the problem. Breaches could include:
- the wrong materials being used in construction or on a project
- materials placed in the wrong area of the worksite and causing a safety risk
- rubbish being disposed of incorrectly and endangering the environment.

You need to ensure that you follow the organisation's reporting protocol when an incident needs to be reported. You should ensure that your supervisor is aware of the situation, and in most circumstances they will put into place the correct actions to address the situation.

LEARNING TASK 1.4   ENVIRONMENTAL REPORTING

Search the internet for your state's Environmental Protection Agency (EPA) website and identify the contact details for reporting suspected illegal waste disposal.

COMPLETE WORKSHEET 5

SUMMARY

In Chapter 1 you have learnt about working effectively and sustainably in the construction industry. In addition, the chapter has:
- identified a general structure of the construction industry
- unpacked how an individual can take responsibility in their own workplace
- explored the different aspects to be considered when working in a team environment, including:
  - having clear goals
  - contributing as an individual
  - mechanisms for improvement
  - potential types of disharmony
- identified different approaches that will assist in the understanding of your own developmental needs
- discussed several aspects relating to resource efficiency.

REFERENCES AND FURTHER READING

Australian Building and Construction Commission, *Building and Construction General On-site Award*, **https://www.abcc.gov.au/your-rights-and-responsibilities/wages-and-entitlements/modern-awards/building-and-construction-general-site-award-2010**

Australian Building and Construction Commission, *Types of Employment*, **https://www.abcc.gov.au/your-rights-and-responsibilities/wages-and-entitlements/types-employment**

Australian Bureau of Statistics, *Australian Industry*, **https://www.abs.gov.au/statistics/industry/industry-overview/australian-industry/latest-release**

Australian Government, Department of Education, Skills and Employment, *Australian Core Skills Framework*, **https://www.employment.gov.au/australian-core-skills-framework**

Australian Government, Department of Education, Skills and Employment, *Training Packages*, **https://www.employment.gov.au/training-packages**. National Careers Institute © Australian Government, Department of Education, Skills and Employment CC BY 4.0 Licence (https://creativecommons.org/licenses/by/4.0/)

Cengage Learning Australia, *Basic Building and Construction Skills*, Chapter 2, Working effectively and sustainably in the construction industry.

NCVER, *Apprenticeship in Australia: An historical snapshot*, **https://www.ncver.edu.au/research-and-statistics/publications/all-publications/apprenticeship-in-australia-an-historical-snapshot**

Training Industry, 2014, *The 70-20-10 Model for Learning and Development*, **https://trainingindustry.com/wiki/content-development/the-702010-model-for-learning-and-development/**

WA Department of Training and Workforce Development 2016, *Work Effectively and Sustainably in the Construction Industry, CPCCCM1012A, Learner's Guide, Certificate II in Building and Construction (Pathway – Trades)*, **https://www.dtwd.wa.gov.au**

GET IT RIGHT

Ryan has begun working with a new company. The company has several teams that specialise in different areas of construction. The team that Ryan has been assigned to has been together for more than five years. The team has been advised that they must now use a different product than they have used in the past. Ryan has been placed with this team as he has experience with the new product and construction methods that are required to use it. The team leader has left Ryan out of many of the team conversations and does not communicate directly with Ryan. Some of the team members have also treated Ryan in the same manner. However, there are other team members who have welcomed Ryan as a part of the group.

The supervisor has noticed that the team's production and quality levels have dropped, and can see that he must take action.

Individually, consider then list your responses to the questions below. Then, as a whole class, work through all the responses that have been generated.

1. What actions could Ryan have taken to become more accepted by the team leader?

2. What actions could the team leader have taken if he was not happy with Ryan becoming a part of the team?

3. What actions could the supervisor have taken and why?

4. What training is required for the team members?

5. What communication protocols would you recommend for the team if you were the supervisor?

 **WORKSHEET 1**

To be completed by teachers
Student competent ☐
Student not yet competent ☐

Student name: _____

Enrolment year: _____

Class code: _____

Competency name/Number: _____

Task
Read through the sections from the start of the chapter up to 'Take responsibility for own workload', then complete the following questions.

1 Identify the three main areas that make up the construction industry.

_____
_____
_____

2 How many people are employed in Australia's construction industry?

_____

3 Describe the types of construction in each of the three areas of the construction industry.

_____
_____
_____
_____
_____
_____

4 Name three areas that technology can change the construction industry and describe how this occurs.

_____
_____
_____

5 What are the four main categories of employment conditions that exist in the construction industry?

_____
_____
_____
_____

WORK EFFECTIVELY AND SUSTAINABLY IN THE CONSTRUCTION INDUSTRY

6 What types of conditions does an award cover?

_____

_____   _____

7 Originations may have additional requirements. Identify two of these and explain what they are intended to cover.

_____

_____

WORKSHEET 2

To be completed by teachers
Student competent ☐
Student not yet competent ☐

Student name: _____

Enrolment year: _____

Class code: _____

Competency name/Number: _____

Task
Read through the sections from the start of 'Take responsibility for own workload' up to 'Work effectively in a team', then complete the following questions.

1 When establishing priorities and deadlines, who should be consulted?

   _____

   _____

2 How does having sound planning processes and communicating outcomes contribute to the project?

   _____

   _____

3 What do you need to develop performance levels through self-awareness?

   _____

   _____

4 How could variations in the workplace occur?

   _____

   _____

5 What skills do you require to be able to report effectively?

   _____

   _____

6 If you ask for additional assistance to complete a task, what will it demonstrate to your supervisor and co-workers?

   _____

   _____

7 Identify at least four reasons why you may ask for assistance.

_____

_____

_____

_____

WORKSHEET 3

Student name: _____

Enrolment year: _____

Class code: _____

Competency name/Number: _____

To be completed by teachers

Student competent ☐

Student not yet competent ☐

Task

Read through the sections from the start of 'Work effectively in a team' up to 'Identify own development needs', then complete the following questions.

1. There is a range of different teams that you will be involved with during the construction of a project. (Circle the correct answer.)

   TRUE   FALSE

   _____

2. What does it take to be an effective team?

   _____
   _____

3. What is the best practice that will assist a team to improve?

   _____
   _____

4. List four reasons that may cause friction within a team environment.

   _____
   _____

5. Why is it necessary to address any friction in a team as quickly as possible?

   _____
   _____

6. What is the first step in resolving a conflict?

   _____
   _____

 **WORKSHEET 4**

To be completed by teachers
Student competent ☐
Student not yet competent ☐

Student name: _____

Enrolment year: _____

Class code: _____

Competency name/Number: _____

Task
Read through the sections from the start of 'Identify own development needs' up to 'Identify and follow environmental and resource efficiency requirements', then complete the following questions.

1 Identify the three skill capability areas required to be effective in the workplace.

   _____
   _____
   _____

2 List four generic skills that are commonly used in the construction industry and provide an example of each skill.

   _____
   _____
   _____
   _____

3 Who will support an apprentice in relation to their training needs?

   _____
   _____

4 How would you keep up to date with the latest training opportunities?

   _____
   _____

WORKSHEET 5

Student name: _____

Enrolment year: _____

Class code: _____

Competency name/Number: _____

To be completed by teachers

Student competent ☐

Student not yet competent ☐

Task

Read through the sections from the start of 'Identify and follow environmental and resource efficiency requirements' up to the end of the chapter, then complete the following questions.

1 Why is it recommended to monitor resource efficiency?

_____

_____

2 What type of items would you list when monitoring resource efficiency?

_____

_____

3 List two types of environmentally hazardous substances and give an example of each.

_____

_____

4 Who would usually make workplace improvement suggestions?

_____

_____

5 Provide an example of where some areas of improvement may occur.

_____

_____

6 What are the three levels of legislation that regulate workplace environmental practices?

_____

_____

_____

7 What types of breaches can occur in relation to workplace environmental practices?

_____

MY SKILLS

EMPLOYABILITY SKILLS

Using the following table, describe the activities you have undertaken that demonstrate how you developed employability skills as you worked through this unit. Keep copies of material you have prepared as further evidence of your skills.

Employability skills	The activities undertaken to develop the employability skill
Communication	
Teamwork	
Planning and organising	
Initiative and enterprise	
Problem-solving	
Self-management	
Technology	
Learning	

PLAN AND ORGANISE WORK

This chapter covers the outcomes required by the unit of competence 'Plan and organise work'. These outcomes are:
- Determine and plan basic work task activities.
- Organise performance of basic work task.

Overview

Planning and organising work on a construction site is necessary to ensure the safe, efficient and high-quality completion of tasks. Many levels of planning and organisation take place from the overall project planning phase through to work tasks and individual workers' daily work schedules.

There are several capabilities that are required to plan and organise work effectively in the construction industry.

Construction workers must be able to communicate well with their co-workers. This is in relation to how each trade area can impact on the other trades on site. This is usually covered by the sequencing of works and trades so there is minimal impact on the overall project. If the communication is strong between the trades then there is a higher chance that quality, efficiencies and environmental considerations can be achieved in a safe manner.

You will also need to review the sequencing of tasks and identify and discuss improvements with your co-workers. This is an ongoing process that results in continuous improvement of the work processes on site. Documenting these improvements ensures all improvements are adopted and implemented as required by the company's policies and procedures.

The ability to plan and organise a construction project can have a significant impact on the profitability of the project. In addition to this, the customer will have expectations about the duration of the project and the estimated completion date. If the completion date is not met, there may be undesirable consequences for the customer.

Employers value people who fit into their workplace, solve day-to-day problems, manage their time and are keen to continue learning. These types of skills are known as employability skills. The employability skills you develop while working through this unit will be assessed at the same time as the skills and knowledge.

It is important to show evidence that you have developed these skills. Your trainer or assessor will discuss with you how to record your employability skills. The following table provides some examples of things you might do to develop employability skills in this unit.

EMPLOYABILITY SKILLS

Employability skill	What this skill means	How you can develop this skill
Communication	Speaking clearly, listening, understanding, asking questions, reading, writing and using body language.	• Listen carefully to work instructions. • Ask questions to clarify task requirements.
Teamwork	Working well with other people and helping them.	• Work individually and as a team to complete tasks in a timely manner. • Liaise with others to avoid conflicting work tasks.
Planning and organising	Planning what you have to do. Planning how you will do it. Doing things on time.	• Plan resources required for each task. • Plan sequence of activities in order to successfully complete work task.
Initiative and enterprise	Thinking of new ways to do something. Making suggestions to improve work.	• Prepare information before you discuss issues. • Suggest improvements to processes.
Problem-solving	Working out how to fix a problem.	• Manage task sequencing to cater for changes to the work environment. • Adapt to cultural differences in communications.
Self-management	Looking at work you do and seeing how well you are going. Making goals for yourself at work.	• Evaluate and improve own methods of communication. • Aim for quality results within specified task time frames.
Technology	Having a range of computer skills. Using equipment correctly and safely.	• Implement safe work procedures for machinery. • Use a range of communication tools such as two-way radios, telephone and email.
Learning	Learning new things and improving how you work.	• Research online product and work practice information. • Observe more experienced construction workers and adopt their practices to improve your work processes.

Determine and plan basic work task activities

Determining and clarifying task requirements is an important skill in the construction industry and impacts on the successful management of people, resources and budget. In order to complete job tasks successfully you need to know exactly what is required of you, how to locate information about your job and how to get the assistance you require if necessary.

Communication is the key factor to finding out what you need to know. Listening to instructions and asking questions ensures you understand your role in completing a given task. Advice and mentoring from more experienced colleagues is also useful. However, there are times when you'll need to source the information yourself. This may require you to access information from manufacturers, work sheets, drawings and online sources to ensure that correct interpretations of the task and material requirements are achieved.

Planning and prior preparation play an important role in the smooth completion of any task in the workplace. Always find out what is required of you in advance and read information or ask questions about the task.

Read and/or interpret a work order

To clearly establish what is required to undertake a task, you will need to refer to several different forms of information. Not all information is in the form of

documentation and the list below is a combination from different sources:
- Read plans and specifications.
- Listen to and read instructions.
- Attend toolbox meetings and safety and information meetings.
- Ask questions.
- Talk to workmates.
- Ask for advice from more experienced tradespeople.

A work order is either a hard-copy document, an app or a software package that provides sufficient information to complete the required body of work. The apps and software packages may be referred to as 'construction management instruments' that can be applied to both small works and large construction projects. If you were to use a software system, you would need to research the most appropriate program that would suit the type of work that is undertaken by your company.

The work order (see Figure 2.1) will have standard information such as the location of the project and the contact details of the client. The work order will also have a description of the works that are to be undertaken. In many instances, there will have been a discussion with the client to establish all the details required to complete the piece of work that is required. This may have a broad range of items depending on the works required. Typically, work orders are broken down into specific trade areas and distributed accordingly.

Determine work task, WHS and equipment requirements for the work

The work tasks need to be broken down into a logical sequence in a step-by-step process. This involves a thorough knowledge of the different construction elements and how they interrelate and depend on each other. This type of sequencing is usually identified in the construction planning stage.

There are several considerations that need to be identified at this stage of the planning cycle. A common instrument that is used is a Gantt chart (see Figure 2.2). The general schedule will not break down to each task, but it will provide a sequence and time frame for the different trade areas.

Once the tasks have been identified, consideration needs to be given to several aspects that will have an impact on the performance of the task. The first consideration that needs to be addressed is work health and safety (WHS). This will require the guidance of any legislation that may be in place. This may be in the form of a code of practice that will guide the best practices for given construction tasks and procedures. The main document that will provide the required outcomes will be a safe work method statement (SWMS) that is filled out in relation to the specifics of the particular project and task.

The construction method and WHS considerations will determine to an extent the different types of tools that should be used. An example of this may be the use of a nail gun compared to the use of a hammer and nails. The use of a nail gun can reduce the time required for a task; however, if for some reason in relation to WHS practices a nail gun cannot be used, this will impact the task being completed in a timely manner.

A systematic approach that can be used to identifying a task and its requirements can be as simple as:
- What has to be done?
- Where will it take place?
- When will it be done?
- Who will be doing it?
- How will it be done?

A consistent approach that is consistently followed will give you a structure that will become second nature and will build your capability to plan and organise projects from small simple activities to large complex projects.

Clarify work task, WHS and equipment requirements with supervisor

Successful completion of any work task is dependent on accurately determining exactly what it is you are required to do. It is worth clarifying all of the job requirements at the beginning of a project in order to minimise delays and wastage due to errors from guessing or making assumptions. There are many communication methods to identify exactly what your work task involves. Communication tips include:
- Listen carefully to your supervisor at daily planning meetings (or kick-off meetings) to ensure you fully understand what is required of you. You will also be provided with information about other subcontractors' requirements and how work arrangements are scheduled so tasks are completed in a logical and efficient manner.
- Ask questions to check you understand the required tasks or to clarify points you are uncertain of. Your direct supervisor will appreciate you clarifying aspects you are unsure of instead of proceeding and wasting time, resources and money through making errors. Sometimes it may be necessary to use a two-way radio or phone to speak to your supervisor if they are not close by.
- Read all instructions, plans, specifications and drawings, job sheets and other relevant documentation as this provides further clarification about the task and what you are expected to do.
- Use any visual cues. Your supervisor can inform you of any formal visual cues, safety or warning signs that you need to be aware of, especially where loud equipment may be operating. Pay particular attention to this information as each workplace might have unique cues that are used on site.

FIGURE 2.1 Typical work order with sufficient information to undertake the required preparation and completion of the activity

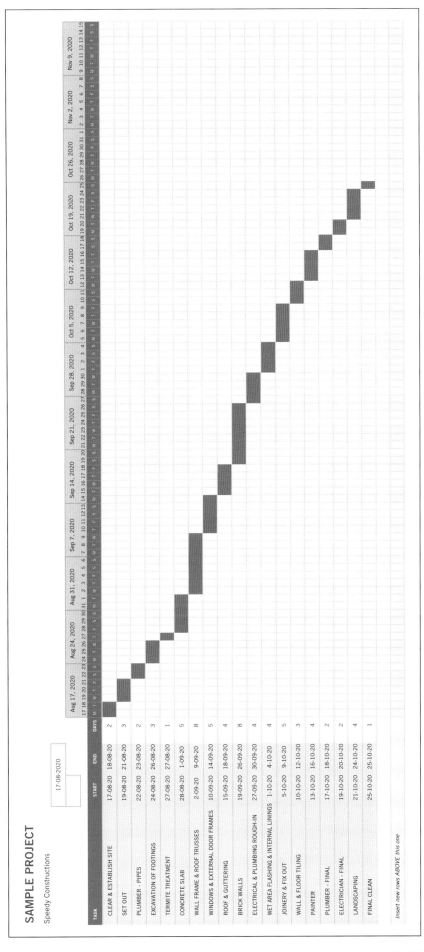

FIGURE 2.2 Sample Gantt chart used to map out sequencing and duration of works

PLAN AND ORGANISE WORK 33

- Use technology such as a mobile phone or computer to assist if your supervisor is not available. Companies may provide fact sheets or information about using their products. Product labels and advertising information may provide a phone number or website address that might be a good source of information.

Misinterpreting information in the construction industry can be very costly in terms of time, money and safety. Your attitude and communication skills with co-workers, subcontractors, suppliers and the client's representatives are a major factor in the success of your job. Always make sure that you choose the most appropriate communication method for each situation and take into consideration cultural, technological, safety and environmental factors.

Clear communication, good organisation and prior preparation will ensure that you do your job well and avoid costly mistakes or injury.

Make sure you are familiar with any specifications, plans and drawings so that the quality of your work is to the required standard and you will be less likely to misinterpret information as you work. If you know what your task will be ahead of time, you can prepare yourself by reading relevant documents or discussing the task with your co-workers. Being familiar with task methods and required resources before you begin a task will cut down on time spent learning on the job.

Safety and environmental protection issues are an essential element to any work carried out on a construction site. Make sure you are familiar with safety and environmental requirements prior to undertaking any task and plan your task stages with this in mind.

Plan steps to complete work task requirements

A construction project is a dynamic work environment where the work conditions continually change as work on the project progresses. The continual change can lead to safety issues and this is why it is essential for workers to receive a site induction upon commencing work on any construction site. In addition, workers must remain aware of, and kept up to date with, any safety aspects that change on the construction site. This may require adjustments to work plans.

It takes time to plan a work task, but the time taken is an investment that will yield results as neglecting to plan will result in missing vital information to complete a task. Missing such information may result in workers having to return to complete a task at a later date, which will consume time that may not have been allowed for in the original quote for the work.

Clear communication with co-workers and other subcontractors is the best way to ensure planned work is completed efficiently. It is good practice to let others know what your priorities are and to listen to theirs, so

LEARNING TASK 2.1 PLANNING SOFTWARE

Research and identify three different software programs that can be used for construction management. Rank the main features of the three software programs in the table below as poor, acceptable or strong.

Based on your review, identify which software program you would recommend for use based on the level of detail in the scheduling function and the automation of the estimating function for both quantities and costing. Use the remaining features to assist in making a final recommendation.

Software program	Features	Poor	Acceptable	Strong
	Scheduling			
	Estimating			
	Budgeting/cost control			
	Accessibility by members of your team			
	Cost of the software			
	Scheduling			
	Estimating			
	Budgeting/cost control			
	Accessibility by members of your team			
	Cost of the software			
	Scheduling			
	Estimating			
	Budgeting/cost control			
	Accessibility by members of your team			
	Cost of the software			

you can work as a team to get tasks achieved safely, on time, on budget and with a quality result.

Analyse task and identify work steps to ensure efficient conduct of work

All tasks require thorough planning if a quality result is to be achieved. Complex construction tasks also require work health and safety and environmental protection issues to be considered, along with quality requirements. You need to be aware of relevant legislation, regulations, codes of practice, organisational safety policies and procedures, and worksite-specific safety plans, and incorporate this information into your daily work routines.

Work health and safety

Before you commence a work task you should be aware of any potential hazards that may be present and how to deal with these should the need arise; for example, power cables, earth leakage boxes, trip hazards, restricted access barriers and faulty equipment.

Make sure you know the exact location of fire extinguishers and continually assess risks associated with tasks. Pay particular attention to safety data sheets (SDS) as they give you the correct procedures for handling resources and for protecting yourself and others against hazards. Identify the best methods for clean-up and disposal of any hazardous materials to help ensure safety and environmental concerns are addressed.

This will be covered by conducting a risk assessment to identify any potential risks that need to be managed. Conducting a risk assessment can be as simple as a standard checklist.

Preparation

Workplace documentation will provide you with information about the tasks and additional requirements. These documents consist of task sheets, work orders, diagrams, plans, specifications and drawings. The documentation contains information such as measurements, materials required and location-specific instructions. The information in the documents will contain sufficient information to undertake the required task; however, there are incidental items that may not be explicitly noted in the documentation. In many instances, it will be necessary to rely on your experience or the experience of co-workers to ensure that missing information is addressed. The use of experience comes into play when problems occur or there is an unintended discrepancy from the documented information. For example, diagrams or specifications will often assume that all existing floors of a building are level, but this might not be the on-site reality.

Once the exact task requirements have been identified, it is time to plan how the task will be completed. For example, if a task will take several days, it is important to ensure that all tools and materials are available. This may be difficult if there is restricted storage space on the construction site. Such situations may require a sequenced delivery of material over the duration of the task. The flow-on effect may also be applicable if different tools and equipment are required for different activities over the duration of the task.

Interpreting tasks and sequencing your work plan

Instruction sheets or job process sheets show the sequence of work tasks to get the job completed. However, if these are unavailable you need to determine the required steps. Think about the desired result and try to work backwards from there to identify the logical required sequence. It is a good idea to write down the sequence to develop your own work plan, which you can discuss with your supervisor. There are several products in the construction industry that have manufacturer's installation instructions (see, for example, Figure 2.3). These instructions provide a guide to the process of installation and, in some instances, if the installation process is not followed then any warranties may be rendered invalid.

Also identify opportunities to boost efficiencies in work processes throughout the completion of the task. It is important continually to reflect on the execution of a task and ensure that you incorporate any improvements the next time that you undertake the same task.

If you have to wait for a product to dry or set before progressing (such as floor tiles that have just been laid), you can plan and prepare resources for the next stage or task or commence a clean-up to protect yourself and others against dust and other hazardous materials.

Creating efficient work practices saves time and money and ensures your safety and the safety of your co-workers, equipment and the environment.

Often there are many labourers working on a site. The site coordinator or manager might have a pre-start meeting to discuss what is required each day and how tasks will be sequenced. However, it is up to you to break down the job you are responsible for into manageable steps and sequence these steps logically to ensure a safe work environment and that quality and environmental concerns are addressed.

Plan task steps in conjunction with team members

There are often many people working on a construction site at any given time. It is therefore important to communicate clearly so everyone can work together to achieve the best possible results. For example, you would not want someone walking through your concreting job before it is set because they were unaware that it was not yet ready.

Poor communication can be costly and cause frustration in work environments as well as increasing safety risks.

LIGHT-WELL CONSTRUCTION
INSTALLATION INSTRUCTIONS

Introduction

There are three basic types of light-well which can be used with your Skydome Skylight:

1. **Skyflex** — **Factory supplied,** flexible light-well kits which are available in a limited number of sizes. Check with our friendly sales staff on available sizes.

2. **Custom Wood** — Timber light-wells which DO NOT come in kit form but are straight forward for trades people or basically equipped handy persons.

3. **Plasterboard** — Plasterboard light-wells which DO NOT come in kit form and are best left to trades people.

Skyflex - Fixing Instructions

Tips For A Successful Installation

> Handle the Skyflex material carefully.
> Before use - store in a clean dry area, (Not Dusty).
> Offset as little as possible for best results.
> Keep the Skyflex taut not tight, to much slack will reduce the light level.

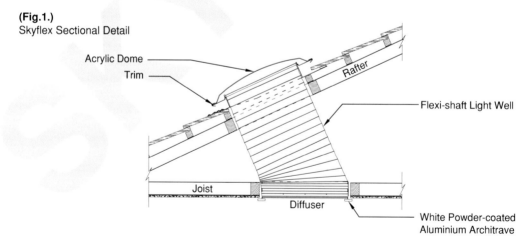

(Fig.1.)
Skyflex Sectional Detail

Step 1.
Fit your Skylight to the roof as per specific instructions.

FIGURE 2.3 Typical installation guide provided for installers from manufacturers

There are several ways you can work with others to ensure your job tasks are completed appropriately; for example:
- Discuss your priorities and job task requirements at the morning kick-off meeting. If you know where others are working and your work does not impact on them, you are free to continue as required. If your work does impact on others, discuss the time you need to complete the task so they can adapt their work tasks as necessary.
- If your work impacts on others, identify ways to lessen the impact. For example, if you are using noisy machinery or equipment that causes dust and vibration, you might be able to conduct that part of your activity away from colleagues.
- Check with others if what you're about to do impacts on them and, if so, either try to minimise the impact or work on a different aspect of the task until a better opportunity arises.
- Ensure that others know when your work task requires special consideration by communicating with them and by using the appropriate signage; even a text message to the team can assist communications on site.

There is a wide range of people you may need to consult with at different times. These people may come from various cultural backgrounds or have different skill and knowledge levels. It is important to respect other people in all interactions and always choose the most appropriate methods of communication. People you may have to consult with could include:
- supervisors and work colleagues
- other subcontractors
- inspectors
- clients or client representatives
- manufacturers and delivery personnel
- members of the general public and visitors.

COMPLETE WORKSHEET 1

Organise performance of basic work tasks

For a project to be successful, it relies on thorough planning and preparation. Work plans also need continual adjustment to ensure that priorities are addressed to support the project progressing at the required level. Often there are many teams working on one site to complete specialised tasks. It is necessary for these teams to communicate effectively to enable the safe, efficient completion of tasks. Individuals within these teams must also be able to sequence their own work tasks to enable a smooth progression of tasks within the group.

Communication is the key to organising workers and teams to prioritise and complete the jobs they have. It is also essential that information is recorded and that these records are maintained so that supervisors and co-workers can identify the progress of the project at any stage. The safety of workers depends on everyone knowing where they can safely focus their activities each day and how long they have to complete tasks in certain areas.

LEARNING TASK 2.2 CONSTRUCTION SEQUENCING

You are required to construct a fence with the following specifications:
- The fence is to be 6.000 m in length and 0.900 m in height.
- Consider the ground to be flat and level.
- Posts are to be concreted into the ground at 2.000 m intervals
- There will be two rows of rails: one at the top of the posts and a rail 0.100 m off the ground between posts.
- The pickets are to be screwed to the rails.
- The fence is to be painted white.
- All material is already on the site.
- Any excess dirt and material need to be removed.

Use the following table to create a schedule of sequential sub-tasks and the tools and materials required to complete them.

Main task	Sub-tasks	Tools and materials
1 Set out fence		
2 Dig holes		
3 Concrete in posts		
4 Install rails		
5 Fix pickets to rails		
6 Clean site		
7 Paint fence		

Sequence work task steps and allocate time and resources to each step

Safety is a priority when planning any activity in the construction industry. It is not uncommon to have many people involved in a single task on a construction site, or for several tasks to be occurring at the same time. For this reason, it is important that roles, responsibilities and task sequencing are discussed and issues clarified prior to commencing the tasks. Communication is particularly important where there are several people working in the same area.

Organising how the task is to be completed is usually managed by the site coordinator or manager at a daily pre-start meeting. Discussions focus on the appropriate sequencing of tasks to enable work progression in a logical, safe and efficient manner. They assign responsibilities to you and your colleagues, making sure that skills and abilities are matched to tasks to obtain the best results.

Your supervisor should give you an indication of how long it should take you to safely complete a task. They may also discuss what other activities depend on the completion of your task. The time needed to complete a task will have been estimated in the planning phase and will be set as a target. It is important you understand the impact that completing your task in a timely manner has on other areas of the project.

All construction projects will have a schedule that will be referred to as the project progresses. Methods such as Gantt charts or critical path analyses will clearly identify any dependencies of each trade area or tasks to be performed. Using such planning techniques will identify which tasks can occur at the same time and which tasks need to be completed for the next task to commence.

A common practice is to have workers who have a high level of experience with a particular task to lead a team, as their experience will provide a sense of certainty of the actions to be taken. In other instances, a higher number of workers may need to be allocated to the task to meet the time requirements. The allocation of human resources is a crucial element to running not only the project but also the business. The time consumed by the workers allocated to a task is an expense that is not static, and if miscalculated can reduce the profit of the project. For example, as a business owner, you would not like to be paying for workers to stand around if the task has been completed. Conversely, if there are not enough workers and they are rushing to complete the task, the quality of the work may be substandard.

Having charts and sequences available will help to ensure that everyone is aware of the project priorities and each other's role in the daily management of tasks. Communication helps to maintain a safe work environment where all personnel are working as a team towards achieving a common quality outcome.

People who may be involved in planning and organising work tasks could include:
- supervisors and work colleagues
- other subcontractors
- inspectors
- clients or client representatives
- manufacturers and delivery personnel.

On construction sites there will be a range of workers from different backgrounds and cultures. It is the responsibility of every worker to ensure they choose the most appropriate methods of communication so all workers fully understand the information being presented. The workforce that makes up the construction industry at times will consist of foreign labour depending on trade agreements with other countries. Diagrams may be very useful in conveying information if there are language differences, particularly when discussing processes or resources.

Another important form of communication is the use of nonverbal communication such as hand signals. These may be used when hearing is restricted due to loud equipment necessitating the use of PPE. This practice is common for tasks that involve the use of a crane.

Complete records of task planning and progress

Documenting activity on a construction site is useful for several reasons. The monitoring of the overall progress of the project against the original schedule will allow supervisors to make adjustments and take appropriate decisions to maintain the targets set for the project. The documentation of tasks or works being completed also allows for inspection of the works to be completed to meet regulatory requirements, and for the claiming of any progress payments depending on the contractual arrangements.

Documentation can be an efficient method of communicating information as it can be located centrally, allowing everyone easy access.

By completing the relevant documentation, supervisors can make decisions about the following:
- What materials need to be ordered or if the supplier provided the correct items and quantities
- Who should be working in specific areas
- What tasks must be completed before others commence
- Safety issues relating to tasks and risk management
- Hazards identified and dealt with
- Who is responsible for job completion and quality.

A range of different documentation is used across the construction industry. These documents will vary due to the variety of construction types and sizes. The documentation will also be set by the organisation and will align to the systems and processes of the particular

organisation. Workplace documentation may include some of the following:
- A general summary of major site and construction activities and achievements during the period
- Progress reports by area or awarded contracts during the period
- Construction hours, projected hours and hours actually taken to complete a task
- Job safety analysis documentation
- List of equipment or resources purchased and received during the project
- Checking material quantity and conformity to requirements
- Construction cost control reports and explanations
- An up-to-date construction schedule
- Reporting and recording hazards and risks.

LEARNING TASK 2.3 REPORTING DOCUMENTS

Research the four different documents listed in the table below that are used for reporting purposes. Give examples of the type of information that you would expect to be in the document.

Document	Types of information
Job safety analysis documentation	1
	2
	3
Construction cost control report	1
	2
	3
Construction progress report	1
	2
	3
Construction schedule	1
	2
	3

Review planning of activities to establish the effectiveness of the process

To improve productivity, you need to ensure that you include time to review how effective your planning has been. Figure 2.4 illustrates the cyclical nature of planning, task completion and review. Each phase is dependent upon the other and, with regular review, will result in process improvement.

Reviews can cover several different aspects of project planning, including:
- work health and safety issues
- environmental protection requirements
- quality assurance
- new technologies
- time efficiencies
- resource cost-effectiveness.

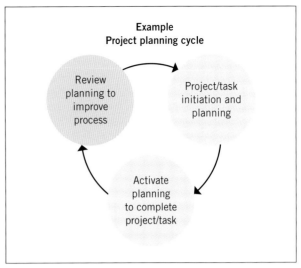

FIGURE 2.4 Project planning cycle of review

A formal review of the project is traditionally undertaken on its completion; however, smaller reviews may be necessary during the project as more information becomes available. For example, if a supplier cannot get an order of materials to the site on time, then the manager will have to review priorities and reschedule tasks to accommodate the changed delivery schedule. The planning process review may include identifying another supplier or allowing more lead time for the order to ensure materials arrive on time.

These decisions impact the planning and organisation for the next project, ideally leading to improvements such as better time management, better resource management and the selection of suppliers who can meet the demands of the project's constraints.

It is also common to have regular reviews of what has been planned for the day or week. This may be due to earlier tasks not being completed, in which case a site manager or foreperson will make such adjustments. It is important that the cause of the reorganisation is documented and fed back into the planning unit so that appropriate allowances can be made in the planning stages of future projects.

All members of the team should have the opportunity to provide feedback about their observations of the progress of the project or the impacts of decisions in the workplace. Such information is fed back through line managers at the conclusion of each phase of the project in discussion with the site manager.

Each worker should review their own work activities to recognise improvements that could be made in their work practices. This may mean making adjustments to your daily work plans or task procedures and should be discussed with your supervisor. In order to do this, you need to ask yourself several questions:
- Did I complete the task within the material and time constraints allowed for the task?

- Was my planning and preparation of the task appropriate?
- What went wrong and why did it go wrong?

Undertaking such practices creates an environment of continuous improvement. The benefits of having continuous improvement practices in place helps to create a higher quality product and working environment for all workers involved in the construction of a project.

Provide feedback and suggestions on improvements for future projects

It is important that all staff have a method of recording or expressing ideas for improvement. Each organisation has its own strategies and/or forms for this purpose. These range from formal recorded methods through to informal discussions over a break or after work.

Formal methods of recording suggestions for improvement are intended to be shared by work teams and might include:

- a job safety analysis form (JSA) – if a worker has recognised a need for a change to a process or the organisation of a work activity from a safety perspective, they can record it on a JSA
- progress reports from teams or general summaries of achievements on site that have an area for feedback
- process improvement forms, which are completed by contractors and subcontractors in some organisations
- minutes of team meetings (more commonly seen in larger companies).

Where staff numbers are small, it is more likely that there will be only informal methods of gaining feedback. Such informal methods may result in a supervisor taking notes. For an individual, this may build their own knowledge of a work task.

It is important that any feedback is not relayed as a criticism or seen as placing blame. You must be aware of how the feedback is provided so that the intent of raising the item is not lost in the way that the message is communicated. When putting forward your suggestion, it is professional to discuss the process itself or refer to job titles rather than personalise the issue by using individual people's names.

Before a suggestion is made, it is best to discuss with a co-worker. This will give another perspective to the matter you are raising, and you may find that there are considerations that you may not be aware of. It is important to also consider the impact of your suggestion on other trades or tasks that occur before or after the task you are considering.

COMPLETE WORKSHEET 2

SUMMARY

In Chapter 2 you have learnt about planning and organising work concepts that are commonly used in the construction industry. In addition, the chapter has:
- explored the different software available that will assist in planning and scheduling a construction project
- identified different aspects that need to be considered when planning a construction project, which include:
  - WHS requirements
  - tools and equipment
  - staffing requirements
- considered a systematic approach that can be used to identify tasks and associated requirements
- considered behaviours to be used when giving feedback

REFERENCES AND FURTHER READING

Skydome, **http://www.skydome.com.au**.

Uher, Thomas E., *Programming and Scheduling Techniques*, UNSW Press, 2003.

GET IT RIGHT

John works for a company that builds pergolas. He has followed the work order supplied to him in preparing for the work, but has found that there is information missing from the work order. Review the work order and identify which elements are not clear or are missing.

Construction Work Order Form		
Organisation:	Project: Simple Pergola	Team: John's Team
Client	**Address**	**Contact**
Mr and Mrs Stone		Fred - 0411 555 789
Work Order Number: 01022021		**Work Order Date:** 1st February 2021
Construction Work Details		
Type of Construction job: Timber pergola		
Plumbing required: Nil		
Electrical Installation required: Nil		
Material required:	Treated pine posts Treated pine beams Treated pine rafters	Treated pine battens Associated hardware
Work Details		
Quantity	**Description**	**Price**
4 /	Treated pine posts	
1 / 6.000	Treated pine beams	
10 / 3.300	Treated pine rafters	
5 / 6.000	Treated pine battens	
2 of	Metal feet	
Hardware		
	Total Price:	
Delivery date of work: 1st April 2021	**Actual date of completion:**	
Site plan	Front elevation	

_____
_____
_____
_____
_____

WORKSHEET 1

To be completed by teachers
Student competent ☐
Student not yet competent ☐

Student name: _____

Enrolment year: _____

Class code: _____

Competency name/Number: _____

Task
Read through the sections from the start of the chapter up to 'Organise performance of basic work tasks', then complete the following questions.

1. Identify three different sources of information you could refer to when preparing for a work task.

   _____
   _____
   _____

2. List five items that you would expect to be listed in a work order.

   _____
   _____
   _____
   _____
   _____

3. List three methods of communication that may be used to gain information before a task is undertaken.

   _____
   _____
   _____

4. Why is it essential for workers to receive a site induction upon commencing work on any construction site?

   _____
   _____

5. What actions can you take from a WHS perspective when planning your work task?

   _____
   _____
   _____

6 List at least three pieces of information you will be looking for when referring to documentation for the task to be undertaken.

_____

_____

_____

7 What documents might manufacturers supply that may guide you with your work task?

_____

_____

8 Identify four different people that you may need to consult with before you undertake a task.

_____

_____

_____

_____

WORKSHEET 2

To be completed by teachers
Student competent ☐
Student not yet competent ☐

Student name: _____

Enrolment year: _____

Class code: _____

Competency name/Number: _____

Task
Read through the sections from the start of 'Organise performance of basic work' to the end of the chapter, then complete the following questions.

1 List four people who may be involved in planning and organising work tasks.

   _____
   _____
   _____
   _____

2 How does completing documentation assist supervisors?

   _____
   _____
   _____

3 Identify three benefits that may result from having a continuous improvement process in the workplace.

   _____
   _____
   _____

4 How often should reviews take place and why?

   _____
   _____

5 Who should be involved in a feedback process?

   _____
   _____
   _____

6 Identify two perspectives that may be considered when undertaking a review and why it is important to consider these perspectives.

   _____
   _____
   _____

7 Why do you need to give consideration to the way feedback is given?

   _____
   _____
   _____

EMPLOYABILITY SKILLS

Using the following table, describe the activities you have undertaken that demonstrate how you developed employability skills as you worked through this unit. Keep copies of material you have prepared as further evidence of your skills.

Employability skills	The activities undertaken to develop the employability skill
Communication	
Teamwork	
Planning and organising	
Initiative and enterprise	
Problem-solving	
Self-management	
Technology	
Learning	

PLAN AND ORGANISE WORK

CARRY OUT MEASUREMENTS AND CALCULATIONS

This chapter covers the outcomes required by the unit of competence 'Carry out measurements and calculations'. These outcomes are:
- Obtain measurements.
- Select and apply methods of obtaining measurements required to meet construction task requirements.
- Use a rule or tape to obtain linear measurements accurate to 1 mm.
- Perform basic calculations.
- Take basic measurements and calculate quantities of materials in a construction environment, using basic formulae for each of: weight, area, volume, perimeter, circumference, ratio and percentage.
- Convert measurements in metres to millimetres and measurements in millimetres to metres.
- Check calculations for accuracy and record calculation workings and results.

Overview

To work effectively in the construction industry, you need to know how to take accurate measurements and make correct calculations. Measurements are used to determine lengths, widths and heights of materials and work areas. Calculations are combinations of figures used to determine the amount of materials needed, such as how much concrete for a slab or how much timber for a frame.

Taking measurements and doing calculations involves literacy and numeracy skills and the ability to use some basic tools. These tools include rulers, tape measures, calculators and possibly computers.

Communication is a critical skill that is needed in all aspects of the construction industry. Communication skills include the ability to share information about measurements and calculations; check the materials that need measuring; and the ability to read and interpret work documents, drawings and specifications.

Employers value people who fit into their workplace, solve day-to-day problems, manage their time and are keen to continue learning. These types of skills are known as employability skills. The employability skills you develop while working through this unit will be assessed at the same time as the skills and knowledge.

It is important to show evidence that you have developed these skills. Your trainer or assessor will discuss with you how to record your employability skills. The following table provides some examples of things you might do to develop employability skills in this unit.

EMPLOYABILITY SKILLS

Employability skill	What this skill means	How you can develop this skill
Communication	Speaking clearly, listening, understanding, asking questions, reading, writing and using body language.	• Liaise with others when organising work activities. • Interpret plans and working drawings.
Teamwork	Working well with other people and helping them.	• Work as part of a team to prioritise tasks. • Offer advice and assistance to team members and also learn from them.
Planning and organising	Planning what you have to do. Planning how you will do it. Doing things on time.	• List tasks in a logical order of action. • Select appropriate tools and materials.
Initiative and enterprise	Thinking of new ways to do something. Making suggestions to improve work.	• Initiate improvements in using resources. • Respond constructively to workplace challenges.
Problem-solving	Working out how to fix a problem.	• Participate in job safety analyses. • Check and rectify faults in tools and equipment if appropriate.
Self-management	Looking at work you do and seeing how well you are going. Making goals for yourself at work.	• Maintain performance to workplace standards. • Set daily performance targets.
Technology	Having a range of computer skills. Using equipment correctly and safely.	• Apply computer skills to report on project work. • Follow manufacturers' instructions.
Learning	Learning new things and improving how you work.	• Use opportunities for self-improvement. • Trial new ideas and concepts.

Obtain measurements

Different methods are needed to obtain measurements for different tasks, so you need to be able to select the appropriate measuring and calculating equipment, and check it for serviceability (and rectify or report any faults). You also need to learn to obtain and record linear measurements accurate to 1 mm.

Select and apply method of obtaining measurements for task requirements

The type of task or work application often determines the method for measurements you need to use. This may also be determined by the type and nature of materials being used in the construction process. In some workplace situations, all measurements must be in line with specifications or standards relating to the job.

The instructions for a particular work task are usually given out by the supervisor. These instructions may be with guidance from site plans, project specifications, material management information, engineering specifications or other related workplace information. It is important that you understand the instructions you are given for the allocated tasks. If you are unsure, ask questions or have the task demonstrated so you are clear about what needs to be done.

There are many examples of where measurement techniques and methods may be required (see Table 3.1).

It is necessary to know the units of measure for the worksite. The general practice in construction is to use the metric system and give measurements for length, width, height or depth in millimetres (mm) and metres (m); for example, a 2400 mm length of timber, a 100 mm PVC pipe, or a hole dug to 750 mm deep.

All weights are given as kilograms (kg) or tonnes (t); for example, a 20 kg bag of cement or a 5 kg bag of lime. Litres (L) and millilitres (mL) are the measurements used for liquids such as water, paint, solvents and liquid chemicals.

Sometimes the tool that you use to measure materials will have directions for use from the manufacturer. In other cases, measure methods may be

TABLE 3.1 Requirements of measurement techniques and specific methods of calculation

What measurements are required for	Examples	How results are expressed
Length of materials	Timber, steel or piping are measured in linear (or straight-line) metres; these are usually measured with a standard tape measure.	The results are generally expressed in millimetres (mm) or metres (m).
Depth of holes or excavations	A measuring device such as a tape measure, a string or specific depth instrument is used to calculate how deep the hole is.	The results may be in millimetres (mm) or metres (m) for depth; or in cubic metres (m^3) if measuring the volume of the excavation.
Working out areas of worksite features	For marking out the area of a concrete slab, there is a width and length measurement; these may be required in millimetres (mm) or metres (m), depending on the task.	The results are generally expressed in square metres (m^2).
Material quantity or volume measures	These measurements have three dimensions: length, width and height or depth. For example, to calculate how much concrete is required for a slab, multiply length by width by depth.	These results are given in cubic metres (m^3).
Weight of materials	The amount of steelwork in a building (steel beams, columns or reinforcements) may be measured by weight. The area in square metres (m^2) or cubic metres (m^3) can be multiplied by a weight factor to give the total weight.	These results are given as kilograms (kg) or tonnes (t).

detailed in workplace procedures or instructions. Ask your supervisor if you are in doubt about what tool or method to use and the requirements for the work.

The method of obtaining measurements will largely be dependent on the timing of taking the measurements in relation to the progress of the construction tasks.

Measurements are initially required at the preconstruction phase of the project. Following this is the actual construction phase, which requires a closer level of accuracy in measurements.

The method used to take any measurements will be determined by the material that is to be used. The manner in which the material is supplied will determine the nature of the calculations that are required. An example of this is the concrete used in footings, which is measured in cubic metres, compared to Gyprock™ sheets used to line the internal walls of a building, which are calculated in square metres. This approach is appropriate at the preconstruction phase when the quantities of the materials are being calculated for the purposes of costing the construction works. However, when it comes to the actual ordering of the material, the Gyprock™ sheets will be ordered in certain sheet lengths to match what is physically in place.

A high level of accuracy is required when estimating and measuring on site, as any miscalculations can affect the timing and completion of a project, as well as its profitability. Underestimation or under-ordering will delay the progress of the project or works and may incur additional costs if workers have to return to finish the job. In the worst case, there may not be additional material available to complete the job. Overestimation is equally undesirable, if excess material is ordered that is not required for the project.

Therefore, it is important that the measurement and ordering of material for a project is accurate – the axiom of measuring twice and cutting once is particularly important in the construction industry.

Select equipment, check for serviceability and rectify or report faults

The pieces of equipment used in the pre-construction phase and the construction phase are very different, reflecting the different tasks required in each phase.

Pre-construction phase

In the pre-construction phase of small jobs, calculators and scale rulers are commonly used for measuring and calculating. However, when you are working from professionally developed drawings for larger projects, there are software programs that have preloaded information that will automatically calculate quantities and costs (see Figure 3.1). These programs include:
- Buildsoft: https://www.buildsoft.com.au
- Buildxact: https://www.buildxact.com.au
- Constructor: https://constructor.com.au

Construction phase

A large range of measuring instruments may be used during the construction phase. These instruments are designed to take specialised readings and include rulers, tape measures and laser levels. If instruments have no moving parts, their accuracy can be relied upon, but when moving parts are a part of the instrument it is

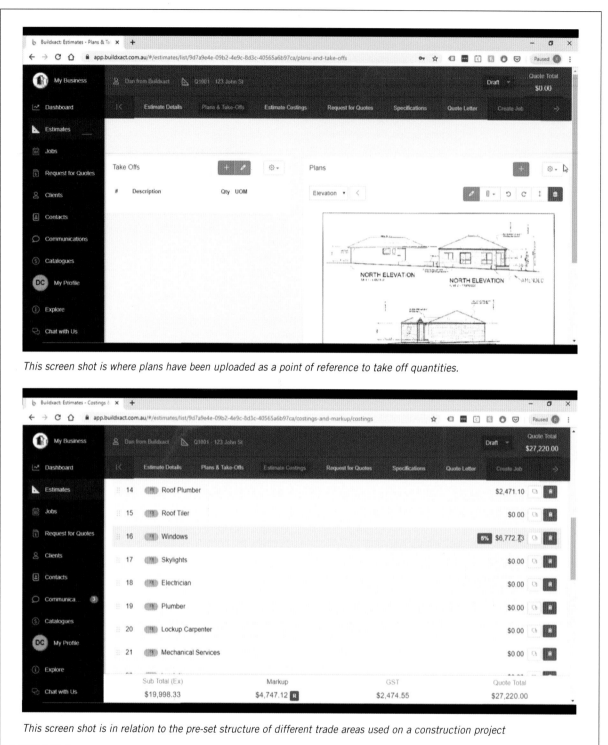

This screen shot is where plans have been uploaded as a point of reference to take off quantities.

This screen shot is in relation to the pre-set structure of different trade areas used on a construction project

FIGURE 3.1 Online estimating programs

essential that the equipment is serviced on a regular basis. This relates mostly to instruments such as laser levels and automatic levels.

Once you have selected the correct equipment, you should check it to ensure it is functional for the task. Check for faults, damage or other factors that could make the equipment unusable. This is especially true with laser and survey equipment. To check for faults or damage, you must follow the equipment instructions or user manuals. These manuals should be kept with the equipment.

Calibrating (checking or adjusting) the equipment used for measuring may need to be done with every use. Instructions to calibrate the equipment vary

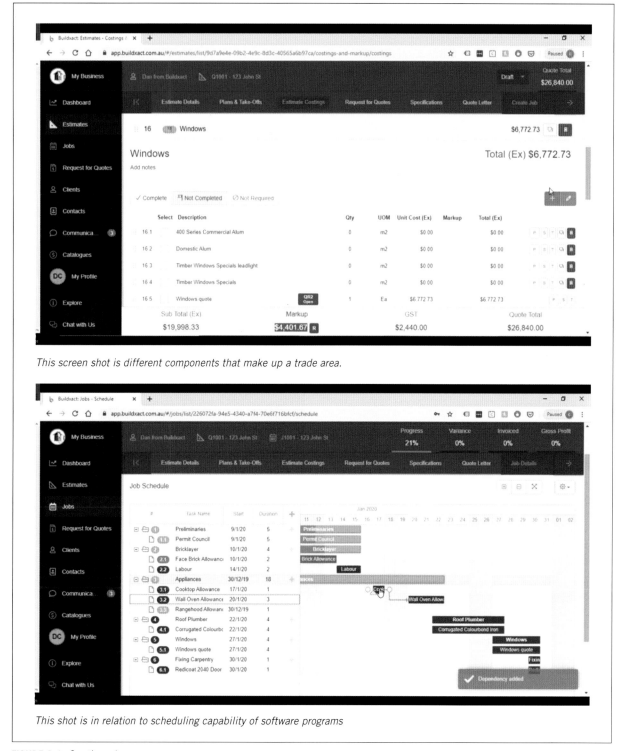

This screen shot is different components that make up a trade area.

This shot is in relation to scheduling capability of software programs

FIGURE 3.1 *Continued*

between brands and types of equipment. Read your instruction or user manual to identify the calibration process.

Checking simple pieces of equipment such as rulers, building squares, tape measures or measuring wheels may seem unnecessary, but always check for parts that are warped, bent, twisted, stretched, splintered or damaged. Even small measurement errors can affect the safety and quality of your work.

If you identify any faults in the equipment, report them to your site supervisor or other designated person. Only fix faulty equipment if you have received the appropriate training or have the relevant qualifications, and always check with your supervisor if you are

CARRY OUT MEASUREMENTS AND CALCULATIONS 55

unsure about what to do. For example, you may be able to rectify a fault when recalibrating laser equipment, but not be able to fix a fault with measuring equipment.

Often it is better to write off the equipment and replace it to ensure measurements are accurate.

Obtain and record linear measurements accurate to 1 mm

To obtain correct measurements it is important to select the appropriate tool for the task. By selecting the most appropriate tool, the chances of incorrectly measuring are greatly reduced. It is also important to check the tool to be used to make sure that it functions properly. Measuring tools that are damaged or broken may result in incorrect measures, which in turn will result in the wrong quantity of materials being ordered.

Measuring equipment

Measuring tools that are used in the construction industry include the following:
- scale rule
- four-fold rule
- retractable metal tape
- long tape measure
- trundle wheel
- laser distance-measuring devices and range finders.

Scale rule

The scale rule is a plastic rule 150 mm or 300 mm in length that is used to scale off dimensions not given on a drawing (see Figure 3.3). There is a range of different scales on the different faces and sides of the rule.

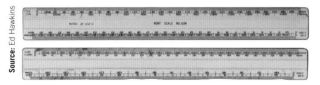

FIGURE 3.2 Typical scale rule

Four-fold rule

The four-fold rule is 1 m long and is made in four hinged sections so that it can be folded for convenience (see Figure 3.3). This rule is best used for measuring and marking out dimensions of less than 1 m. If the rule is used to measure or mark out a dimension greater than 1 m, an error is likely to compound each time the rule is moved.

Depending on whether the rule is made from boxwood or plastic, the rule blades are approximately 4 mm or 5 mm thick. Because of this thickness, parallax error (discussed later in this section) is likely to occur when measuring or marking out with the rule laid flat on a surface. To overcome this problem, the rule should be used on its edge so that the graduations marked on

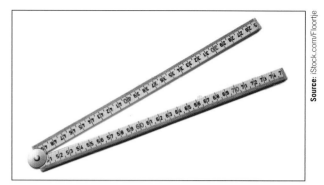

FIGURE 3.3 Standard four-fold rule

the blades of the rule are in contact with the surface of the material being measured or marked out.

Retractable metal tape

The retractable metal tape (see Figure 3.4) is available in lengths from 1 m up to 10 m and is most suitable for measuring and/or marking out dimensions over 1 m and up to 10 m. Metal retractable tapes have a hook on the end, which adjusts depending on whether the measurement to be taken or marked out is internal or external.

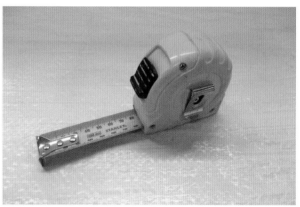

FIGURE 3.4 Typical retractable metal blade tape measure

The hook slides a distance equal to its own thickness so that an internal measurement (for example, between two walls) begins from the outside of the hook, while an external measurement (for example, from the end of a piece of timber) begins from the inside of the hook. Care must be taken when returning the blade to the case so that the hook does not slam against the case, which will stretch or distort the hook, making the tape inaccurate. This can also tear the hook from the end of the tape, making it useless and needing replacement.

Long tape measures

For measuring distances greater than 10 m there are long tape measures available. They are constructed from either steel or fibreglass and are manually wound back into a closed case or open carriage (see Figure 3.5). These tape measures range in length from 30 m up to 100 m and are used on construction sites where

FIGURE 3.5 Open-reel and closed-case long tape measures

measurement of longer distances is required. These tape measures generally have a specialist hook on the end so that they may be hooked onto a nail for simple one-person operation.

Trundle wheel

A trundle wheel is simply a lightweight wheel constructed from timber or plastic and has a circumference of exactly 1 m (see Figure 3.6). It has a handle attached to allow a person to walk it along easily and a counter so that every rotation (1 m) is counted. These tools are not accurate and are mainly used to obtain a quick estimate of distances over 100 m up to several kilometres.

FIGURE 3.6 A trundle wheel

Laser distance-measuring devices and range finders

Laser distance-measuring devices and range finders can be used to accurately measure distances up to 30 m or 40 m, with an error margin of around 3 mm (see Figure 3.7). They are simply placed against the object or surface you wish to measure from and pointed at the object you wish to measure to.

FIGURE 3.7 A laser distance-measuring device

Technological change for these devices moves at a fast pace and like all electronic items they need to be looked after as moisture and rough handling will damage them.

Parallax error

Parallax error is a function of the way that measurements are read. When reading a tape measure it needs to be in alignment of sight perpendicular to the material being measured. If you read a tape measure on an angle, the length you see may be longer or shorter than the actual length. In the example in Figure 3.8, the true length is 1.002 m, but if you look at the measurement from an angle on either side, it could be incorrectly read as 0.907 m or 1.007 m.

 When using any tool that has moving components, care must be taken by the operator. The blade of a tape measure can retract at a high speed and the operator must take care not to be cut.

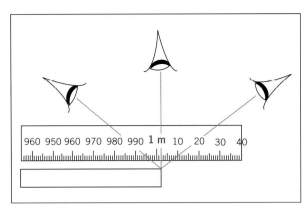

FIGURE 3.8 Parallax error when reading a tape measure incorrectly

LEARNING TASK 3.1
USING MEASURING EQUIPMENT

Select a room. In groups of two, take sufficient measurements of the floor and record the measurements appropriately to calculate the area of the floor in square metres.

Next, measure and calculate the area of each wall. Ensure that you deduct any openings in each wall such as doors and windows.

As a class, compare each group's results and identify any discrepancies. If there are discrepancies, break down the measurements of each area to find the points of difference. This may require comparing the lineal metres that were initially measured.

COMPLETE WORKSHEET 1

Perform basic calculations

The calculation of material will be determined by the type, shape and application of the material. Material can be supplied in lineal metres, square metres and cubic metres, which are all standard units of measurement. Another unit that is used in construction is 'number of'; that is, the number of pieces.

In some situations multiple calculations will be needed. For example, if you are calculating volume for a footing, several calculations will be required to acquire all the measurements needed to undertake the final calculation of volume for the job. However, the first step is to have a basic understanding of the different types of measurements.

Types of measurements

There are certain types and dimensions of measurement that you need to be familiar with. These include:
- length
- area
- volume
- mass
- perimeter
- circumference.

Length

Measuring length is the most common task that you will undertake as a tradesperson. Generally, in the construction industry length is measured in millimetres (mm). Of course, this is not used on all occasions, especially when there are longer lengths required such as the boundary of a building site. Where longer measurements are required they are expressed in metres (m).

The measurements of all dimensions of a material to be used are required, so it is vitally important that you have the capability to take a single measurement with accuracy and an understanding of the dimensions required for a specific task.

In some circumstances, the determination of lineal metres can require additional calculations beyond just adding lengths together. An example of this is when circular work is required for formwork to prepare for a concrete path. In this case you would also need to be able to determine the circumference of a circle (discussed later in this section).

Area

Area is expressed in square metres (m^2). When the area of any item is required for a construction task, it is vital that the correct dimensions are recorded so that the appropriate calculation can take place. For example, if you need to determine the floor area of a room that is rectangular in shape, then you will require only the length and the width (see Figure 3.9). This kind of shape is referred to as a regular shape.

If the room is an 'L' shape, then you will need to obtain more measurements to be able to calculate the total floor area (see Figure 3.10). An L shape is an example of an irregular shape. Where there is an irregular shape that needs to be measured, the approach is to 'cut up' the irregular shape into regular shapes. In this case, it means splitting the smaller section off so you're left with two regular rectangles, then calculate the area of each and add the amounts together to come up with a final overall amount.

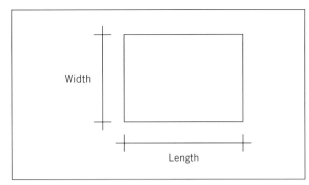

FIGURE 3.9 Dimensions required to calculate the area of a rectangle

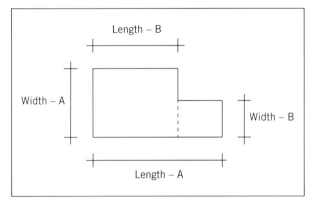

FIGURE 3.10 Dimensions required to calculate the area of an L-shaped area

Volume

Volume is expressed in cubic metres (m³), which means that there are three dimensions that need to be measured to perform the calculation and determine the quantity (see Figure 3.11).

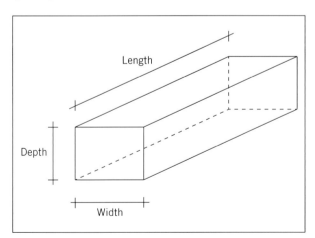

FIGURE 3.11 Dimensions required to calculate volume

The base formula is Length × Width × Depth (or L × W × D). In the construction industry this formula is used to calculate materials such as concrete and also spoil from excavations. In some situations, the use of volume is also a part of the calculation process for mass.

Again, there may be situations where the shapes to be calculated are irregular and need to be broken down into regular shapes. The typical approach in these situations is to find the area of the end section and then multiply this by the length.

Mass

Mass is more commonly known as the weight of a particular item. The calculation for mass is determined by the volume (usually expressed in m³), which is then multiplied by a factor that is applicable to the particular material. For example, the mass of a cubic metre of water is different from the mass of a cubic metre of concrete: 1 m³ water is 1000 kg whereas 1 m³ concrete is 2500 kg (see Figure 3.12). The mass of material is important to know in the construction industry as the weight needs to be considered in transporting, lifting and placing material, and the weight of the material needs to be supported once it is in place.

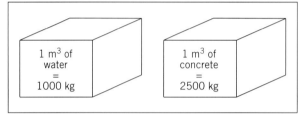

FIGURE 3.12 Comparison of volume and mass of different materials

Perimeter and circumference

The perimeter is the length around the outside of a shape. This type of information is gathered to calculate, for example, the length of a fence that follows the complete boundary of a block of land.

The circumference is similar to a perimeter, except that it expresses the distance or length around a circle. There are many circumstances when such calculations are required on a construction site. For example, it is fashionable to have curved staircases and these will not only require calculations for the structure of the staircase, but also for the internal framework and linings that support the staircase.

Useful formulas

There are several standard formulas that are worth committing to memory for future reference. Table 3.2 lists common shapes that you will come across in the construction industry and the formulas used to calculate their area or perimeter. Table 3.3 lists the formulas used to calculate volume.

Types of calculation

There are certain types of calculations that you must be familiar with. The four basic actions are addition, subtraction, division and multiplication. These can be further reduced to two groups of opposites: addition is

TABLE 3.2 Formulas commonly used in the construction industry for area and perimeter calculations

	Area	Perimeter
Rectangles	$A = a \times b$	$P = 2(a + b)$
Squares	$A = a \times a$	$P = 4a$
Circles	$A = \pi r^2$	$C = \pi d$
Triangles	$A = \frac{1}{2} bh$	$P = a + b + c$
Trapeziums	$A = \frac{a + b}{2} \times h$	$P = a + b + c + d$

Source: Based on *Illustrated Maths Dictionary*, 3rd Edition, Judith de Klerk, 1999.

TABLE 3.3 Formulas commonly used in the construction industry for volume calculations

	Volume
Cubes	$V = a \times a \times a$ or $V = a^3$
Cylinders	$V = \pi r^2 h$

Source: Based on *Illustrated Maths Dictionary*, 3rd Edition, Judith de Klerk, 1999.

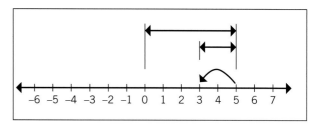

FIGURE 3.14 A number line demonstrating subtraction

the opposite of subtraction, and division is the opposite of multiplication.

Addition
Addition is the mathematical operation of joining numbers. This action is indicated by a plus symbol ('+'). Another form of expressing addition is using a number line (see Figure 3.13). Addition is constantly used in the construction industry both on and off the job site.

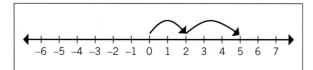

FIGURE 3.13 A number line demonstrating addition

Subtraction
Subtraction is the mathematical operation taking something away and finding what is left. This action is indicated by a minus symbol ('−'). Figure 3.14 is an expression of subtraction on a number line. This form of calculation is as commonly used in the construction industry as addition.

Division
Division is the mathematical operation of taking several items and grouping them into the number of groups required. This action is indicated by a division symbol ('÷'). This function is commonly used in the construction industry to work out the number of items required. For example, if you have 15 bricks and you need to split them into three groups, then the calculation is 15 ÷ 3 = 5 bricks in each group (see Figure 3.15).

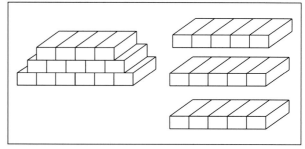

FIGURE 3.15 A diagrammatical demonstration of division

Another example is when you have a wall 4.700 m in length that requires studs spaced at 0.600 m intervals.

How many studs are required? The calculation is 4.700 ÷ 0.600 = 7.833, which rounds up to eight spaces. Then you add one more stud for the end, which means nine studs are required (see Figure 3.16).

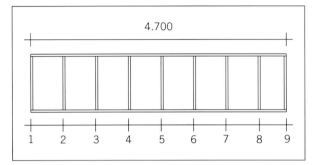

FIGURE 3.16 An example of division applied to a wall frame

Multiplication

Multiplication is the mathematical operation of adding together a number of equal groups. This action is indicated by a multiplication symbol ('×'). For example, if you had four boxes that each contained 12 tiles it would be 4 × 12 = 48 tiles (see Figure 3.17).

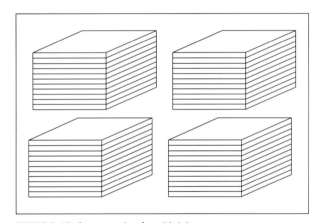

FIGURE 3.17 An example of multiplying

Percentages

A percentage is a part of a certain amount – a ratio expressed as a fraction of 100. In the construction industry, percentages are commonly used when calculating wastage. For example, concrete poured directly into a footing in the ground will have a higher waste percentage than concrete being poured into a building element that is completely contained by formwork. Percentages are also used to work out profit margins and to make allowances for additional material or amounts required.

Commonly, to add 10% to a figure, you would multiply it by 1.1, and to deduct 10% you would multiply it by 0.9. However, there are different methods for calculating percentages depending on the functions of the calculator you are using.

Conversions

Conversion is where items are expressed differently. This is applicable when amounts increase to higher quantities, such as converting kilograms to tonnes. For example, trucks have a tonne load limit. If you're loading a truck with 20 kg bags of cement you need to be able to identify the total amount you can load in tonnes before you exceed the limit.

The most common conversion that takes place in the construction industry is from millimetres to metres (see Table 3.4).

Conversions also may be expressed as different units of measure for length, area, mass and liquid volume (see Table 3.5).

TABLE 3.4 Converting millimetres to metres

100s of millimetres in metres	10s of millimetres in metres	Individual millimetres in metres
1000 mm = 1.0 m	90 mm = 0.09 m	9 mm = 0.009 m
900 mm = 0.9 m	80 mm = 0.08 m	8 mm = 0.008 m
800 mm = 0.8 m	70 mm = 0.07 m	7 mm = 0.007 m
700 mm = 0.7 m	60 mm = 0.06 m	6 mm = 0.006 m
600 mm = 0.6 m	50 mm = 0.05 m	5 mm = 0.005 m
500 mm = 0.5 m	45 mm = 0.045 m	4 mm = 0.004 m
400 mm = 0.4 m	40 mm = 0.04 m	3 mm = 0.003 m
300 mm = 0.3 m	35 mm = 0.035 m	2 mm = 0.002 m
200 mm = 0.2 m	30 mm = 0.03 m	1 mm = 0.001 m
100 mm = 0.1 m	25 mm = 0.025 m	0.5 mm = 0.0005 m

TABLE 3.5 Units of measure

Length			
10 millimetres	mm	= 10 centimetres	cm
10 centimetres	cm	= 1 metre	M
1000 millimetres	mm	= 1 metre	m
1000 metres	m	= 1 kilometre	km
Area			
100 square millimetres	mm²	= 1 square centimetre	cm²
10 000 square centimetres	cm²	= 1 square metre	m²
10 000 square metres	m²	= 1 hectare	ha
100 hectares	ha	= 1 square kilometre	km²
1 square kilometre	km²	= 1 000 000 square metres	m²
Mass			
1000 milligrams	mg	= 1 gram	g
1000 grams	g	= 1 kilogram	kg
1000 kilograms	kg	= 1 tonne	t
Liquid volume			
1000 millilitres	mL	= 1 litre	l
1000 litres	L	= 1 kilolitre	kl
1 mL (for liquids)		= 1 cm³ (for solids)	
1 kL (for liquids)		= 1 m³ (for solids)	

Source: Based on *Illustrated Maths Dictionary*, 3rd Edition, Judith de Klerk, 1999.

Ratios

Ratios are used in the construction industry when different materials are mixed together. An example of this is mortar, which has elements mixed in different portions to make up the desired strength of the mortar. For example, cement mortar consists of one part cement to one part hydrated lime (or lime putty) to six parts sand. This can be expressed as a ratio of 1:1:6.

Ratios are also applicable to other scenarios in the construction industry, such as when using solvents and resins.

Pythagoras' theorem

Pythagoras' theorem is an important element of trigonometry (discussed in the next section) that is applicable where there is a right-angled triangle. The theorem states that the square of the hypotenuse is equal to the sum of the squares of the sides.

This concept is used in building when setting out the position of walls on a concrete slab or setting out the footings that need to be excavated. Pythagoras' theorem, which is commonly known as the 3-4-5 method (see Figure 3.18 and Figure 3.19), can also be used when a tradesperson is checking existing structures to confirm that the elements are square.

Trigonometry

Trigonometry is used where you need to find the length of the longest side of a triangle and you have only been

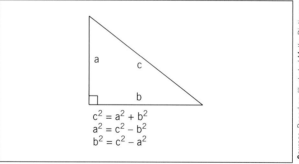

FIGURE 3.18 Right-angle triangle

provided with one length of the triangle and the angle for one point of the triangle. This situation occurs in the construction industry for situations such as determining the length of roof rafters (see Example 3.1). To undertake calculations for trigonometry you will need to use a scientific calculator that has the appropriate functions of sin, cos and tan.

If one angle of a triangle is 90 degrees (a right angle) and one of the other angles is known, then you can work out the angle of the third, because the angles of any triangle add up to 180 degrees.

This means that once one of the other angles is known, the ratios of the other sides are always the same, no matter the overall size of the triangle.

These ratios – sin, cos and tan – are the trigonometric functions of the known angle A, where a,

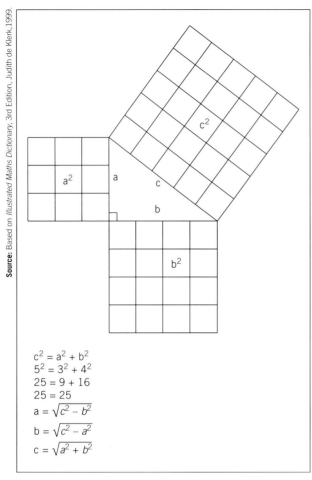

$c^2 = a^2 + b^2$
$5^2 = 3^2 + 4^2$
$25 = 9 + 16$
$25 = 25$
$a = \sqrt{c^2 - b^2}$
$b = \sqrt{c^2 - a^2}$
$c = \sqrt{a^2 + b^2}$

FIGURE 3.19 Right-angle triangle with expanded view of Pythagoras' theorem

b and c refer to the lengths of the sides of the triangle (see Figure 3.20).

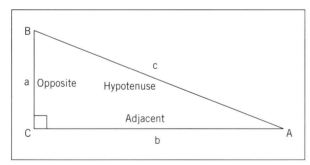

FIGURE 3.20 Right-angle triangle identifying elements for trigonometry calculations

The *sine* function (sin) is the ratio of the side opposite the angle to the hypotenuse:

$$\sin A = \frac{Opposite}{Hypotenuse} = \frac{a}{c}$$

The *cosine* function (cos) is the ratio of the adjacent leg to the hypotenuse:

$$\cos A = \frac{Adjacent}{Hypotenuse} = \frac{b}{c}$$

The *tangent* function (tan) is the ratio of the opposite leg to the adjacent leg:

$$\tan A = \frac{Opposite}{Adjacent} = \frac{a}{b} = \frac{\sin A}{\cos A}$$

The *hypotenuse* (c) is the side opposite to the 90-degree angle in a right triangle; it is the longest side of the triangle, and is adjacent to angle A. The *adjacent leg* (b) is the other side that is adjacent to angle A. The *opposite side* (a) is the side that is opposite to angle A.

EXAMPLE 3.1

USING TRIGONOMETRY

You can use trigonometry to work out the length of the angled section of a rafter if you know the length of the adjacent side (b) and one of the acute angles.

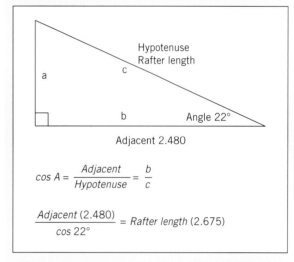

$$\cos A = \frac{Adjacent}{Hypotenuse} = \frac{b}{c}$$

$$\frac{Adjacent\ (2.480)}{\cos 22°} = Rafter\ length\ (2.675)$$

Therefore, the length of the rafter is 2.675 m.

Calculate, confirm and record areas and volumes

The calculation of quantities is used for several different purposes in the construction industry, including the amount of spoil from an excavation, the amount of concrete required for a pour, the number of bricks needed for a wall and the number of tiles needed for a floor. These are not a straightforward processes, and are dependent on the type of material that will be used, and the way it will be used in the construction process.

Excavation

In calculating the volume of spoil to be excavated, the centre line length of the trench is required (see Figure 3.21). In addition, if you do not 'cut' each length off at each external corner then you risk doubling up on these areas when calculating the quantity. As indicated in Figure 3.22, a line is drawn across the diagram to indicate that the triangular area has been flipped from one section to the other, and this will need to be reflected in the calculations.

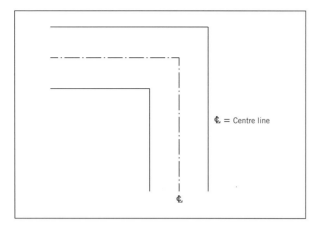

FIGURE 3.21 Centre line length of trench

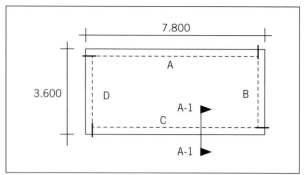

FIGURE 3.23 Diagram of plan with footing elements labelled for easy identification

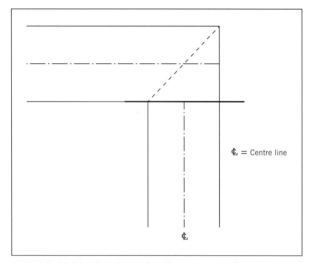

FIGURE 3.22 Line to indicate length to external edge of footing

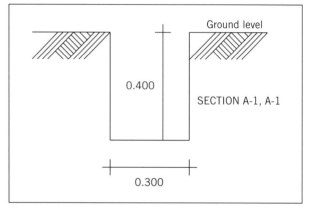

FIGURE 3.24 Cross-sectional view of footing system

By labelling the different footings you have a way of identifying the elements that have already been included in the calculation (see Figure 3.23). Arrows are also used to show the direction in which you are working – in this case, from the top left-hand corner in a clockwise direction. The use of letters is preferred so that there is no confusion with numerical measurements. The remaining dimensions to complete the required calculations, such as depth of the footings, can be obtained from a cross-section view of the footing system (see Figure 3.24).

Concrete

To calculate the quantity of concrete for the area illustrated in Figure 3.25 and Figure 3.26 involves several steps that must be undertaken in a methodical manner.

Brickwork

Calculations for a skin of brickwork around the perimeter of the floor plan requires not only the dimension of the floor plan (see Figure 3.27), but also the thickness and

EXAMPLE 3.2

CALCULATING THE VOLUME OF SPOIL

Refer to Figure 3.23 for these calculations.

Step 1 Calculate the length of the edge beam.

Length − width of footing = centre line length of footing

A → 7.800 − 0.300 = 7.500
B ↓ 3.600 − 0.300 = 3.300
C ← 7.800 − 0.300 = 7.500
D ↑ 3.600 − 0.300 = 3.300
Total lineal metres = 21.600

Step 2 Calculate the volume.

Volume = Length × Width × Depth
= 21.600 × 0.300 × 0.400
= 2.592 m³

height of the brickwork – in this case, 110 mm thick and 2.400 m high. This calculation in Example 3.4 goes a step further to provide the number of bricks required, but this is only possible when there is a known square metre rate for the bricks – in this case, 50 per square metre.

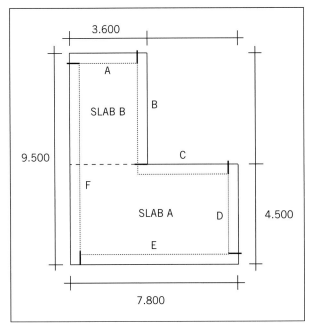

FIGURE 3.25 Floor plan of typical concrete floor slab

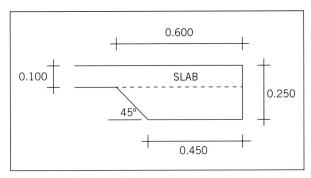

FIGURE 3.26 A section through the concrete floor slab and edge beam

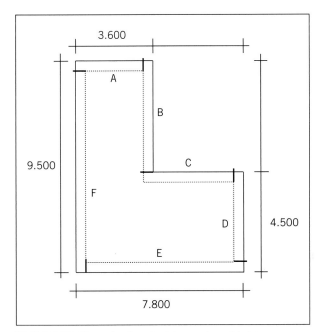

FIGURE 3.27 Floor plan to calculate number of bricks

EXAMPLE 3.3

CALCULATING CONCRETE QUANTITIES

Refer to Figure 3.25 and Figure 3.26 for these calculations.

Step 1 Break the irregular shape up into regular shapes and label the slabs.

Step 2 Calculate volume of each section.

$$\text{Area A} = \text{Length} \times \text{Width} \times \text{Depth}$$
$$= 7.800 \times 4.500 \times 0.100$$
$$= 3.510 \text{ m}^3$$
$$\text{Area B} = \text{Length} \times \text{Width} \times \text{Depth}$$
$$= 5.000 \times 3.600 \times 0.100$$
$$= 1.800 \text{ m}^3$$
$$\text{Area A} + \text{Area B} = \text{m}^3$$
$$3.510 \text{ m}^3 + 1.800 \text{ m}^3 = 5.310 \text{ m}^3$$

Step 3 Calculate average width of edge beam.

$$\frac{\text{Average}}{\text{width}} = \frac{\text{Large measurement} + \text{small measurement}}{2}$$
$$= \frac{0.600 + 0.450}{2}$$
$$= 0.525 \text{ m}$$

Step 4 Calculate the length of the edge beam.

Length − width of footing = centre line length of footing

A → 3.600 − 0.525 = 3.075
B ↓ 5.000 − 0 = 5.000
C → 4.200 − 0 = 4.200
D ↓ 4.500 − 0.525 = 3.975
E ← 7.800 − 0.525 = 7.275
F ↑ 9.500 − 0.525 = 8.975
Total lineal metres = 32.500

Step 5 Calculate the volume.

$$\text{Volume} = \text{Length} \times \text{Width} \times \text{Depth}$$
$$= 32.500 \times 0.525 \times 0.150$$
$$= 2.559 \text{ m}^3$$

Step 6 Calculate total of all concrete required.

Slab + edge beams
5.310 m³ + 2.559 m³ = 7.869 m³

Tiling

Calculating the number of floor tiles to cover a concrete slab that is not a regular shape (see Figure 3.28) will require breaking down the shape into regular shapes and then calculating the area in square metres. In Example 3.5, there is an additional step to allow for 10% per square metre for cuts and breakage. To work out the number of tiles required, the size of each tile is required – in this case 400 mm × 400 mm, which equals five tiles per 0.8 m².

EXAMPLE 3.4

CALCULATING THE NUMBER OF BRICKS

Refer to Figure 3.27 for these calculations.

Step 1 Calculate the face area of the wall in square metres.

Area = Length × Height

A → 3.600 − 0.110 = 3.490
B ↓ 5.000 − 0 = 5.000
C → 4.200 − 0 = 4.200
D ↓ 4.500 − 0.110 = 4.390
E ← 7.800 − 0.110 = 7.690
F ↑ 9.500 − 0.110 = 9.390
Total lineal metres = 34.160
Length × Height = Area
34.160 × 2.400 = 81.984 m^2

Step 2 Multiply the result by the square metre rate.

Area × 50 per m^2
81.984 × 50 = 4099 bricks

This answer gives the number of bricks required for one skin of brickwork. If there are two skins of brickwork, the answer will have to be multiplied by 2.

Further calculations will be needed for deductions due to any openings, such as doors and windows. In addition to this, an allowance will need to be made for wastage.

EXAMPLE 3.5

CALCULATING THE NUMBER OF FLOOR TILES

Refer to Figure 3.28 for these calculations.

Step 1
Break the irregular shape up into regular shapes and label for the slabs (Slab A and Slab B).

Step 2
Calculate area of each section.

Area = Length × Width
Slab A = 7.800 × 4.500
    = 35.100 m^2
Slab B = 5.000 × 3.600
    = 18.000 m^2

Slab A + Slab B = total m^2
35.100 m^2 + 18.000 m^2 = 53.100 m^2

Step 3
Allow 10% for wastage.
53.100 m^2 × 1.1 = 58.410 m^2

Step 3
Calculate number of tiles
Tiles 400 mm × 400 mm = 5 tiles per 0.8 m^2
58.410 m^2 ÷ 0.8 m^2 = 73.013
73.013 × 5 tiles = 365.065

Rounded up = 366 tiles.

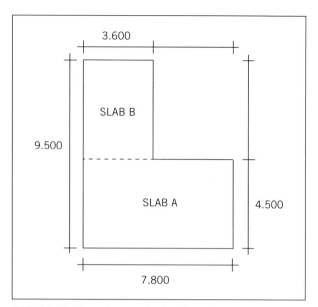

FIGURE 3.28 Floor plan to calculate floor tiles

LEARNING TASK 3.2 CALCULATION ACTIVITY

You are required to calculate the amount of concrete required for the floor slab and footing beam in Figure 3.29. Allow 10% for wastage.

Show all steps required to complete the task.

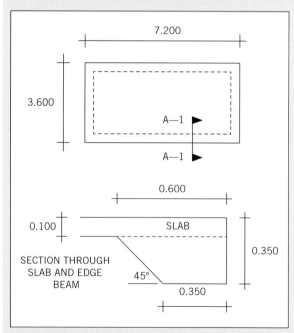

FIGURE 3.29 Floor plan of concrete floor slab with section through slab and edge beam

 COMPLETE WORKSHEET 3

 COMPLETE WORKSHEET 4

SUMMARY

In Chapter 3 you have learnt about the tools used to take measurements and the basic measurements that are commonly used in the construction industry. In addition, the chapter has:
- identified the different measuring tools used in the construction industry
- reviewed the standard formulas used in trade calculations
- explored the different calculations that are used to determine the quantities of material
- identified different approaches that are required to perform calculations
- applied simple and complex calculations for a range of materials and construction elements

REFERENCES AND FURTHER READING

de Klerk, J. 1999, *Illustrated Maths Dictionary*, 3rd Edition, Pearson Education Australia.

Department of Training and Workforce Development 2016, Government of Western Australia, *Carry Out Measurements and Calculations, Certificate II in Building and Construction, (Pathway – Trades) CPCCCM1015A, Learners Guide*.

GET IT RIGHT

The photo below shows a person using a tape measure to determine the required size of the opening. Identify the incorrect practices and how this could affect the accuracy of the final measurement.

WORKSHEET 1

To be completed by teachers
Student competent ☐
Student not yet competent ☐

Student name: _____

Enrolment year: _____

Class code: _____

Competency name/Number: _____

Task
Read through the sections from the start of the chapter up to 'Perform basic calculations', then complete the following questions.

1 Identify the two distinct phases during which measurements are performed in a construction project.

   _____
   _____

2 What impacts may occur if inaccurate measurements and calculations are performed?

   _____
   _____

3 If equipment has been identified as faulty, who should you advise of the equipment's status?

   _____
   _____

4 Identify at least two tools in each construction phase that are used to take measurements or perform calculations.

   _____
   _____
   _____
   _____

WORKSHEET 2

To be completed by teachers
Student competent ☐
Student not yet competent ☐

Student name: _____

Enrolment year: _____

Class code: _____

Competency name/Number: _____

Task

Read through the sections from the start of 'Perform basic calculations' up to 'Ratios', then complete the following table.

Convert from millimetres to metres			Convert from metres to millimetres		
86 mm	=		45 m	=	
12300 mm	=		2.450 m	=	
345 mm	=		0.089 m	=	
89 mm	=		0.09 m	=	
10 mm	=		1.200 m	=	
458 mm	=		3.450 m	=	
2400 mm	=		28.500 m	=	
820 mm	=		0.435 m	=	
2040 mm	=		28.500 m	=	
45 mm	=		0.008 m	=	
240 mm	=		2.111 m	=	

WORKSHEET 3

To be completed by teachers

Student competent ☐

Student not yet competent ☐

Student name: _____

Enrolment year: _____

Class code: _____

Competency name/Number: _____

Task

Read through the sections from the start of 'Ratios' to the end of the chapter, then complete the following task.

You are required to calculate the number of tiles required to cover the floor plan below. The floor tiles are 500 mm × 500 mm, which equates to four tiles per square metre. A 12.5% allowance needs to be made for cuts and breakages. You are required to clearly set out and show all working out for the calculations required.

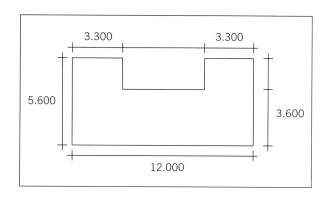

WORKSHEET 4

To be completed by teachers
Student competent ☐
Student not yet competent ☐

Student name: _____

Enrolment year: _____

Class code: _____

Competency name/Number: _____

Task

Read through the sections from the start of 'Ratios' up to the end of the chapter, then complete the following task.

Calculate the amount of spoil that is to be excavated from the floor plan and section below.

You are also required to calculate the amount of concrete needed for the footings, given that the concrete will be 0.400 m in depth. Allow 10% for spillage and unevenness of the excavation.

Show all steps required to complete the task.

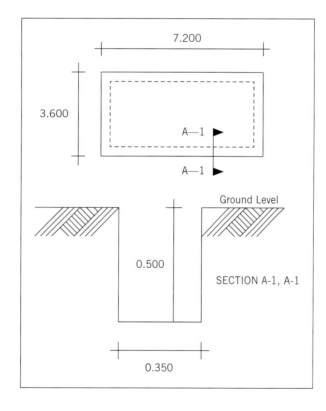

CARRY OUT MEASUREMENTS AND CALCULATIONS

EMPLOYABILITY SKILLS

Using the following table, describe the activities you have undertaken that demonstrate how you developed employability skills as you worked through this unit. Keep copies of material you have prepared as further evidence of your skills.

Employability skills	The activities undertaken to develop the employability skill
Communication	
Teamwork	
Planning and organising	
Initiative and enterprise	
Problem-solving	
Self-management	
Technology	
Learning	

PREPARE TO WORK SAFELY IN THE CONSTRUCTION INDUSTRY

This chapter covers the outcomes required by the unit of competency 'Prepare to work safely in the construction industry'. These outcomes are:
- Identify health and safety legislative requirements of construction work.
- Identify construction hazards and risk-control measures.
- Identify health and safety communication and reporting processes.
- Identify incident and emergency response procedures.

Overview

The term 'work health and safety' (WHS) – also referred to as occupational health and safety (OHS) – is used to describe a broad range of workplace practices covered under federal and state laws. These aim to improve the standards of workplace health and safety. The ultimate goal of these practices is to reduce the number of work-related injuries and deaths, and to bring about a healthier and safer working environment for everyone.

Building construction workers must be familiar with these workplace practices and must understand their responsibilities, whether they are workers, employers or persons conducting a business or undertaking (PCBUs).

Employers value people who fit into their workplace, solve day-to-day problems, manage their time and are keen to continue learning. These types of skills are known as employability skills. The employability skills you develop while working through this unit will be assessed at the same time as the skills and knowledge.

It is important to show evidence that you have developed these skills. Your trainer or assessor will discuss with you how to record your employability skills. The following table provides some examples of things you might do to develop employability skills in this unit.

EMPLOYABILITY SKILLS

Employability skill	What this skill means	How you can develop this skill
Communication	Speaking clearly, listening, understanding, asking questions, reading, writing and using body language.	• Liaise with others when organising work activities. • Interpret plans and working drawings.
Teamwork	Working well with other people and helping them.	• Work as part of a team to prioritise tasks. • Offer advice and assistance to team members and also learn from them.
Planning and organising	Planning what you have to do. Planning how you will do it. Doing things on time.	• List tasks in a logical order of action. • Select appropriate tools and materials.
Initiative and enterprise	Thinking of new ways to do something. Making suggestions to improve work.	• Initiate improvements in using resources. • Respond constructively to workplace challenges.
Problem-solving	Working out how to fix a problem.	• Participate in job safety analyses. • Check and rectify faults in tools and equipment if appropriate.
Self-management	Looking at work you do and seeing how well you are going. Making goals for yourself at work.	• Maintain performance to workplace standards. • Set daily performance targets.
Technology	Having a range of computer skills. Using equipment correctly and safely.	• Apply computer skills to report on project work. • Follow manufacturers' instructions.
Learning	Learning new things and improving how you work.	• Use opportunities for self-improvement. • Trial new ideas and concepts.

Health and safety legislative requirements of construction work

To fully understand any piece of legislation, it is important to appreciate the origins and reasons for the law first being introduced. There were five main reasons for the development of WHS laws and these are outlined in Table 4.1.

In Australia, prior to the implementation of the model *Work Health and Safety Act 2011*, each state and territory had separate responsibility for making and enforcing laws concerning WHS.

- ACT introduced the *Occupational Health and Safety Act* in 1989. The current Act is the *Work Health and Safety Act 2011*.
- New South Wales proclaimed the *Occupational Health and Safety Act* in 1983. It was enacted following the Williams Inquiry into health and safety practices in the workplace. In 1987, major changes were made to that Act. In 2001, that legislation gave way to the *Occupational Health and Safety Act 2000*,

TABLE 4.1 Five reasons for the development of WHS Laws

Self-regulation not working	Allowing organisations to regulate their own WHS programs was not working; at one stage, over 500 people nationally were dying each year due to work-related accidents, injuries and diseases.
National and overseas legal developments	There were efforts internationally as well as nationally to produce and then update existing WHS legislation to bring it into line with the 20th-century working environment.
Many workers not covered	When WHS legislation began, only approximately one-third of the workforce was covered. Today all workers are covered.
Too much legislation	Although only one-third of the workforce was covered by any WHS legislation at all, there were up to 26 different Acts in one state alone relating to occupational/work health and safety. Enforcement procedures for these Acts were a legal nightmare.

Cost of compensation	Every year billions of dollars were spent on workplace compensation and rehabilitation for those who had been injured at work. WHS legislation was introduced in conjunction with workers' compensation to provide for injured workers who were unable to work.

No 40. Later, New South Wales enacted the *Work Health and Safety Act 2011*.

- Northern Territory first introduced legislation in 1989. The current Act is the *Work Health and Safety (National Uniform Legislation) Act 2011* (in force 1 May 2016).
- Queensland first introduced legislation in 1989. The current Act is the *Work Health and Safety Act 2011*.
- South Australia was the first state to introduce legislation. In 1972, it introduced the *Industrial Safety and Welfare Act*. In 1986, it enacted the *Occupational Health, Safety and Welfare Act*, following the general form of Victorian state legislation. Finally, on 1 January 2013, South Australia adopted the model *Work Health and Safety Act 2012*.
- Tasmania first introduced legislation in 1977 known as the *Industrial Safety, Health and Welfare Act*. The current Act is the *Work Health and Safety Act 2012*.
- Victoria introduced legislation in 1985 that became the model for the rest of the country at the time. Far-reaching social and industrial concepts were incorporated into the legislation. Victoria currently uses the *Occupational Health and Safety Act 2004*.
- Western Australia introduced the *Occupational Health, Safety and Welfare Act in 1984*. In 2021 the Work Health and Safety Bill 2014 was before parliament and this bill will result in a new Act being created.

Current WHS Acts

Today the industrial workplace in Australia is governed by both federal and state legislation, followed by regulations and then codes of practice (see Table 4.2). The main piece of legislation for all of Australia is the model *Work Health and Safety Act 2011* (WHS Act). Laws and regulations provide a set of minimum standards for the protection and the health and safety of workers. Codes of practice demonstrate practical methods for undertaking the work safely. By and large, each principal WHS Act sets out the requirements of different groups of people who play a role in workplace health and safety.

Safe Work Australia is the national body that works to coordinate and develop policy, and assists in the implementation of the model Act. Safe Work Australia seeks to build cooperation between the three groups involved – governments, business and unions – to bring them together to create strategies and decide on policy.

TABLE 4.2 Current state and territory WHS Acts and Regulations

State/ territory	Current WHS act	Current WHS regulation	WHS regulating authority	Website and contact number
ACT	Work Health and Safety Act 2011	Work Health and Safety Regulations 2011	WorkSafe ACT	Website: http://www.worksafe.gov.au Contact: 02 6207 3000
NSW	Work Health and Safety Act 2011	Work Health and Safety Regulations 2017	SafeWork NSW	Website: http://www.safework.nsw.gov.au Contact: 13 10 50
NT	Work Health and Safety (National Uniform Legislation) Act 2011 (in force 1 May 2016)	Work Health and Safety (National Uniform Legislation) Regulations 2011 (in force 21 August 2019)	NT WorkSafe	Website: http://www.worksafe.nt.gov.au Contact: 1800 019 115
QLD	Work Health and Safety Act 2011	Work Health and Safety Regulations 2011	WorkCover Queensland	Website: http://www.worksafe.qld.gov.au Contact: 1300 362 128
SA	Work Health and Safety Act 2012	Work Health and Safety Regulations 2012	WorkCover SA	Website: http://www.workcover.com Contact: 13 18 55
TAS	Work Health and Safety Act 2012	Work Health and Safety Regulations 2012	WorkCover TAS	Website: http://www.workcover.tas.gov.au Contact: 1300 366 322
VIC	Occupational Health and Safety Act 2004	Occupational Health and Safety Regulations 2017	WorkSafe Victoria	Website: http://www.worksafe.vic.gov.au Contact: 1800 136 089
WA	Occupational Safety and Health Act 1984	Occupational Safety and Health Regulations 1996	WorkCover WA	Website: http://www.commerce.wa.gov.au/WorkSafe Contact: 08 9327 8777

Regulations, codes of practice and guidelines

Some workplace hazards have the potential to cause so much injury or disease that specific regulations or codes of practice are warranted. These regulations and codes, adopted under state and territory WHS Acts, explain the duties of particular groups of people in controlling the risks associated with specific hazards.

Note that:
- regulations are legally enforceable
- codes of practice and guidelines provide advice on how to meet regulatory requirements. As such, codes are not legally enforceable, but they can be used in courts as evidence that legal requirements have or have not been met.

The basic purpose of these codes and guidelines is to provide workers in any industry with practical, common sense, industry-acceptable ways by which to work safely.

They are now generally written and published by Safe Work Australia as 'Model Codes of Practice'. They are adopted by each state and territory's regulating authority, and cover such areas as 'Managing the risk of falls at workplaces', 'How to safely remove asbestos', 'Labelling of workplace hazardous chemicals', 'Demolition work' and 'Excavation work', to name just a few. A typical example is shown in Figure 4.1.

FIGURE 4.1 Typical code of practice – *How to safely remove asbestos* (front cover)

LEARNING TASK 4.1 LICENCES

Using the information in Table 4.2, visit the website of the regulating authority for your state or territory and search for 'Code of practice – asbestos'. Using this document, complete the following three questions.
1. What type of licence is required to remove 'any amount of friable and non-friable asbestos or Asbestos-Containing Material (ACM)'?
2. What type of licence is required to remove 'any amount of non-friable asbestos ONLY or ACM'?
3. If your state or territory allows: What is the maximum amount in square metres of non-friable asbestos or associated ACM that may be removed without a licence?

General construction induction training

The Work Health and Safety Regulations 2011 (Commonwealth) requires workers to complete general construction induction training before they can carry out construction work.

General construction induction training provides basic knowledge of construction work, the WHS laws that apply, common hazards likely to be encountered in construction work and how the associated risks can be controlled.

Source: Comcare © Commonwealth of Australia 2018. CC BY 4.0 International (https://creativecommons.org/licenses/by/4.0/)

On completion of a WHS general construction induction training session, a worker will be issued with a statement that outlines the training they have received, and identifies the training body, the training assessor and the date of the assessment.

Although introducing a nationally consistent construction induction card has been discussed, each jurisdiction or state/territory differs and provides its own particular card. All cards must show the cardholder's name, the date training was completed, the number of the registered training organisation (RTO) providing the training, the jurisdiction in which it was issued and a unique identifying number.

As each state/territory provides its own card, there has been a mutual agreement between the applicable state and territory regulating authorities to accept WHS induction cards from other states and territories, as long as the training meets existing standards for currency. A sample of the White Card used in South Australia is shown as an example in Figure 4.2.

The card should be carried on site at all times and produced on demand for inspection.

 COMPLETE WORKSHEET 1

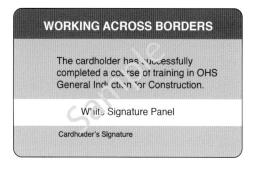

FIGURE 4.2 White Card sample as issued in South Australia

Basic roles, responsibilities and rights of duty holders

Each state and territory has its own specific requirements, which may include any or all of the following.

PCBUs/employers

Persons conducting a business or undertaking (PCBUs)/employers have a duty of care and must provide for the health, safety and welfare of their employees/workers at work. To do this, PCBUs/employers must:
- identify any foreseeable hazard that may arise from the work carried out by the company or partnerships
- provide and maintain equipment and systems of work that are safe and without risks to health
- make arrangements to ensure the safe use, handling, storage and transport of equipment and substances
- provide the information, instruction, training and supervision necessary to ensure the health and safety of employees at work
- maintain places of work under their control in a safe condition and provide and maintain safe entrances and exits
- make available adequate information about research and relevant tests of substances used at the place of work.

PCBUs/employers must not require employees/workers to pay for anything done or provided to meet specific requirements made under the Acts or associated legislation. They must also ensure the health and safety of people visiting their places of work who are not employees/workers.

Employees/workers

Employees/workers must take reasonable care of the health and safety of themselves and others. They must cooperate with their PCBU/employers and comply with WHS requirements. Employees/workers must not:
- interfere with or misuse any item provided for the health, safety or welfare of people at work
- obstruct attempts to give aid or attempts to prevent a serious risk to the health and safety of a person at work
- refuse a reasonable request to assist in giving aid or preventing a risk to health and safety.

Subcontractors

The PCBU/principal contractor must ensure that each subcontractor provides a written safe work method statement (SWMS) for the work to be carried out by the subcontractor, which should include an assessment of the risks associated with that work.

Duty of care requirements

The Work Health and Safety Acts and Regulations require both you and your business owner to do everything that is reasonably practicable to protect yourself and others from harm in the workplace.

The term duty of care means to act in a way that does not put others at risk of potential harm. Depending on the role that you have in the workplace, this can cover many different situations. The range of considerations is not just limited to the potential for physical harm but also for harm to mental health. To ensure that such matters are being managed, a range of mechanisms need to be implemented, such as ensuring that staff are correctly trained and licensed for high-risk activities in the workplace.

Safe work practices in construction

Safe work practice means working in a way that minimises risk to yourself, other people, equipment, materials, the environment and work processes.

Specific high-risk areas in construction

Since the introduction of the national model WHS regulations, high-risk construction work legally requires a SWMS. High-risk construction work is defined as work carried out:
- on a telecommunications tower
- in or near a shaft or trench with an excavated depth greater than 1.5 m, or in a tunnel
- on or near pressurised gas distribution mains or piping
- on or near chemical, fuel or refrigerant lines
- on or near energised electrical installations or services

- in an area that may have a contaminated or flammable atmosphere
- on, in or adjacent to a road, railway, shipping lane or other traffic corridor that is in use by traffic other than pedestrians
- in an area at a workplace in which there is any movement of powered mobile plant
- in an area in which there are artificial extremes of temperature
- in or near water or other liquid that involves a risk of drowning.

  It further includes tasks that involve:
- the risk of a person falling more than 2 m
- tilt-up or precast concrete
- the use of explosives
- the demolition of an element of a structure that is load-bearing
- the removal of asbestos
- structural alterations or repairs that require temporary support to prevent collapse
- confined spaces work
- diving work.

  Source: Adapted from *Work Health and Safety Management Systems and Auditing Guidelines* (edition 5) May 2014 © NSW Government.

Even though a SWMS is only required for 'high-risk construction work', it is still advisable to prepare a SWMS for all work (see https://www.safeworkaustralia.gov.au/search/site?search = safe + work + method + statement), thus controlling all potential hazards.

COMPLETE WORKSHEET 2

Construction hazards and risk-control measures

There are many different types of hazards on a construction site. It is important to have a systematic approach to identifying hazards and instituting the appropriate risk-control measures. Having a consistent approach will ensure that there is a high level of safety on construction sites and that this high level is maintained as the workplaces of construction sites change.

Basic principles of risk management

Risk management is designed to save you or someone else from being seriously injured or killed. In applying risk management, you aim do one of two things:
1. eliminate the risk of harm, injury or death altogether
2. if elimination of the hazard is not possible, then reduce the risk of harm, injury or death as much as possible.

According to Safe Work Australia, undertaking risk management involves four fundamental steps:

1. Identify hazards – find out what could cause harm.
2. Assess risks if necessary – understand the nature of the harm that could be caused by the hazard, how serious the harm could be and the likelihood of it happening.
3. Control risks – implement the most effective control measure that is reasonably practicable in the circumstances.
4. Review control measures to ensure they are working as planned.

Step 1 – how to identify hazards

Safe Work Australia classifies hazards into four different categories. These categories can stand alone or several of them can occur together. The categories are:
- physical work environment
- equipment, materials and substances used
- work tasks and how they are performed
- work design and management.

How to find hazards

On large construction sites there are regular scheduled inspections, but on smaller sites inspections are usually undertaken by the subcontractor when they begin on the site. The inspections are usually done by walking around the workplace and observing how things are being done and what plans there are for upcoming work. For an inspection to be effective, the people undertaking the inspection need to have a good understanding of the type of work that will be undertaken.

Once work has commenced, work practices need to be observed in relation to how people actually work, how plant and equipment is used, what chemicals are being used, and the general state of housekeeping.

According to Safe Work Australia, things to look out for include the following:

- Does the work environment enable workers to carry out work without risks to health and safety (for example, space for unobstructed movement, adequate ventilation, lighting)?
- How suitable are the tools and equipment for the task and how well are they maintained?
- Have any changes occurred in the workplace which may affect health and safety?

... Make a list of all the hazards you can find, including the ones you know are already being dealt with, to ensure that nothing is missed. You may use a checklist designed to suit your workplace to help you find and make a note of hazards.

... Analyse the site records of health monitoring, workplace incidents, near misses, worker complaints, sick leave and the results of any inspections and investigations to identify hazards.

Source: Safe Work Australia © Commonwealth of Australia.

Step 2 – How to assess risks

A risk assessment involves considering what could happen if someone were exposed to a hazard and the likelihood of it happening. Safe Work Australia states that a risk assessment can help you determine:

- how severe a risk is
- are any of the existing control measures effective
- what action is required to control the risk
- how urgently the action needs to be taken.

... A risk assessment can be undertaken with varying degrees of detail depending on the type of hazards and the information, data and resources that you have available.

Source: Safe Work Australia © Commonwealth of Australia.

Figure 4.3 provides an example of an approach for managing risk management that is used in many workplaces.

Consequence \ Likelihood		Rare — The event may occur in exceptional circumstances	Unlikely — The event could occur at some time.	Moderate — The event will probably occur at some time.	Likely — The event will occur in most circumstances.	Certain — The event is expected to occur in all circumstances.
		Less than once in 2 years	At least once per year	At least once in 6 months	At least once per month	At least once per week
	Level	1	2	3	4	5
Negligible — No injuries, Low financial loss	0	0	0	0	0	0
Minor — First-aid treatment, Moderate financial loss	1	1	2	3	4	5
Serious — Medical treatment required, High financial loss, Moderate environmental implications, Moderate loss of reputation, Moderate business interruption	2	2	4	6	8	10
Major — Excessive multiple long-term injuries, Major financial loss, High environmental implications, Major loss of reputation, Major business interruption	3	3	6	9	12	15
Fatality — Single death	4	4	8	12	16	20
Multiple — Multiple deaths and serious long-term injuries	5	5	10	15	20	25

Legend

Risk Rating	Risk Priority	Description
0	N	No Risk: The costs to treat the risk are disproportionately high compared to the negligible consequences.
1–3	L	Low Risk: May require consideration in any future changes to the work area or processes, or can be fixed immediately.
4–6	M	Moderate: May require corrective action through planning and budgeting process.
8–13	H	High: Requires immediate corrective action.
15–25	E	Extreme: Requires immediate prohibition of the work, process and immediate corrective action.

FIGURE 4.3 Risk matrix diagram

Source: Small Business Development Corporation, Government of Western Australia

A risk assessment should be done when:
- there is insufficient knowledge of how the hazard will result in an injury or an illness
- there are several different hazards that have been identified and it is not clear how the different hazards impact or interact with each other
- there are changes to the workplace that may render the control measures inadequate for their original intended use.

A risk assessment should consider the following factors:
- the potential severity of harm from the hazard
- how the hazard may cause harm
- the likelihood of harm occurring.

Potential severity of harm

The potential harm from a single hazard can range from a minor injury to death. For example, a circular saw can cause muscle and joint strain through repetitive use. However, it can also cause cuts when in use, from minor lacerations to severe cuts that could maim the operator or result in the operator's death.

Factors that may increase the severity of harm also need to be considered. For example, the greater the height that a person is working at, the greater the potential for more severe harm if they fall.

The potential breadth of the harm needs to be considered as well. This means the potential number of people who may be exposed to the harm. Every precaution needs to be taken to limit the number of people who could be injured by a hazard.

There is the potential for a cascade effect, where one failure leads to another possible larger failure if not controlled.

How the hazard may cause harm

There are several ways that hazards may cause harm, even if control measures have been put into place; for example, not adhering to best practices in the workplace, taking short cuts, deviating from manufacturers' instructions and ignoring work orders.

The likelihood of harm occurring

The likelihood that harm will occur can be estimated by using the criteria in Figure 4.3 where the likelihood is graded at five different levels from rare up to certain.

Step 3 – How to control risks

Safe Work Australia states that the best way to control any risk is to remove the risk. If this is not possible then the next best thing is to minimise the risk. This can be achieved by using hierarchy of control measures.

Under this model, risks are ranked from the highest level of protection and reliability to the lowest, as shown in Figure 4.4.

The hierarchy of control measures should be applied against any risk. Working through a structure of the hierarchy of control will produce the most appropriate

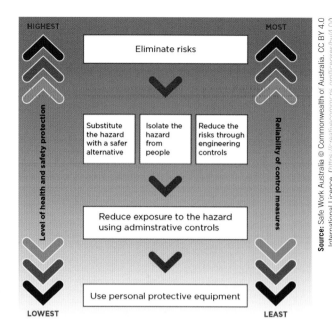

FIGURE 4.4 The hierarchy of control measures

measures without causing unnecessary disruption to the activity that is to take place. You need to keep in mind that the lower the level of control in the hierarchy that is used then the less effective the control will be.

Administrative controls and personal protective equipment (PPE) are the least effective at minimising risk because they do not control the hazard at the source and rely on human behaviour and supervision.

Elimination

The most effective control measure involves eliminating the hazard and associated risk.

The only way to do this is to not have the activity occur in the workplace. This is not always possible, but the planning stage of the work gives you the opportunity to put into place alternatives that may prevent the risk occurring on the job site. It is at this point that you may choose to have as much production done off site as on the job. This may also assist with the time frames of the on-site activity.

Simple ways to eliminate hazards on site can be as simple as each worker progressively keeping their immediate work space clean and free of debris.

Substitution, isolation and engineering controls

There are several different approaches that can be used to minimise hazards. It is possible that not all of these measures will be able to be used, though all approaches need to be investigated.

Substitute the hazard with something safer

This may be as simple as using a different type of material that does not expose the worker to any harm. The removal of asbestos-based building products is a measure that has occurred at a broad level across the whole nation.

Isolate the hazard from people
This particular measure is used on all sites. An example is the use of temporary fencing around a building site. Similar measures are the use of barriers and hand rails. Many road works companies use heavy-duty barriers made of concrete, while in areas of pedestrian traffic plastic barriers may be used.

Use engineering controls
One engineering control is the use of a crane. The use of cranes on domestic construction sites is a fairly new practice, but the introduction of prefabricated house frames and smaller-sized cranes means this engineering control measure has become commonplace.

Use administrative controls
Administrative controls include practices such as having SWMSs and JSAs in place that all workers are aware of and understand. The ongoing training of workers in the correct use of control measures is another common administrative control. An example of this is the requirement for all workers on a construction site to have induction training at an industry level as well as site-specific safety inductions.

Personal protective equipment
The use of PPE is the last line of defence to protect against a hazard. The use of PPE relies on the individual having a sound understanding of the appropriate PPE as it relates to the hazard and understanding the correct use of the PPE.

How to develop and implement control options
Information about suitable controls for many common hazards and risks can be found in:
- codes of practice and guidance material
- manufacturers and suppliers of plant, substances and equipment used in your workplace
- industry associations and unions.

In the instances of materials, the supplier will have SDSs that will provide recommendations of the types of controls that can be used for different materials.

Developing specific control measures
There may be instances where specific control measures may need to be developed based on the unique circumstances of a particular task or site. The best way for this to happen is through the process of developing a SWMS or JSA. Working through the process and identifying the sequence of events will provide you with an opportunity to come up with ideas that may provide broader and more effective solutions than the use of several controls in isolation. This will also allow you to prioritise the controls that are required for the hazards that are of the highest risk.

Implementing controls
When implementing any controls, especially if the controls are specifically identified, it is necessary to support the measures through avenues such as work procedures; training, supervision and maintenance.

Work procedures
Develop work procedures that describe the task and the hazards associated with that task. If this is a new or different approach, then a clear understanding is needed of the best practice for the situation.

Training, instruction and information
If new work practices are required, it is essential for all workers to have a full understanding of the skills and knowledge to be able to perform the tasks in a safe manner that produces a high-quality outcome. For example, a worker may be required to use a different material and associated materials to perform a task, such as when a bricklayer is required to install ACC blocks. Even though the basic principles of the task are the same, the materials used are very different from usual bricks and blocks, and behave in a different manner.

Supervision
If the tasks are of a different or unique nature, closer supervision will be required to ensure that any inexperienced workers are supported when undertaking the task. A second set of more experienced eyes can pick up potential hazards that the worker may not notice.

Maintenance
Any physical control measures that are put in place need to be maintained. In some situations, the control measures can be in place for extended periods of time and possibly subjected to different weather conditions. In these cases, there should be regular checks on the equipment being used. Depending on the size of the site, it is recommended that there is a clear understanding of who is responsible for the maintenance of any physical control measures. This includes controls such as PPE.

Step 4 – How to review controls
There should be a culture of consistent review of the control measures that have been put in place. It is too late if an accident occurs and only then a review of the measures is undertaken. According to Safe Work Australia, you should review the controls:
- when the control measure is not effective in controlling the risk
- before a change at the workplace that is likely to give rise to a new or different health and safety risk that the control measure may not effectively control
- if a new hazard or risk is identified
- if the results of consultation indicate that a review is necessary
- if a health and safety representative requests a review.

Managing work health and safety risks is an ongoing process that needs attention over time, but particularly when any changes affect work activities.

> **COMPLETE WORKSHEET 3**

Construction hazards

In the construction industry, electricity, falls, collapsing trenches and melanoma can kill. Chemicals, corrosives, noise and dust inhalation can result in blindness, deafness, burns and injuries to lungs. Back problems or other serious strains or sprains can slow workers down and put them out of action for weeks or even permanently.

It is therefore important to be able to identify workplace hazards and implement a process of identifying, assessing and controlling risks (see Figure 4.5).

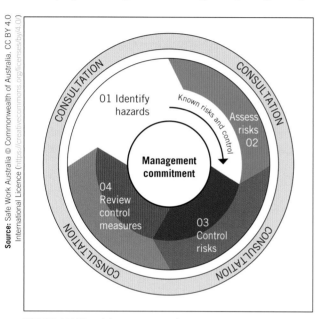

FIGURE 4.5 The risk-management process

Hazards

A hazard is anything in the workplace that has the potential to injure, harm, damage, kill or destroy people, property or plant and equipment. Note the important word is 'potential'; this means that a hazard is not only something that has caused damage or injury but also includes anything that can possibly do damage or injury.

Hazards can include objects in the workplace, such as machinery or chemicals; they also include work processes where manual handling, excessive noise and fatigue may cause long-term damage. Hazards are divided into two categories – acute and chronic.

An acute hazard is one where short-term exposure to the hazard will cause an injury or sickness (for example, being burnt in an explosive fire).

A chronic hazard is one where long-term exposure to the hazard will cause an injury or sickness (for example, melanoma from extended exposure to the sun, or slow poisoning from chemicals building up in the body's system over a long period of time).

Risk

Risk is the likelihood or probability of a hazard causing damage or injury. The level of risk will depend on factors such as how often the person is exposed to the hazard and how serious any potential injuries.

Incident/accident

An incident is where a hazard actually results in injury, damage or death. It may be random or it may be an intentional act or action due to neglect or purposeful intention.

By definition, an accident is when something happens unexpectedly, without design, or by chance. An accident may cause injury, death, damage or destruction.

To reduce incidents and accidents we need to first recognise the hazard and then rate the level of risk (risk assessment). The most common way of achieving this is by using a risk matrix as shown in Figure 4.3.

The following example applies the risk matrix from Figure 4.3 to a real-life context. On the job site, the foreperson sets up the compound mitre saw on the floor and asks you to cut timber to length; approximate time on the saw is four hours.

What are the hazards? Consider the following:
- hearing damage
- foreign objects being thrown into the eyes
- back strain
- breathing in timber dust
- amputating a hand or fingers.

Each of the above needs to be considered using the matrix diagram; however, we will look at 'Amputating a hand or finger' for our example.

So, ask the first question: how *serious* will an accident be if I put my hand through the saw? Multiple, Fatality, Major, Serious, Minor or Negligible? The answer is most likely 'Serious' or 'Major'.

Now the second question: how *likely* is an accident to occur? Well, with no training, impaired visibility because you are squatting over the top of the saw and the fact that you are concerned about your back hurting all the time, then it all adds up to being 'Certain' that you will do damage.

So, using the matrix and plotting the result, the answer would be either Gold (level 10) 'High requiring immediate corrective action' or Red (level 15) 'Extreme – requires immediate prohibition of the work and immediate corrective action'.

So, what can be done to control the hazard and risk? Very simply, place the saw on a saw stand and be trained on how to use the saw correctly. This will reduce the likelihood of harm to 'Rare'. Now check the result on the matrix and it is a level 2 or 3.

Purpose and use of PPE

PPE is the last line of defence to protect your health and safety from workplace hazards. It is the PCBU/

employer's responsibility to provide the PPE and training to protect the worker. It is your responsibility to wear and look after the equipment provided.

PPE must be appropriate to each particular hazard. It is important to consider the following when assessing workplace PPE requirements:

- The workplace – could it be made safer so that you don't need to use PPE?
- PPE selection – is the PPE designed to provide adequate protection against the hazards at your workplace?
- PPE comfort and fit – is the PPE provided comfortable to wear? Even if the equipment theoretically gives protection, it won't do the job if it doesn't fit properly or is not worn because it is too uncomfortable. For example, close-fitting respirators give protection only if the person is clean-shaven. Those with a beard or 'a few days' growth' will need to use a hood, helmet or visor-type respirator. The respirator must also be properly cleaned and maintained.

PPE can be grouped according to the part of the body it will protect; for example:

- head – safety helmets and sun hats
- eyes/face – safety spectacles, goggles and face shields
- hearing – ear muffs and ear plugs
- airways/lungs – dust masks and respirators
- hands – gloves and barrier creams
- feet – safety boots, safety shoes and rubber boots
- body – clothing to protect from sun, cuts, abrasions and burns; high visibility safety garments; and fall protection harnesses.

To decide what PPE and clothing is required, you must first be able to identify the hazards involved. Types of hazards commonly identified where PPE and clothing are a suitable means of protection include:

- physical hazards – noise, thermal, vibration, repetitive strain injury (RSI), manual and radiation hazards
- chemical hazards – dusts, fumes, solids, liquids, mists, gases and vapours.

Once the hazards have been identified, suitable equipment and clothing must be selected to give the maximum protection.

Safety helmets

Wearing safety helmets on construction sites may prevent or lessen a head injury from falling or swinging objects, or through striking a stationary object.

Safety helmets must be worn on construction sites when:

- it is possible that a person may be struck on the head by a falling object
- a person may strike his/her head against a fixed or protruding object
- accidental head contact may be made with electrical hazards
- carrying out demolition work
- instructed by the person in control of the workplace.

Safety helmets (see Figure 4.6) must comply with AS/NZS 1801 Occupational protective helmets, and must carry the AS or AS/NZS label, and must be used in accordance with AS/NZS 1800 Occupational protective helmets – Selection, care and use.

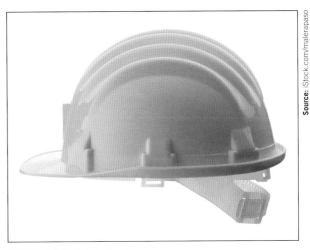

FIGURE 4.6 Safety helmet

When wearing a helmet, the harness should be adjusted to allow for stretch on impact. No contact should be made between the skull and the shell of the helmet when subjected to impact.

Sun shade

The awareness of skin cancer for building workers is increasing. The neck, ears and face can be particularly exposed. Workers should wear sun protection at all times when working outdoors (including in the winter).

Sun shades include wide-brimmed hats and foreign legion-style sun shields fixed to the inner liner of safety helmets, or safety helmet 'foreign legion sun brims' (see Figure 4.7).

FIGURE 4.7 Fabric sun brim accessory for a safety cap and bucket hat

Eyes/face protection

The design of eye and face protection is specific to the application. It must conform to AS/NZS 1337.1 Eye protectors for industrial applications. The hazards to the eyes are of three categories:

1. physical – dust, flying particles or objects, and molten metals
2. chemicals – liquid splashes, gases and vapours, and dusts
3. radiation – sun, lasers and welding flash.

The selection of the correct eye protection to protect against multiple hazards on the job is important (see Figure 4.8 and Figure 4.9). Most eyewear is available with a tint for protection against the sun's UV rays, or may have radiation protection included.

FIGURE 4.8 Clear wide-vision goggles

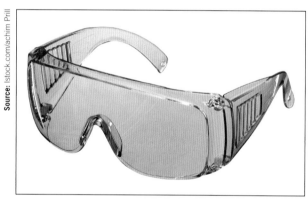

FIGURE 4.9 Clear-framed spectacles

Face shields

Face shields give full face protection, as well as eye protection. They are usually worn when carrying out grinding and chipping operations, and when using power tools on timber. Shields are also worn for full-face protection when welding (see Figure 4.10). The shield may come complete with head harness (see Figure 4.11) or be designed for fitting to a safety helmet.

Hearing protection

You should always wear ear protection in areas where loud or high-frequency noise operations are being carried out, or where there is continuous noise. Always wear protection when you see a 'Hearing protection

FIGURE 4.10 Full-face welding mask

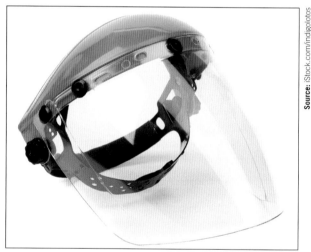

FIGURE 4.11 Face shield

must be worn' sign, and when you are using or are near noisy power tools.

The two main types of protection available for ears are:
1. ear plugs – semi- and fully disposable
2. ear muffs – available to fit on hard hats where required (see Figure 4.12).

Choose the one that best suits you and conforms to AS/NZS 1270 Acoustic – Hearing protectors.

Disposable dust masks

Dust masks are available for different purposes and it is important to select the correct type. If the work that you are undertaking is mowing or general sweeping, a nuisance-dust mask is appropriate and is designed to filter out nuisance dusts only. A nuisance-dust mask is easily recognised as it has only one strap to hold it onto the face (see Figure 4.13). If, on the other hand, you are working with toxic dusts (such as bonded asbestos), you will require greater protection. A mask with two straps and labelled with either P1 class particle dust

FIGURE 4.12 Hearing protection

FIGURE 4.14 P1/P2 disposable mask

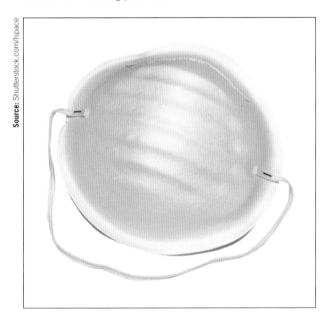

FIGURE 4.13 Mini dust mask, for nuisance dust only

FIGURE 4.15 Half-face respirator with P2 class dust filters fitted

filter (minimum protection) or P2 class particle dust filter (mid-range protection) (see Figure 4.14) will be required.

Respirators

Half-face and full-face respirators have cartridge type filters that are designed to keep out dusts, smoke, metal fumes, mists, fogs, organic vapours, solvent vapours, gases and acids, depending upon the combination of dust and gas filters fitted to the respirator. Cartridge type filters can be identified by the classification ratings from AS/NZS 1716 Respiratory protective devices.

Respirators fitted with P2 class particle dust filter (see Figure 4.15) are suitable for use with the general low-toxic dusts and welding fumes that are commonly found on construction sites.

Some full-face or half-face respirators may also be connected to an air supply line or bottle that provides clean filtered air to the user. These are generally used for loose (friable) asbestos removal or in contaminated or low-oxygen environments, where face and eye protection is also required.

Further information on respirators and dust masks should be obtained from the manufacturers. It is very important to be trained in the correct methods of selecting, fitting, wearing and cleaning of the equipment in accordance with AS/NZS 1715 Selection, use and maintenance of respiratory protective equipment. When selecting, it is important that tight-fitting respirators and masks must have an effective seal on the face to ensure that all air entering your respiratory passages has been fully filtered.

Gloves

Gloves are used to protect hands and arms from both physical and chemical hazards. Stout leather gloves are required when handling sharp or hot materials.

Rubberised chemical-resistant short or long gloves are used when handling hazardous chemical substances. Gloves should conform to AS/NZS 2161.1 Occupational protective gloves – Selection, use and maintenance (see Figure 4.16).

FIGURE 4.16 Gloves

Creams

Barrier creams may be used to protect the hands from the effects of cement and similar low-toxic hazards (see Figure 4.17) when gloves are too restrictive.

FIGURE 4.17 Barrier cream

Foot protection

It is mandatory to wear protective footwear in the workplace at all times. Thongs are not permitted at any time. Footwear should conform to AS/NZS 2210.1 Safety, protective and occupational footwear guide to selection, care and use.

All safety footwear must have:
- stout oil-resistant, non-slip soles or steel midsoles to protect against sharp objects and protruding nails
- good uppers to protect against sharp tools and materials
- reinforced toecaps to protect against heavy falling objects.

Safety boots should be worn in preference to safety shoes on construction sites to give ankle support over the rough terrain. Safety joggers may be required when carrying out roof work or scaffold work; they must have reinforced toecaps. Rubber boots should be worn when working in wet conditions, in wet concrete, or when working with corrosive chemicals. They must have reinforced toecaps. See Figure 4.18 for examples of foot protection.

FIGURE 4.18 Foot protection

Clothing

Good-quality tough clothing is appropriate for construction work. It should be kept in good repair and cleaned regularly. If the clothing has been worn when working with hazardous substances it should not be taken home to launder but sent to a commercial cleaning company; this will prevent the hazards from contaminating the home and the environment.

A good fit is important, as loose-fitting clothing is easily caught in machine parts or on protruding objects.

Work pants should not have cuffs or patch pockets, as hot materials can lodge in these when worn near welding or cutting operations.

Clothing should give protection from the sun's UV rays, cuts, abrasions and burns.

Industrial clothing for use in hazardous situations should conform to AS/NZS 4501.2 Occupational protective clothing – General requirements.

Fall protection harnesses

In some instances when working at heights, it may be necessary to wear a harness. These are specialist items that need to be fitted to the individual. A harness must be correctly fitted or serious injury may result if or when the person falls. All harnesses must comply with AS 2626 Industrial safety belts and harnesses – Selection, use and maintenance.

Cleaning and maintenance

All PPE must be cleaned and maintained on a regular basis. This must be done by someone who has been trained in inspection and maintenance of such equipment. Remember your life and **wellbeing** depends on this PPE, and if it is faulty or damaged or simply not functioning properly, you are at risk.

LEARNING TASK 4.2 JSA/SWMS DEVELOPMENT

In groups of two, complete a JSA/SWMS for an identified practical project that has been identified by your teacher, using the pro forma provided.

Ensure that you identify each logical step of the project, which may be provided to you in a work instruction from your teacher.

Identify any potential hazards and then associated measures to assist in managing the hazard.

Once the JSA/SWMS has been completed, discuss each step of the project with the class and establish the most appropriate actions to manage any hazards.

COMPLETE WORKSHEET 4

Health and safety communication and reporting processes

Construction sites feature many different kinds of signs and messages to alert workers to potential hazards and to prevent injury or death. If an incident does occur, the reporting process is not only crucial to ensuring the worker is properly supported, but is also part of the continuous improvement process to create full levels of awareness so similar situations can be better managed in the future.

Safety signs and symbols

Safety signs are placed in the workplace to:
- warn of health and safety hazards
- give information on how to avoid particular hazards, thereby preventing incidents and accidents
- indicate the location of safety and fire protection equipment
- give guidance and instruction in emergency procedures.

Standards Australia has three Standards covering the use of safety signs in industry:
- AS 1216 Class labels for dangerous goods
- AS 1318 Use of colour for the marking of physical hazards
- AS 1319 Safety signs for the occupational environment.

AS 1319 identifies three main types of safety signs:
- Picture signs – these use symbols (pictures) of the hazard, equipment or the work process being identified, as well as standardised colours and shapes, to communicate a message (see Figure 4.19).
- Word-only signs – these are written messages using standardised colours and shapes to communicate the required meaning (see Figure 4.20).

FIGURE 4.19 Example of a picture sign

FIGURE 4.20 Example of a word-only message

- Combined picture and word signs – these are clearer to understand with a picture and a short written message (see Figure 4.21).

FIGURE 4.21 Example of a combined picture and word sign

To make sure the message reaches everyone at the workplace, including workers from non-English-speaking backgrounds and workers with low literacy skills, picture signs should be used wherever possible.

When this is not the case, it may be necessary to repeat the message in other languages.

Colour and shape

There are seven categories of safety signs and they are identified by colour and shape, as shown in Table 4.3.

Prohibition signs

These signs indicate that this is something you must not do. They feature a red circular border with cross bar through it, a white background and a black symbol (see Table 4.3, Figure 4.22 and Figure 4.23).

Mandatory (must do) signs

These signs tell you that you must wear some special safety equipment. They feature a blue solid circle and a white symbol, with no border required (see Table 4.3, Figure 4.24 and Figure 4.25).

Restriction signs

These signs tell of the limitations placed on an activity or use of a facility. They feature a red circular border, with no cross bar and a white background. Limitation or restriction signs normally have a number placed in them to indicate a limit of some type (for example, a speed limit or weight limit; see Table 4.3).

TABLE 4.3 The seven categories of signs

Sign category	Picture	Sign category	Picture
1. Prohibition (must not do) signs	 Fire, naked flame and smoking prohibited	5. Mandatory (must do) signs	 Head protection must be worn
2. Restriction signs	 40 km/hr restriction sign	6. Hazard warning signs	 Fire, naked flame and smoking prohibited
3. Danger (hazard) signs	 Danger warning sign – asbestos removal in progress	7. Emergency information signs	 Emergency exit signs
4. Fire-fighting equipment signs	 Fire safety – fire hose reel		

FIGURE 4.22 Digging prohibited

FIGURE 4.25 Hearing protection must be worn

FIGURE 4.23 No pedestrian access

FIGURE 4.26 Fire hazard

FIGURE 4.27 Toxic hazard

FIGURE 4.24 Eye protection must be worn

Hazard warning signs

These warn you of a danger or risk to your health. They feature a yellow triangle with a black border and a black symbol (see Table 4.3, Figure 4.26, Figure 4.27 and Figure 4.28).

Danger hazard signs

These signs warn of a particular hazard or hazardous condition that is likely to be life-threatening. They

FIGURE 4.28 Electric shock hazard

PREPARE TO WORK SAFELY IN THE CONSTRUCTION INDUSTRY **95**

feature a white rectangular background, white/red DANGER text, and a black border and wording (see Figure 4.29).

FIGURE 4.29 Danger signs

Emergency information signs

These show you where emergency safety equipment is kept. They feature a green solid square with a white symbol (see Table 4.3, Figure 4.30 and Figure 4.31).

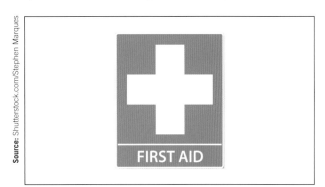

FIGURE 4.30 First aid

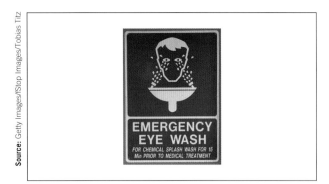

FIGURE 4.31 Emergency (safety) eye wash

Fire signs

These tell you the location of fire alarms and firefighting facilities. They feature a red solid square and a white symbol (see Table 4.3 and Figure 4.32). Sometimes a safety sign is required to be placed on a piece of equipment to indicate a problem or potential hazard if

FIGURE 4.32 Fire alarm call point

the machine is switched on. This is usually in the form of a tag.

Safety and accident prevention tags for electrical equipment

Electrical wires and equipment that are being worked on or are out of service, or are live or may become live, must have Warning or DANGER safety tags fixed to them to help prevent accidents. Once the hazard has been removed, only the person who put the tag in place should remove it or authorise its removal.

Any standard safety sign may be made smaller and used as an accident protection tag (see Figure 4.33). If words are to be used, this will generally be in the form of a danger sign. A tag should be at least 80 mm × 50 mm, plus any area required for tying or fixing the tag in place.

FIGURE 4.33 Electrical safety signs and tags

The background colour of the tag should be yellow for warning signs and white for danger signs.

For more picture safety signs and their meanings, see:
- Standards Australia: AS 1319 Safety signs for the occupational environment
- SafeWork South Australia: 'Work Safety Signs'
- Safety sign manufacturers' catalogues.

Placement of safety signs

Signs should be located where they are clearly visible to all concerned, where they can easily be read and where they will attract attention. If lighting is not adequate, illuminated signs can be used.

Signs should not be located where materials and equipment are likely to be stacked in front of them, or where other obstructions could cover them (for example, doors opening in front of them). They should not be placed on movable objects such as doors, windows or racks so that when the object is moved they are out of sight or the intention of the sign is changed.

The best height for signs is approximately 1500 mm above floor level. This is within the normal line of sight of a standing adult. The positioning of the sign should not cause the sign itself to become a hazard to pedestrians or machine operators.

Regulation and hazard-type signs should be positioned in relation to the hazard to allow a person plenty of time to view the sign and take notice of the warning. This distance will vary; for example, signs warning against touching of electrical equipment should be placed close to the equipment, whereas signs on construction work may need to be placed far enough away to permit the warning to be understood before the hazard is reached.

Care should be taken where several signs are intended to be displayed close together. The result could be that so much information is given in one place that little or no notice is taken of it, or that it creates confusion.

 COMPLETE WORKSHEET 5

Procedures for reporting hazards, incidents and injuries

There are five different health and safety personnel roles that are usually in place on a construction site. Each role has its own set of responsibilities that together safeguard the WHS of all staff in the workplace.

First-aid officers

First-aid officers are people who have completed a course and are qualified first aiders. They are responsible for helping people who are sick or injured and in many instances are the first responders to an incident.

If the injury they are treating is minor, they might be able to solve the problem themselves with basic treatment.

If the injury is more serious, the first-aid officer's job is to help the sick or injured person until professional medical help arrives.

Supervisors

On most construction sites, the supervisor is the first person to contact if you need to talk about safety. In most circumstances the supervisor will have ensured that the appropriate control measures are in place. The supervisor is the first point of contact if you have any concerns or if clarification is required about safety on the site.

Wardens

A warden makes sure everyone evacuates properly if there is an emergency. In the event of an emergency the warden will give clear instructions and coordinate with any emergency services if required. The all-clear to return to the site will be given by the warden.

Health and safety representatives

Health and safety representatives (HSRs) represent the safety interests of workers on the construction site. They are elected by other workers. A construction site might have one or more HSRs. The person in this role can help workers, supervisors and PCBUs communicate about safety on the site.

Health and safety committee members

Health and safety committee members are people who meet as a group to talk about safety on the construction site. Committee members often come from different areas of the site. They can also be a mix of workers, supervisors and HSRs.

Accident/incident reporting

All workers have a responsibility to report any illness, accident or near miss they are involved in or see at the workplace. They should report such incidents to their immediate supervisor, WHS committee, health and safety representative (HSR), a union delegate or first aid officer. If there is any plant or equipment involved they must make the operator aware of the problem immediately.

Once the relevant personnel become aware of an illness, accident or near miss, it must be determined whether it is a reportable incident. If so, the necessary forms must be completed and forwarded to the appropriate authorities (see Figure 4.34). This reporting procedure allows for steps to be taken on a state and national level to help reduce workplace accidents and

Accident and Injury Report

Details of injury (e.g., to a worker or visitor) and treatment			
Date of incident	Time of incident		am pm
Nature of incident	Near miss	First aid	Medical treatment
Name of injured person			
Home address			
Date of birth			
Telephone			
Email			
Employer			
Activity in which the person was engaged at the time of injury			
Site address and location on site of incident			
Nature of injury – e.g., fracture, burn, sprain, foreign body in eye			
Body location of injury (indicate location of injury on the diagram)	Right Left Right Front view Rear view		
Treatment given on site			
Referral for further treatment? Yes No			
Witness to incident (each witness may need to provide an account of what happened)			
Witness name		Witness contact	
Witness name		Witness contact	
Completed by			
Name		Position	
Signature		Date	

FIGURE 4.34 Sample accident report form

illnesses in the future. It also means inspections of the workplace at which the accident or illness occurred will be made by the state's WHS regulating authority to ensure that steps are taken to prevent similar incidents from happening again.

Reportable accidents/incidents

AS 1885.1 Measurement of occupational health and safety performance (known as the National Standard for Workplace Injury and Disease Recording) provides a template for describing and reporting occupational injuries and disease. From this national standard, each state and territory developed an accident reporting register and reporting procedure to match its own WHS legislation.

Your state/territory regulating authority may require serious work-related illnesses, injuries or dangerous occurrences to be reported on an accident report form.

Accidents that must be reported by law are known as reportable accidents.

Accident/incidents

An accident/incident is any event that causes human injury or property damage. Such events may occur as a result of unsafe acts that may include the following:
- practical jokes gone wrong
- using tools and/or equipment in a manner for which they were not designed
- rushing and taking short cuts
- not using PPE or other safety devices
- throwing materials/rubbish from roofs or upper floors.

Other causes of accidents and incidents include:
- little or no training in safety and proper use of tools and equipment
- poor housekeeping
- poor management of site safety issues
- poorly maintained tools and/or equipment
- damaged tools and/or equipment
- inadequate PPE or PPE not supplied or not used
- poor site conditions and congestion due to lack of preparation.

An example of an illness that must be reported is:
- an employee/worker has a medical certificate stating that he/she is suffering from a work-related illness that prevents the employee from carrying out his/her usual duties for a continuous period of at least seven days.

An example of an injury that must be reported is:
- one resulting from a workplace accident where a person dies or cannot carry out his/her usual duties for a continuous period of at least seven days.

Examples of dangerous occurrences that must be reported (even if no one is injured) are:
- damage to any boiler, pressure vessel, plant, equipment or other item that endangers or is likely to endanger the health and safety of any person at a workplace
- damage to any load-bearing member or control device of a crane, hoist, conveyor, lift, escalator, moving walkway, plant, scaffolding, gear, amusement device or public stand
- any uncontrolled explosion, fire or escape of gas or dangerous goods or steam, or where there is a risk that any of these events is likely
- a hazard exists that is likely to cause an accident at any moment and may cause death or serious injury to a person (for example, an electric shock) or substantial damage to property.

PCBUs/employers in control of a workplace are normally required to send an accident report form to their state or territory regulating authority even if the person injured or killed is not one of their employees (for example, is a subcontractor or visitor to the site).

Near misses

These are accidents/incidents that didn't quite happen. They occur when the conditions are right for an accident but people have not been hurt and equipment is not damaged.

Near misses usually indicate that a procedure or practice is not being carried out correctly or site conditions are unsafe. By reporting these near misses, the problems may be looked at by management and rectified before someone is hurt or killed or equipment is seriously damaged.

Company accident report forms

As well as completing a state or territory accident/incident notification form, a pro-forma company form or Worker's Compensation Insurance Accident form must also be completed. This accident report normally requires the following information to be given.

Information about the PCBU/employer:
- name of company
- office address
- address of site where accident happened
- main type of activity carried out at the workplace (for example, building construction)
- major trades, services or products associated with this activity
- number of people employed at the workplace and whether there is a WHS committee/HSR at the workplace.

Information about the injured or ill person:
- name and home address
- date and country of birth
- whether the person is an employee of the company
- job title and main duties of the injured person.

Information about the injury/illness:
- date of medical certificate
- type of illness as shown on the medical certificate
- whether the injury resulted in death
- particulars of any chemicals, products, processes or equipment involved in the accident.

Information about the injury or dangerous occurrence:
- time and date that it happened
- exact location of the event
- details of the injury
- type of hazard involved
- exactly how the injury or dangerous occurrence was caused
- how the injury affected the person's work duties
- details of any witnesses
- details of the action taken to prevent the accident from happening again.

Details of the person signing and the date of signing the accident report are also required. When the report is completed, copies are generally sent to the company's worker's compensation insurance organisation and a copy is kept by the PCBU/employer.

Injury management

The loss or disruption a company can experience as a result of a hazardous incident may be multiplied tenfold when that incident leads to a worker being injured.

A comprehensive risk-management system should include a well thought-out plan to maximise the opportunity for injured workers to remain at work. This allows the worker to be productive in some capacity and assists with the recovery and rehabilitation process.

Therefore, the risk-management system should cover the following points:
- early notification of the injury
- early contact with the worker, his or her doctor and the builder's insurance company
- provision of suitable light duties as soon as possible to assist with an early return to work
- a written plan to upgrade these duties in line with medical advice.

Payment of compensation

Worker's compensation insurance is a system that provides payment and other assistance for workers injured through work-related accidents or illnesses. It may also provide their families with benefits where the injury is very serious or the worker dies.

PCBU/employers must take out worker's compensation insurance to cover all workers considered by law to be their employees/workers.

Eligibility for compensation

To be able to claim worker's compensation, a worker must have suffered an injury or disease. The injury or disease must be work-related: that is, it must have happened while working, during an allowed meal break, or on a work-related journey.

The injury or disease must result in at least one of the following:
- death of the worker
- the worker being totally or partially unable to perform work
- the need for medical, hospital or rehabilitation treatment
- the worker permanently losing the use of some part of the body.

A claim for compensation is made by:
- informing the PCBU/employer and lodging a claim as soon as possible
- seeing a doctor and obtaining a medical certificate.

If any problems arise with the compensation claim the worker should contact their state or territory compensation board or authority, or their own union.

A record of all injuries that occur at a workplace must be entered in an injuries register kept at the job site.

COMPLETE WORKSHEET 6

Incident and emergency response procedures

It is never desirable to be in an incident or emergency situation, but to ensure the best outcome for everybody, there needs to be a structured approach. It is essential to have a clear procedure to follow in an emergency and that everybody has an understanding of their role. The ability to provide first aid, if required, can greatly enhance the recovery for any person who may be injured. The ability to take the appropriate action if the situation is still dangerous is also a high priority.

Procedures for responding to incidents and emergencies

An emergency may develop due to any number of reasons, such as a fire, gas or toxic fumes leak, improper use of flammable materials, partial collapse of a building, bomb threat, crane overturning, unstable ground, materials improperly stored or a trench collapse. Therefore, every organisation must have an emergency procedure in place and personnel appointed to control the safe exit of people from the workplace as well as to assist anyone injured or seriously hurt.

It is the responsibility of the PCBU/employer to ensure that in the event of an emergency the following arrangements have been made:
- safe and rapid evacuation of people from the place of work to a designated assembly area on the site
- emergency communications, such as a landline phone or mobile with emergency phone numbers, are clearly visible and accessible at all times
- provision of appropriate medical treatment of injured people by ambulance, medical officer or access to a suitable first aid kit.

Note: If a person is seriously injured the relevant state or territory regulating authority must be notified. If a person is killed, the relevant state or territory regulating authority and the police must be notified.

Roles of designated health and safety personnel

On all large sites there are specialised roles that need to be performed in the case of an emergency. It is essential that everybody has a clear understanding of the roles and responsibilities of staff on the site so there is no confusion when giving directions is required.

Responsible personnel

On a large building site the responsible personnel may include the:
- head contractor/PCBU
- safety officer
- head foreperson or site supervisor.

Small building site personnel may include the:
- builder
- foreperson
- leading hand or a nominated tradesperson.

These people are responsible for following set procedures to get all other people on site out of the danger area to a predetermined emergency evacuation point, so that during an incident all people may easily be accounted for. The nominated responsible personnel are indemnified against liability resulting from practice evacuations or emergency evacuations from a building, where the people act in good faith and in the course of their duties.

On large sites, these people are identified by a coloured helmet they wear, which would be determined as part of the organisation's emergency plan.

Roles

Many companies will have selected and trained competent individuals to take charge during an emergency; as a worker it is important to recognise the various roles that these responsible personnel perform.

The emergency coordinator:
- determines the nature of the emergency and the course of action to be followed
- sets off any alarm or siren to warn people of an emergency
- contacts appropriate emergency services such as police, fire or ambulance
- initiates the emergency procedure and briefs the emergency services when they arrive.

The warden or controller:
- assumes control of the occupants/workers until the emergency services arrive
- notifies all people regarding the nature of the emergency
- gives clear instructions and makes a record of what was carried out
- reports all details to the emergency coordinator as soon as possible.

Casualty control:
- attends to casualties and coordinates first aid
- coordinates the casualty services when they arrive
- arranges for further medical or hospital treatment.

Evacuation process

It is vital that PCBU/employers prepare an emergency plan for the workplace, and that each worker is aware of what to do in the event of an emergency evacuation.

An emergency plan is a written set of directions that outlines what workers and others at the workplace must do in an emergency. An emergency plan must provide the following information:
- emergency procedures, including:
  - an effective response to an emergency
  - evacuation procedures
  - notifying emergency service organisations at the earliest opportunity
  - contact details for medical treatment and assistance
  - effective communication between the person authorised to coordinate the emergency response and all people at the workplace
- testing of the emergency procedures – including the frequency of testing
- information, training and instruction to relevant workers on implementing the emergency procedures.

Source: Safe Work Australia © Commonwealth of Australia. CC BY 4.0 Licence (https://creativecommons.org/licenses/by/4.0/)

Another important part of the evacuation process is the evacuation diagram. This diagram shows important information, such as the location of evacuation assembly points, fire hose reels, fire extinguishers and where the Fire Indicator Board (FIB) is. These diagrams should be placed at strategic locations in the structure, allowing all workers easy access to the information shown on them (see Figure 4.35).

COMPLETE WORKSHEET 7

Procedures for accessing first aid

A PCBU/employer must ensure that employees have access to first aid facilities that are adequate for the immediate treatment of common medical emergencies. In addition, if more than 25 people are employed at the workplace, there should be trained first aid personnel. Trained first aid personnel may include a person with a current approved first aid certificate, a registered nurse or a medical practitioner.

The following first aid kit recommendations should be referred to as a guideline only. It is highly recommended that you refer to your own state or territory WHS legislation in relation to this. Alternatively, visit the St John Ambulance Australia website at http://www.stjohn.org.au. The following,

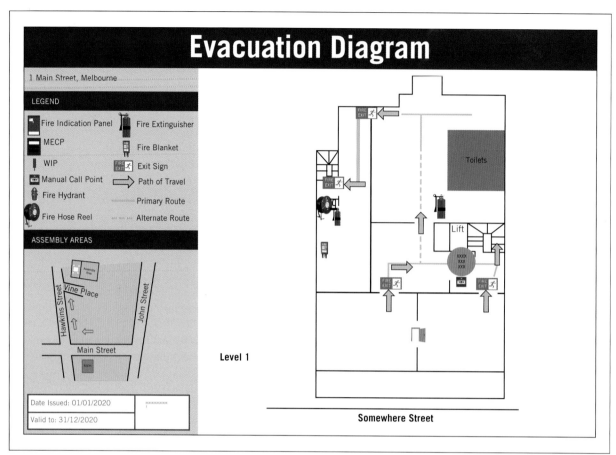

FIGURE 4.35 Evacuation diagram

therefore, is to be referred to as an example only; requirements may change from jurisdiction to jurisdiction:
- First aid kit A – for construction sites at which 25 or more people work or other places of work at which 100 or more people work
- First aid kit B – for construction sites at which fewer than 25 people work or other places of work at which fewer than 100 and more than 10 people work
- First aid kit C – for places of work (other than construction sites) at which 10 or fewer people work.

Most residential building sites would fall into the 'kit B' category, as it would be rare to have 25 or more people on the site at any one time.

Note: 'Kit C' is not suitable for construction sites of any kind, as it lacks many of the items required for first aid treatment of common building site injuries.

Contents of a first aid kit B

The contents of a type B kit are listed in Table 4.4 and the first aid kit and stores are shown in Figure 4.36 and Figure 4.37.

Types and purpose of fire safety equipment

One of the biggest hazards on any construction site is fire – it produces heat that can burn and smoke that can

TABLE 4.4 Contents of a first aid kit B

Item	Quantity
Adhesive plastic dressing strips, sterile, packets of 50	1
Adhesive dressing tape, 25 mm × 50 mm	1
Bags, plastic, for amputated parts:	
Small	1
Medium	1
Large	1
Dressings, non-adherent, sterile, 75 mm × 75 mm	2
Eye pads, sterile	2
Gauze bandages:	
50 mm	1
100 mm	1
Gloves, disposable, single	4
Rescue blanket, silver space	1
Safety pins, packets	1
Scissors, blunt/short-nosed, minimum length 125 mm	1
Sterile eyewash solution, 10 mL single-use ampoules or sachets	6
Swabs, prepacked, antiseptic, packs of 10	1

Item	Quantity
Triangular bandages, minimum 900 mm	4
Wound dressings, sterile, non-medicated, large	3
First aid pamphlet (as issued by the St John Ambulance or the Australian Red Cross Society or as approved by WorkCover)	1

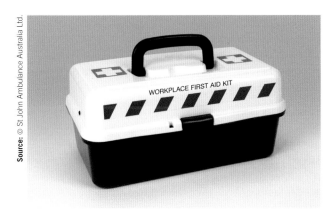

FIGURE 4.36 Type B first aid kit

FIGURE 4.37 Stored contents of the kit

asphyxiate people, and fire can have such a high degree of heat as to bring down steel or brick structures.

Being able to prevent and fight fires is part of every worker's responsibility. It is important for the safety of every worker on a job to understand the procedures to follow in the event of a fire.

Large construction sites and buildings should have firefighting teams responsible for each floor or the whole building. The firefighting team must be specially trained staff members who can direct the evacuation and firefighting operation until the fire brigade arrives.

It is therefore important to have a fire plan set up on the construction site. It may be quite simple; for example, a person discovering a fire should:
- rescue anyone in immediate danger, if it is safe to do so
- alert other people in the immediate area
- if equipment is available, take action to extinguish the fire before it takes hold
- confine the fire by closing any doors
- dial 000 for the fire brigade to attend
- contact the emergency coordinator, or warden, as soon as possible.

Understanding fires and what causes them

To understand how a fire can start and how to fight a fire, you must first understand the three basic elements necessary for combustion to begin (see Figure 4.38):
- Fuel can be any combustible material; that is, any solid, liquid or gas that can burn. Flammable materials are any substances that can be easily ignited and will burn rapidly.
- Heat that may start a fire can come from many sources; for example, flames, welding operations, grinding sparks, heat-causing friction, electrical equipment or hot exhausts.
- Oxygen comes mainly from the air. It may also be generated by chemical reactions.

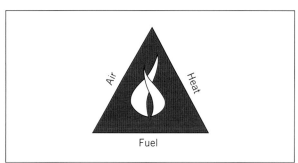

FIGURE 4.38 The elements necessary for a fire

If any one of the three elements is taken away, the fire will be extinguished.

Some of the more common reasons for fires on building sites are:
- burning off rubbish – site rubbish should be cleared away regularly and no burning off should take place, in accordance with EPA requirements
- electrical fires – caused by overloading equipment, faulty equipment and faulty wiring; all equipment should be carefully checked, maintained and used correctly
- contractors using naked flames – such as plumbers and structural steel workers; these contractors must ensure that they do not carry out naked flame operations within the vicinity of stored rubbish, paints, sawdust or any other highly flammable material
- smokers – carelessly disposing of cigarettes and matches; butane lighters may also be a source of ignition and should not be exposed to naked flames or other situations where ignition could occur.

Preventing and fighting fires

The first rule and probably the best rule is: don't give fires the chance to start.

Remember, fires need fuel, heat and oxygen; remove any one of these and the fire will go out. So look for and act on the following items, as prevention is better than cure:

- Remove unwanted fuel from the workplace (for example, rubbish and waste materials).
- Store fuels and combustible materials carefully; use safe carrying and pouring cans.
- Use only approved electrical fittings, and keep them in good order.
- Do not overload electrical circuits.
- Do not smoke at the workplace.
- Take special care if working with flammable liquids or gases.
- Be careful of oily rags, which can ignite from spontaneous combustion (for example, turps-soaked or linseed-oil-soaked rags).
- Avoid dust hazards. Many types of dust are so highly flammable that they can explode when mixed with air or when they are exposed to flame or sparks.

In the event of a fire

- Don't panic – keep calm and think.
- Warn other people in the building or on the worksite.
- Workers who are not needed should leave the building at once and assemble at the designated emergency evacuation point.
- Arrange for someone to phone the fire brigade.
- Have the power and gas supplies turned off if appropriate (some lighting may still be required). Close doors where possible to contain the fire.
- Stay between a doorway and the fire.
- Be aware of containers of explosive or flammable substances. Only remove them from the area if it is safe to do so.
- If it is safe to fight the fire, select the correct type of extinguisher, and have others back you up with additional equipment.
- Know how to use the extinguisher; be confident and attack the fire energetically.
- If the fire is too large for you to extinguish, get out of the building and close all doors. Assemble at the designated area.

Classes of fires and extinguishers

In the event of a fire you may be called on to fight the fire. Fire extinguishers have been grouped according to the class of fire on which they should be used. The class of fire is determined by the type of material or equipment involved in the fire. There are five main classes of fires:

1. Class A – ordinary combustible materials; for example, wood, paper, plastics, clothing and packing materials
2. Class B – flammable and combustible liquids; for example, petrol, spirits, paints, lacquers, thinners, varnishes, waxes, oils, greases, petrol or diesel-driven motor vehicles, and many other chemicals in liquid form
3. Class C – flammable gases; for example, LPG, acetylene and gas-powered forklifts
4. Class E – electrically energised equipment; for example, electric motors, power switchboards and computer equipment
5. Class F – cooking oils and fats.

Each class of fire must be fought using a different extinguisher; each type of extinguisher has its own colour code, as seen in Figure 4.39.

Extinguishers to use on a Class A fire

Water-type extinguishers are the best type to use. There are two types of water extinguisher: gas pressure and stored pressure. Water-type extinguisher details:

- Range of operation: up to 10 m.
- Methods of activating the extinguishers are different for each type; it is therefore important to read the instructions on the container before attempting to use it.
- Once activated, the jet of water should be directed at the seat of the fire.
- Any other type of extinguisher may be used except B(E) dry chemical powder.

Extinguishers to use on a Class B fire

Foam-type extinguishers are the best type to use. There are two types of foam extinguisher: gas pressure and stored pressure. Foam-type extinguisher details:

- Range of operation: up to 4 m.
- Methods of activating the extinguishers are different for each type; it is therefore important to read the instructions on the container before attempting to use it.
- Once activated, the jet of foam is directed to form a blanket of foam over the fire. This stops oxygen from getting to the fire long enough to allow the flammable substance to cool below its re-ignition point.
- Dry chemical and carbon dioxide (CO_2) may also be used but are not as effective as foam.

Extinguishers to use on a Class C fire

ABE dry chemical powder extinguishers are the best type to use. Dry chemical powder extinguisher details:

- Small sizes have a range of 3 m, with larger types effective up to 6 m.
- Powder is discharged through a fan-shaped nozzle. It should be directed at the base of the fire with a side-to-side sweeping action.
- Carbon dioxide (CO_2) extinguishers are not as effective on Class C fires so their use should be limited.

Portable Fire Extinguisher Guide

Two colour schemes exist for the extinguishers		Extinguishant	Class A	Class B	Class C	Class E	Class F	Red text indicates the class or classes in which the agent is most effective.
Pre 1999	Post 1999		Wood, Paper, Plastic	Flammable and combustible liquids	Flammable gases	Electrically energised equipment	Cooking oils and fats	
		WATER	YES	NO	NO	NO	NO	Dangerous if used flammable liquid energised electrical equipment and cooking oils and fat.
		FOAM	YES	YES	NO	NO	LIMITED	Dangerous if used on energised electrical equipment.
		POWDER	YES (ABE) / NO (BE)	YES (ABE) / YES (BE)	YES (ABE) / YES (BE)	YES (ABE) / YES (BE)	NO (ABE) / LIMITED (BE)	Look carefully at the extinguisher to determine if it is a BE or ABE unit as the capability is different.
		CARBON DIOXIDE	LIMITED	LIMITED	LIMITED	YES	LIMITED	Limited outdoor use.
		VAPORISING LIQUID	YES	LIMITED	LIMITED	YES	LIMITED	Check the characteristics of the specific extinguishing agent.
		WET CHEMICAL	YES	NO	NO	NO	YES	Dangerous if used on energised electrical (power on) equipment.

LIMITED indicates that the extinguishant is not the agent of choice for the class of fire, but may have a limited extinguishing capability

Source: Fire Gears Australia

FIGURE 4.39 Portable fire extinguisher guide

Extinguishers to use on a Class E fire

Carbon dioxide (CO_2) extinguishers are the best type to use. Carbon dioxide extinguisher details:

- Small sizes have a range of 1 m, with larger types effective to 2.5 m.
- They are useful for penetrating fires that are difficult to access.
- Use as close to the fire as possible. Aim the discharge first to the rear edge of the fire, moving the discharge horn from side to side, progressing forward until flames are out.
  Warning:
- Never use carbon dioxide in a confined area as this could cause suffocation. Move clear of the area immediately after use and ventilate the area once the fire is out.
- Never direct the nozzle at a person as carbon dioxide gas under pressure will freeze their skin causing frostbite-like burns.
- Never use water or foam extinguishers on Class E fires. AB and ABE dry chemical powder extinguishers will have limited effectiveness.

Note: Class F fires usually occur in kitchens and have not been covered here.

Operating and using fire extinguishers

In general, all fire extinguishers operate on the same principle, and you can use the acronym PASS to help you remember how to use them (see Figure 4.40).

FIGURE 4.40 Using a fire extinguisher – PASS

Pull (pin)

Pull the pin at the top of the extinguisher, breaking the seal. When in place, the pin keeps the handle from being pressed and accidentally operating the extinguisher. Immediately test the extinguisher (aiming away from the operator). This is to ensure the extinguisher works and also shows the operator how far the stream travels.

Aim

Approach the fire, stopping at a safe distance. Aim the nozzle or outlet towards the base of the fire.

Squeeze

Squeeze the handles together to discharge the extinguishing agent inside. To stop the discharge, release the handles.

Sweep

Sweep the nozzle from side to side as you approach the fire, directing the extinguishing agent at the base of the flames. After a Class A fire is extinguished, probe for smouldering hot spots that could reignite the fuel.

Other firefighting equipment

Fire blankets can be used on most types of fires (see Figure 4.41 and Figure 4.42). The function of the blankets is to smother the fire. Blankets are very useful on burning oils and electrical fires. Keep the blanket in place until the fire is out and for long enough to allow the flammable substance to cool below its re-ignition point.

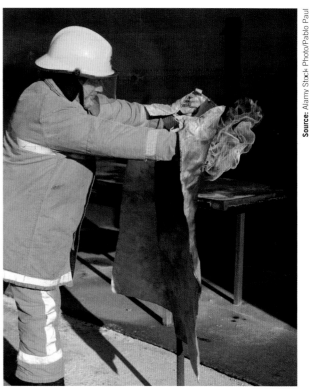

FIGURE 4.42 A fire blanket in use

Hose reels (see Figure 4.43) are the most effective means of fighting Class A fires. Extreme care is required to ensure that no live electrical equipment is in the area. Hose reels should not be used on liquid fires, as the water may spread the burning liquid.

FIGURE 4.41 A typical fire blanket packet

FIGURE 4.43 Hose reel

LEARNING TASK 4.3 EMERGENCY EVACUATION ACTIVITY

As a class you are required to conduct a rehearsal of an emergency evacuation for the site you are located on. Observe the activity and then discuss as a group the emergency rehearsal.

Before the rehearsal, you need to decide who will be taking on the role of warden and what procedures are required for the activity.

At the end of the rehearsal, undertake a debriefing that includes the following questions:
- Did all of the participants follow instructions of the warden?
- Was the procedure followed?
- What were the key activities that were crucial for the warden to ensure the safety of the participants?

COMPLETE WORKSHEET 8

SUMMARY

In Chapter 4 you have learnt about health and safety legislative requirements of construction work, construction hazards and risk control methods, health and safety communication and reporting processes, and incident and emergency response procedures. In addition, the chapter has:

- identified the general induction training requirements for the construction industry
- explored a four-step process to conduct a risk assessment
- covered the different types of personal protective equipment used in the construction industry
- discussed different types of safety signs and symbols
- identified the different fire safety equipment that is available.

REFERENCES AND FURTHER READING

Acknowledgement

Reproduction of the following Resource List references from DET, TAFE NSW C&T Division (Karl Dunkel, Program Manager, Housing and Furniture) and the Product Advisory Committee is acknowledged and appreciated.

Comcare (2015), *General Construction Induction Training*, **http://www.comcare.gov.au/preventing/prevention/work_health_and_safety_authorisations,_registrations_and_licences/construction_workgeneral_construction_induction_training**.

Graff, D.M. & Molloy, C.J.S. (1986), *Tapping Group Power: A practical guide to working with groups in commerce and industry*, Synergy Systems, Dromana, Victoria.

Ministry of Health (n.d.), *Building and Industry Hazards*, NSW Government, Sydney, retrieved from **http://www.health.nsw.gov.au/environment/hazard/Pages/default.aspx**.

National Centre for Vocational Education Research (2001), *Skill Trends in the Building and Construction Trades*, National Centre for Vocational Education Research, Melbourne.

NSW Department of Education and Training (1999), *Construction Industry: Induction & training: Workplace trainers' resources for work activity & site OH&S induction and training*, NSW Department of Education and Training, Sydney.

NSW Department of Industrial Relations (1998), *Building and Construction Industry Handbook*, NSW Department of Industrial Relations, Sydney.

Safe Work Australia (2012), *Emergency Plans Fact Sheet*, Safe Work Australia, Canberra.

Safe Work Australia (2014), *Safe Work Method Statement For High Risk Construction Work: Information sheet*, Safe Work Australia, Canberra.

Safe Work Australia (2017), *Work-related Traumatic Injury Fatalities, Australia 2016*, Safe Work Australia, Canberra.

Safe Work Australia, *Emergency Plans Fact Sheet*, **https://www.safeworkaustralia.gov.au/doc/emergency-plans-fact-sheet**.

SafeWork SA, **https://www.safework.sa.gov.au**.

TAFE Commission/DET (1999/2000), *Certificate 3 in General Construction (Carpentry) Housing*, course notes (CARP series).

WorkCover Authority of NSW (1996), *Building Industry Guide*, WorkCover NSW, Sydney.

WorkSafe Victoria (n.d.), *What is a Safe Work Method Statement?*, retrieved from **https://www.worksafe.vic**.

Relevant Australian Standards
AS 1216 Class labels for dangerous goods
AS 1318 Use of colour for the marking of physical hazards
AS 1319 Safety signs for the occupational environment
AS 1885.1 Measurement of occupational health and safety performance
AS 2626 Industrial safety belts and harnesses – Selection, use and maintenance
AS/NZS 1270 Acoustic – Hearing protectors
AS/NZS 1337.1 Eye protectors for industrial applications
AS/NZS 1715 Selection, use and maintenance of respiratory protective equipment
AS/NZS 1716 Respiratory protective devices
AS/NZS 1801 Occupational protective helmets
AS/NZS 2161.1 Occupational protective gloves
AS/NZS 2210.1 Safety, protective and occupational footwear guide to selection, care and use
AS/NZS 4501.2 Occupational protective clothing – General requirements.

GET IT RIGHT

The photo below shows a person using a drop saw unsafely. Identify the unsafe practices and provide reasoning for your answer.

Source: Richard Moran

_____
_____
_____
_____
_____

WORKSHEET 1

To be completed by teachers

Student competent ☐

Student not yet competent ☐

Student name: _____

Enrolment year: _____

Class code: _____

Competency name/Number: _____

Task

Read through the sections from the start of the chapter up to 'Basic roles, responsibilities and rights of duty holders', then complete the following questions.

1 Identify three reasons for the development of WHS laws at a national level.

_____

_____

_____

2 What is the purpose of one main piece of legislation for all of Australia in relation to WHS?

_____

_____

3 Name the national body that works to coordinate and develop policy, and assist in the implementation of the model WHS Act.

_____

_____

4 Identify the three groups that the national body seeks to build cooperation between.

_____

_____

_____

5 What is the purpose of regulations and codes?

_____

_____

6 What is the difference between regulations and codes?

_____

_____

PREPARE TO WORK SAFELY IN THE CONSTRUCTION INDUSTRY

7 What is the purpose of general construction induction training?

   _____

   _____

8 What sort of card do you need to hold before undertaking any construction works?

   _____

WORKSHEET 2

To be completed by teachers
Student competent ☐
Student not yet competent ☐

Student name: _____

Enrolment year: _____

Class code: _____

Competency name/Number: _____

Task

Read through the sections from the start of 'Basic roles, responsibilities and rights of duty holders' up to 'Construction hazards and risk-control measures', then complete the following questions.

1. As a duty of care, PCBU/employers must provide for the health, safety and welfare of their employees at work. List three ways this could occur.

   _____
   _____
   _____

2. What does the term 'duty of care' mean?

   _____
   _____

3. What obligation do employees have to PCBU/employers in relation to WHS?

   _____
   _____

4. There are several standard safe work practices on a construction site. List five safe work practices that you find to be the most important and why.

   _____
   _____
   _____
   _____
   _____

5 Identify five high-risk areas in the construction industry.

6 Why is it advisable to complete a SWMS for all tasks on a construction site?

7 There are four levels in the WHS framework. Identify each level and provide a brief explanation of the purpose of each document.

WORKSHEET 3

To be completed by teachers
Student competent ☐
Student not yet competent ☐

Student name: _____

Enrolment year: _____

Class code: _____

Competency name/Number: _____

Task
Read through the sections from the start of 'Construction hazards and risk-control measures' up to 'Construction hazards', then complete the following questions.

1 What two outcomes are you working towards when you undertake risk-management processes in the workplace?

   _____

   _____

2 Identify the four fundamental steps when managing risk in the workplace. Provide an explanation of each step.

   _____

   _____

   _____

   _____

   _____

   _____

   _____

   _____

3 When undertaking a workplace inspection, what should you look out for? Identify at least two things.

   _____

   _____

4 Once a hazard has been assessed through a ranking process, which risk levels require immediate action?

   _____

   _____

5. If you are unable to eliminate a risk, what should you be aiming to achieve?

6. When should PPE and safety signs be used as the main method of protection against a hazard?

7. Where could you find information about suitable controls for many common hazards?

8. How could you ensure that control measures are maintained in the workplace?

9. Why is it important for regular WHS reviews to occur?

10. List two instances that may trigger a WHS review to occur.

WORKSHEET 4

To be completed by teachers

Student competent ☐
Student not yet competent ☐

Student name: _____

Enrolment year: _____

Class code: _____

Competency name/Number: _____

Task

Read through the sections from the start of 'Construction hazards' up to 'Health and safety communication and reporting processes', then complete the following questions.

1. What is a hazard?

   _____

   _____

2. What is a chronic hazard?

   _____

   _____

3. What is a risk?

   _____

   _____

4. When should PPE and safety signs be used as the main method of protection against a hazard?

   _____

   _____

5. List the body parts that may be protected by using PPE.

   _____

   _____

6. List four physical hazards that PPE may be used to protect the wearer from.

   _____

   _____

   _____

   _____

7. List four chemical hazards that PPE may be used to protect the wearer from.

___

8. List two reasons why a safety helmet must be worn on a construction site.

___

9. When wearing a safety helmet, is it acceptable to wear a helmet without the helmet harness fitted? Why or why not?

___

10. How is the back of the neck protected from sunburn when wearing a safety helmet?

___

11. List the three hazard categories that eye protection is designed for.

___

12. List the two main PPE methods used to protect your hearing.

___

13. How can you recognise a disposable nuisance-dust mask?

___

14. Write down a suitable use for a P2 class dust filter.

___

15 What hazards do rubberised chemical-resistant long gloves protect the wearer from?

_____

_____

16 List the main requirements that safety footwear must have to provide maximum protection.

_____

_____

WORKSHEET 5

To be completed by teachers
Student competent ☐
Student not yet competent ☐

Student name: _____

Enrolment year: _____

Class code: _____

Competency name/Number: _____

Task

Read through the sections from the start of 'Health and safety communication and reporting processes' up to 'Procedures for reporting hazards, incidents and injuries', then complete the following questions.

1 Briefly describe the three main categories of common safety signage used in the construction industry.

   a   Picture signs

   b   Word-only signs

   c   Combined picture and word signs

2 Briefly describe the shape, colours and the detail/symbol found on a sign used to indicate a toxic hazard warning.

3 Briefly describe the background, detail/symbol and border found on a sign used to indicate an emergency exit sign.

4 What is the purpose of fitting a Warning or DANGER safety tag/accident prevention tag to an electric tool machine?

5 State the preferred position for safety signs.

PREPARE TO WORK SAFELY IN THE CONSTRUCTION INDUSTRY

6 Identify the following signs, stating what they represent.

WORKSHEET 6

Student name: _____

Enrolment year: _____

Class code: _____

Competency name/Number: _____

To be completed by teachers

Student competent ☐

Student not yet competent ☐

Task

Read through the sections from the start of 'Procedures for reporting hazards, incidents and injuries' up to 'Incident and emergency response procedures', then complete the following questions.

1 All workers have a responsibility to report any illness, accident or near miss they are involved in or see at the workplace. To whom should you report such incidents in the workplace?

_____

_____

2 Which two pieces of information are required when completing an incident report form?

_____

_____

3 Identify five unsafe acts that may lead to an accident or an incident in the workplace.

_____

_____

_____

_____

_____

4 What does it mean when a near miss occurs in the workplace?

_____

_____

5 What should a risk-management system include?

_____

_____

_____

_____

6 How do you become eligible for a workers' compensation claim?

WORKSHEET 7

To be completed by teachers
Student competent ☐
Student not yet competent ☐

Student name: _____

Enrolment year: _____

Class code: _____

Competency name/Number: _____

Task

Read through the sections from the start of 'Incident and emergency response procedures' up to 'Procedures for accessing first aid', then complete the following questions.

1 What arrangements should be in place to deal with an emergency?

   _____
   _____

2 Why is it important to have specific roles in the event of an emergency?

   _____
   _____

3 Which two pieces of information must be provided in an emergency plan?

   _____
   _____

4 What items should be identified in an evacuation diagram?

   _____
   _____

WORKSHEET 8

To be completed by teachers
Student competent ☐
Student not yet competent ☐

Student name: _____

Enrolment year: _____

Class code: _____

Competency name/Number: _____

Task
Read through the sections from the start of 'Procedures for accessing first aid' up to the end of the chapter, then complete the following questions.

1 What are the three basic elements necessary for combustion to begin?

   _____

   _____

   _____

2 List three common causes of fires on a site.

   _____

   _____

3 List three ordinary combustible materials that may be used as fuel for a Class A fire.

   _____

   _____

4 State the source or fuel and most suitable types of fire extinguishers for use on the following classes of fires.

   Class A

   _____

   _____

   Class B

   _____

   _____

   Class E

   _____

   _____

5. What type of extinguishers should never be used for Class E fires?

6. Fire extinguishers come in a variety of easily identifiable colours. State the common colour used to identify a foam-type extinguisher.

7. Relating to fire extinguishers, what does the acronym PASS mean?

   P
   A
   S
   S

8. Name two other items that can be used to fight a fire.

EMPLOYABILITY SKILLS

Using the following table, describe the activities you have undertaken that demonstrate how you developed employability skills as you worked through this unit. Keep copies of material you have prepared as further evidence of your skills.

Employability skills	The activities undertaken to develop the employability skill
Communication	
Teamwork	
Planning and organising	
Initiative and enterprise	
Problem-solving	
Self-management	
Technology	
Learning	

APPLY WHS REQUIREMENTS, POLICIES AND PROCEDURES IN THE CONSTRUCTION INDUSTRY

This chapter covers the outcomes required by the unit of competence 'Apply WHS requirements, policies and procedures in the construction industry'. These outcomes are:
- Identify and assess risks.
- Identify hazardous materials and other hazards on worksites.
- Plan and prepare for safe work practices.
- Apply safe work practices.
- Follow emergency procedures.

Overview

In Chapter 4, we covered preparation requirements to work safely in construction. This is essential for understanding your responsibilities in conducting safe work practices. The application of WHS practices is the next step from just having an awareness of your responsibilities. As a construction worker, from both a legal and moral perspective you are required to implement WHS policies in the workplace. Developing and applying an action-oriented approach to WHS will save lives, from taking action to save your own life to ensuring that no one else is put at risk.

You will need to build a broad knowledge of the many aspects of a construction site so that you are able to identify risks on areas of the site that do not directly involve the tasks you are required to perform.

Employers value people who fit into their workplace, solve day-to-day problems, manage their time and are keen to continue learning. These types of skills are known as employability skills. The employability skills you develop while working through this unit will be assessed at the same time as the skills and knowledge.

It is important to show evidence that you have developed these skills. Your trainer or assessor will discuss with you how to record your employability skills. The following table provides some examples of things you might do to develop employability skills in this unit.

EMPLOYABILITY SKILLS

Employability skill	What this skill means	How you can develop this skill
Communication	Speaking clearly, listening, understanding, asking questions, reading, writing and using body language.	• Liaise with others when organising work activities. • Interpret plans and working drawings.
Teamwork	Working well with other people and helping them.	• Work as part of a team to prioritise tasks. • Offer advice and assistance to team members and also learn from them.
Planning and organising	Planning what you have to do. Planning how you will do it. Doing things on time.	• List tasks in a logical order of action. • Select appropriate tools and materials.
Initiative and enterprise	Thinking of new ways to do something. Making suggestions to improve work.	• Initiate improvements in using resources. • Respond constructively to workplace challenges.
Problem-solving	Working out how to fix a problem.	• Participate in job safety analyses. • Check and rectify faults in tools and equipment if appropriate.
Self-management	Looking at work you do and seeing how well you are going. Making goals for yourself at work.	• Maintain performance to workplace standards. • Set daily performance targets.
Technology	Having a range of computer skills. Using equipment correctly and safely.	• Apply computer skills to report on project work. • Follow manufacturers' instructions.
Learning	Learning new things and improving how you work.	• Use opportunities for self-improvement. • Trial new ideas and concepts.

Identify and assess risks

The process of identifying and assessing risk is a necessity on all construction sites. The environment that is produced through the construction process creates situations that are highly dangerous and have led to many deaths. Having practices in place to manage any risk and improve on the management of the risks will lead to saving lives.

In Chapter 4, we looked at the principles and processes involved in risk management and the processes involved in reporting risks. Here, we will look more closely at assessing these risks on site and your responsibilities in reporting these.

Identify, assess and report hazards in the work area to designated personnel

Identifying a hazard is the act of recognising situations and items that have the potential to cause harm. In Chapter 4, we looked at the four fundamental steps of the risk-management process. We learned how to find hazards, and that hazards can come from a range of different facets of the workplace. The main area is the physical work environment, particularly any work that needs to take place at height. This is challenging on many construction sites, and over time has been the cause of many deaths in the construction industry. According to Safe Work Australia, there were 73 deaths between 2014 to 2018 from falls from height and by being hit by falling objects.

Another hazardous area is the use of equipment, materials and related substances. This area relates to tasks that construction workers engage with on a daily basis, irrespective of the size of the construction project.

Further to this the other area that is a concern is how the work is performed. In many instances all protective measures can be in place though if not utilised correctly or procedures followed there is the potential for an incident to occur.

Table 5.1 is an extract from Safe Work Australia's *Code of Practice: How to manage work health and safety*

TABLE 5.1 Examples of common hazards

Hazard	Example	Potential harm
Manual tasks	Tasks involving sustained or awkward postures, high or sudden force, repetitive movements or vibration	Musculoskeletal disorders such as damage to joints, ligaments and muscles
Gravity	Falling objects, falls, slips and trips of people	Fractures, bruises, lacerations, dislocations, concussion, permanent injuries or death
Psychosocial	Excessive time pressure, bullying, violence and work-related fatigue	Psychological or physical injury or illness
Electricity	Exposure to live electrical wires	Shock, burns, damage to organs and nerves leading to permanent injuries or death
Machinery and equipment	Being hit by moving vehicles, or being caught in moving parts of machinery	Fractures, bruises, lacerations, dislocations, permanent injuries or death
Hazardous chemicals	Acids, hydrocarbons, heavy metals, asbestos and silica	Respiratory illnesses, cancers or dermatitis
Extreme temperatures	Heat and cold	Heat can cause burns and heat stroke or injuries due to fatigue. Cold can cause hypothermia or frost bite
Noise	Exposure to loud noise	Permanent hearing damage
Radiation	Ultra violet, welding arc flashes, microwaves and lasers	Burns, cancer or blindness
Biological	Micro-organisms	Hepatitis, legionnaires' disease, Q fever, HIV/AIDS or allergies

Source: Based on Safe Work Australia © Commonwealth of Australia

risks, and provides examples of the hazards and their potential harm that they may cause.

Once the hazard has been identified, you need to assess the risk associated to the hazard. Many workplaces will have a checklist that will assist in identifying a hazard and then determining the likelihood of a related incident occurring. This will include identifying the potential severity of such an incident. Working through the process of conducting a risk assessment will also require you to identify if any control measures will be effective and what action is required to control the risk. The last perspective is to determine the urgency for any of the control measures to be actioned. This is an important step because if a hazard has been identified, but the control is yet to be put in place, the health and safety of people may be jeopardised.

Report safety risks in the work area based on identified hazards

The outcome of completing the specified checklist and risk assessment process will result in several outcomes that need to be reported to the appropriate people on the construction site. It is important at this stage to understand the difference between a hazard and a risk. A hazard is something that can cause harm such as a brick falling off a scaffold; risk is the likelihood or probability of a hazard causing damage or injury.

In the first instance, your immediate supervisor is the appropriate person to report any safety hazard or risk to. A hazard report form should be completed during the inspection of the hazard (see Figure 5.1). Depending on the structure of the templates that are used by the organisation, there should be clear outcomes or recommendations from the activity. The outcomes and recommendations provided in the hazard report form will give guidance for senior staff to implement in the workplace.

On large construction sites there will be a designated safety officer who will collate the outcomes of the inspections and recommendations and lodge them in a register for the construction site. On smaller sites this will be the supervisor or the direct line manager.

Follow safe work practices and duty of care requirements for controlling risks

As indicated in Chapter 4, there are legal obligations for all workers on a construction site that need to be followed or penalties may be incurred. It is common to see workers not follow the appropriate work practices or work instructions. It is in such instances that an accident will often occur. There is often pressure to complete tasks as quickly as possible and many workers knowingly risk their lives and the lives of others for the sake of getting the job done. However, just as you do not compromise the quality of your work, you should never place your safety at risk.

HAZARD REPORT FORM			
Job Location:			Date identified:
Name:			Reported to:
Contact Number:			
DESCRIPTION OF HAZARD			
Location of hazard on site:			
Corrective action taken Yes / No		Corrective action required Yes / No	
DESCRIPTION OF ACTION TAKEN			
Has the hazard been accepted by Management? Yes / No			
Has the hazard been addressed? Yes / No			
Hazard reporter sign off	Name:	Signature:	Date:
Management sign off	Name:	Signature:	Date:

FIGURE 5.1 Typical hazard report form

There is no excuse for not being aware of work practices or work instructions as there are multiple layers of information relating to safety on a construction site. This includes the WHS general induction for all construction sites and a site-specific induction process. Best practice also suggests a toolbox talk at the beginning of each work day and task.

Contribute to WHS, hazard, accident or incident reports

When you are given the opportunity to be involved in any processes that identify hazards, it is important that you participate from two perspectives: first, so that you become aware of hazards on the construction site and the control mechanisms that are put into place; and second, so that you build your own capability to conduct hazard identification processes such as checklists.

You can also become involved with any safety committees that are formed in your workplace. The *Work Health and Safety Act 2011* indicates that a business must establish a health and safety committee within two months of it being requested by staff. Many organisations will already have a committee in place where workers and management can contribute to the overall safety requirements within the organisation.

If there is an incident or an accident in your workplace, the main action that you should take is to be involved with any investigations regarding the incident. Your involvement may assist in preventing future incidents occurring. Further, best practice is not to wait for an incident to occur, so you should call out any situation that raises any concern when it comes to WHS and hazards.

Identify hazardous materials and other hazards on worksites

The construction industry has a wide variety of materials that are made up of many different compounds. These materials in different forms may be harmful to humans (recall the relevant signage used in identifying and communicating these hazards discussed in Chapter 4). It has been found that some materials that have been used in the past are no longer safe to use. An example of this is asbestos, which is covered in more detail in Chapter 10. When any demolition and refurbishment work is to be performed workers may be exposed to such materials and appropriate measures need to be put in place to minimise any harmful effects.

There are several pieces of legislation that are designed to provide guidance when it comes to handling hazardous materials on a construction site. Again, a good example of this is where asbestos has been used in the past.

Identify and use hazardous materials in line with legislative and workplace requirements

For large construction projects there should be a hazardous material management plan. The purpose of such a plan is to ensure that there are appropriate measures in place to safely handle and transport such material, in line with relevant legislative requirements.

Common hazardous materials found in existing structures include the following:
- asbestos (friable and non-friable) – found in wall and roof linings, pipes and insulation
- polychlorinated biphenyls (PCBs) – organic chemicals that have been used in paints, plastics and hydraulic equipment.
- synthetic mineral fibres (SMFs) – a broad term that covers glass, rock, alumina and silica
- lead-based paint – products that have a high lead content
- copper chromium arsenate (CCA) – used to treat timber to prevent wood decay and protect against insect attacks
- cement and lime
- fuels – such as petrol and diesel as these items are very flammable
- gases – there are different types of gas used on a construction site on a regular basis.

Apply measures to control risks and hazards effectively and immediately

Controlling risks and hazards on a construction site is the first priority of all workers on site. It is important that recommendations from safety data sheets (SDSs) are implemented immediately. If you see or become aware that the appropriate controls have not been implemented, or if they have been tampered with, you need to immediately take steps to make the situation safe. Such actions may be only temporary until further support or material is acquired to complete the required control measures. When completing a JSA or SWMS, you will usually be able to identify risks and hazards. During the process of completing the documentation, you will have a chance to identify and put in place appropriate control measures. In some circumstances this may mean that specialists come to the workplace to implement the identified control measures. An example of this may be where it has been identified that scaffolding is required to work at heights. To install scaffolding above a certain height, you need to have a licence.

Use appropriate signs and symbols to secure hazardous materials

Undertaking a risk assessment will assist in determining the appropriate signage required for the situation. In most instances the construction site will already have many of the required safety signs in place, but if you are setting up a new site this may be a task that needs to be performed.

At times additional signage may be required due to a particular task that is being performed, often for a short period of time. An example of this is when an explosive powered tool is going to be in use (see Figure 5.2). If you are performing such tasks, you need to ensure that all appropriate signage is in place before the task commences.

FIGURE 5.2 Sign indicating an explosive power tool is in use

Identify asbestos-containing materials on a worksite and report to designated personnel

To become proficient in identifying material containing asbestos is a skill that is developed over time. A key aspect of identifying such materials is to look beyond the material itself – you need to take in other factors that help to conclude that there may be asbestos, such as the age and style of the construction, the component of the building that is being considered and any typical physical features of the material. Then professionally trained and licenced staff can inspect the suspicious material.

As with any hazardous material, you will need to complete a hazard report form (see Figure 5.1) and submit that to your supervisor, the appropriate tradesperson or the leading hand.

> **LEARNING TASK 5.1 HAZARDOUS MATERIALS**
>
> Research one of the eight hazardous materials identified in this chapter and gather the following information:
> - handling and storage
> - disposal requirements (if any)
> - firefighting measures
> - first aid measures
> - PPE requirements.
>
> Note that safety data sheets will hold most of the information required.

COMPLETE WORKSHEET 1

Plan and prepare for safe work practices

Safe work practices do not just happen on their own – you need to make sure that the appropriate actions are taken to make a workplace safe to operate in. These may be requirements for a particular type of mobile scaffolding, harnesses, signage or PPE.

Identify, wear, correctly fit, use and store appropriate PPE and clothing

Identifying the appropriate PPE requirements is straightforward when there are numerous signs on the job site. However, you may need to determine the PPE required before you arrive on the job site. If the nature of the work task is known before you turn up on site, this will assist in determining the correct PPE. Many tradespeople have a PPE kit that will travel from job to job with them.

Being aware of the appropriate PPE to use is only one aspect of its proper use – the correct application and fitting of PPE is just as important (see the discussion in Chapter 4). Too many times PPE is not used in the correct manner or fitted incorrectly, to the point that the intended benefits of the PPE are not realised. Fitting PPE correctly is a skill that needs to be developed very quickly when starting to work in the construction industry.

Many manufacturers provide instructions on the correct way to fit and wear the PPE that they produce. An example of this is the instructions to fit ear plugs (see Figure 5.3).

The use of specific work wear is essential. There are different requirements for the different types of work that are to be performed. As an example of this, the protective clothing while welding, is different from the protective clothing used when pouring concrete. You will need appropriate eye protection and a respirator as well as other PPE shielding your body from stray sparks and dangerous fumes when undertaking welding activities. You will not need many of these items for pouring concrete, although you will need PPE over some of the same parts of your body for different purposes; for example, you will need waterproof boots, gloves and safety glasses to protect yourself from skin irritation and even potential burns from the chemicals present in wet concrete. Return to Chapter 4 for more detail about selecting suitable PPE.

Performing a task in specialised clothing can prove challenging. This is why work practices need to be planned and considered based not only on the task itself but also the implications of the approaches used to manage an identified hazard.

Select equipment and materials, and organise tasks in accordance with workplace procedures

The selection of the correct tools, equipment and materials is an essential step when managing safety on a construction site. The use of the correct tools and equipment will make the performance of a task safer for all workers on site. For example, the incorrect use of tools may result in jamming of cutting blades or drill bits overheating. This will also lead to potentially poor-quality outcomes.

The correct selection of material may have potential impacts after the construction of the project has occurred. There are many instances where inferior material has been substituted for the specified material on site. This becomes a risk at the stage of installation and may remain an ongoing risk once the project has been completed.

The coordination of tasks with other tradespeople who are working on the same job site has the potential for safety issues to eventuate. This could range from having material delivered on site to working in confined spaces. Proper planning processes (discussed in Chapter 2; see Figure 2.2) will help tasks to be completed in sequence and avoid overlap tasks. Many trades and tasks rely on each other to be completed before the next task can be performed. Many sites do attempt to separate tasks into a lock step sequence to minimise any tasks that may cross over each other.

Determine required barricades and signage, and erect at the appropriate location

Barricades and signage are used in many construction projects. The purpose of the barricades and signage is to provide warning of potential danger. The signage will provide direction as to what to do for both vehicles and pedestrians. As standard practice the signage should be positioned where there is a clear line of sight. The barricades that are typically used on construction sites consist of temporary fencing. The fencing is a physical

Earplugs Fitting Instructions

Keys to Successful Hearing Protection with Earplugs

① Disposable
Roll-Down Foam   MAX®

With clean hands, roll the entire earplug into narrowest possible crease-free cylinder.

Reach over your head with a free hand, pull your ear up and back, and insert the earplug well inside your ear canal.

Hold for 30 – 40 seconds, until the earplug fully expands in your ear canal. If properly fitted, the end of the earplugs should not be visible to someone looking at you from the front.

② Reusable
Push-In Foam   Trust Fit™

While holding the stem, reach a hand over your head and gently pull the top of your ear up and back.

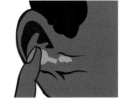

Insert earplug so foam tip is well inside the ear canal. Use a gentle rocking motion while pushing earplug into ear canal to ensure a deep fit.

If properly fitted, the tip of the earplug stem may be visible to someone looking at you from the front.

③ Reusable
Pre-Molded Push-In   SmartFit®

While holding the stem, reach a hand over your head and gently pull top of your ear up and back.

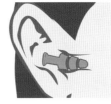

Insert the earplug so all flanges are well inside your ear canal.

If properly fitted, the tip of the earplug stem may be visible to someone looking at you from the front.

④ Banded
Tension Fit   QB2®HYG

Position band under your chin as shown above. Use your hands to press the ear pods well into the ear canal using an inward motion.

Protection levels are improved by pulling your ear up and back when fitting as shown.

In a noisy environment, lightly press the band inward with your fingertips as shown. You should not notice a significant difference in noise level.

DOs and DON'Ts of Howard Leight Earplugs

Proper Fit
If either or both earplugs do not seem to be fitted properly, remove the earplug and reinsert.

Removal
Gently twist earplug while slowly pulling in an outward motion for removal.

Acoustical Check
In a noisy environment, with earplugs inserted, cup your hands over your ears and release. Earplugs should block enough noise so that covering your ears with your hands does not result in a significant noise difference.

WEAR
- Read and follow all earplug fitting instructions.

SELECTION
- Avoid overprotection in minimal noise environments – in selecting the best earplug for your situation, consider noise levels and your need to communicate with co-workers or hear warning signals on the job.

MAINTENANCE
- Prior to fitting, inspect earplugs or ear tips for dirt, damage, or hardness. Discard & replace immediately if compromised.
- For proper hygiene discard disposable earplugs after use.
- With proper maintenance reusable earplugs should be replaced every 2-4 weeks. Push In Foam earplugs should be replaced every 5 days.
- Wash with mild soap/water, pat dry or air dry, and store in a case when not in use.
- Clean regularly and replace ear tip pods every 2-4 weeks on banded earplugs.

FIGURE 5.3 Manufacturer's instruction for fitting earplugs correctly

barrier that prevents any members of the public wandering onto a job site, and is often used to barricade construction sites where excavations are taking place. Typical excavations create a falling hazard, which is one of the leading causes of death on construction sites (see Figure 5.4). Temporary fencing is a combination of fence panels, infills, counterweight support systems and a base. All of these components need to meet AS 4687-2007 Temporary fencing and hoardings.

FIGURE 5.4 Temporary fencing and signage around deep excavation site

At the planning stage of the project, you will need to determine the extent of the fencing that will be required. This is usually taken from site plans, which will consider the access and egress requirements for the site to function. There are standard checklists that can assist with undertaking such a process (see Figure 5.5).

Depending on the site circumstances, other temporary barriers may be required, such as during delivery of material and equipment. When a roadway needs to be blocked, portable barriers can be set up to manage pedestrian traffic. When it comes to managing vehicles, then spotters and signallers can be used:

- A spotter's job is to be the eyes and ears of the workers. They will direct the workers to stop or wait if a vehicle is approaching.
- A signaller directs traffic; for example, so that scaffolding can be carried safely across a road.

Apply safety data sheets, job safety analyses and safe work method statements

Safety data sheets (SDSs), also known as material safety data sheets, will be in place for all materials that are used on the construction site. These are supplied by the manufacturer of the material (see Figure 5.6). The content of the SDS is information relating to:

- the actual material or chemical and the different ingredients that make up the product
- any hazards that relate to the material
- the physical properties of the chemical and ingredients
- safe handling and storage requirements
- emergency procedures if there are any injuries related to the chemical
- any first aid instructions
- how the material should be transported.

Safe Work Australia has a code of practice for SDSs that covers all aspects of materials and their use. As some materials and goods are imported, SDSs should be checked against WHS legislation.

The information in the SDS should be understood by all workers who will be handling the material. Staff who are responsible for the site and business also need to be familiar with the SDS, as they will be responsible for ensuring the appropriate management and storage facilities are available for the material.

A job safety analysis (JSA) and a safe work method statement (SWMS) are used to identify any hazards and risks associated with a particular type of task. It is recommended that a SWMS or JSA is completed for any activity conducted on a construction site.

Again, any workers who are involved in performing the identified task should have a full understanding of the content of the SWMS or JSA. (See Figure 4.3 in Chapter 4 for an example of a standard SWMS.) Variations of these documents may be used, but they will all follow the same principles to assist in managing safety on a construction site.

LEARNING TASK 5.2

SDS PLANNING INFORMATION

Search the internet for an SDS on the materials listed in the following table. Identify the manufacturer of the material and list at least three PPE items that are identified for use in the SDS.

Material	Manufacturer	PPE requirements
GP cement		
Silicon		
Treated pine		
Ceramic tiles		

As a class, identify which pieces of PPE are unique to the different materials and discuss why this may be the case.

COMPLETE WORKSHEET 2

SAFETY AND SECURITY FENCING FOR CONSTRUCTION SITES

SITE-SPECIFIC RISK ASSESSMENT (SSRA)

The principal contractor (builder) should complete this SSRA before starting work, and keep a copy on file.

January 2021

SCOPE OF WORKS	Residential construction				
COMPANY		DATE		TIME	
SITE ADDRESS					
PERSON COMPLETING THIS ASSESSMENT (Print)		SIGNATURE			

PART 1

This checklist will help you to determine if you need safety and security fencing for your construction site. If you tick 'yes' to any question, you may need to install fencing.

KNOWN HAZARDS AND POTENTIAL RISKS	Yes	No
Is the site near to other residential buildings where children may play out of site working hours?	☐	☐
Is the site near to local schools or routes that children may take home?	☐	☐
Are there excavations (confined spaces) on site – e.g. open trenches, utilities, pier holes or swimming pools?	☐	☐
Are there any impaling materials left on site – e.g. reo bars, nails in timber or form work materials?	☐	☐
Is there a potential for collapse – e.g. of concrete formwork or battered areas of soil from site cuts?	☐	☐
Are there deliveries of building materials to site?	☐	☐
Is there a potential for the public to be hit by falling objects related to the construction process on the site?	☐	☐
Is the public exposed to high-risk construction work, or plant and equipment?	☐	☐
Is there a potential for construction building debris (including wind-blown debris) to leave the site – i.e. inadequate waste containment?	☐	☐
Is there a potential for unauthorised access on this site by any individuals? Consider access: • to hazardous materials • to plant and materials • overnight or when the site is unsupervised • by climbing over or under the existing fence • to electricity or services • to scaffolding, upper storey floor levels, ladders or penetrations.	☐	☐
Are there any other hazards on site that need to be isolated?	☐	☐

If you tick 'yes' to any question but choose not to install fencing, explain here how you will secure the site:

_____
_____
_____
_____

Note:
Consider installing safety signage – e.g. 'Danger – excavation' or 'Danger – no unauthorised access'.
If it is not practical for you to use safety and security fencing (e.g. if space is restricted or the fencing itself could pose an additional hazard), you may be able to barricade off a danger area with a suitable hi-visibility mesh barrier, using 1.0–1.2 m high star posts.

Source: Based on Work Cover NSW

FIGURE 5.5 Standard checklist assessing the fencing requirements of a construction site

FIGURE 5.6 Typical safety data sheet supplied by a manufacturer

---

SAFETY DATA SHEET

1. IDENTIFICATION OF THE MATERIAL AND SUPPLIER

1.1 Product identifier

Product name BLENDED CEMENT

Synonyms BLUE CIRCLE BUILDERS CEMENT • BLUE CIRCLE SPECIAL PURPOSE CEMENT • BUILDERS CEMENT • FA BLEND • FLY ASH BLEND • GB CEMENT • GENERAL BLEND • GENERAL PURPOSE BLENDED CEMENT • GP SLAG BLEND • LOW HEAT CEMENT • MARINE CEMENT • POZZOMENT • SLAG BLEND • SLAGMENT • SPECIAL PURPOSE CEMENT • SULFATE RESISTING CEMENT • TRIPLE BLEND • TYPE GB • TYPE LH • TYPE SR

1.2 Uses and uses advised against

Uses BINDER - REFRACTORIES • CONCRETE • CONSTRUCTION • GROUT • INDUSTRIAL APPLICATIONS • MANUFACTURE OF CEMENTS • MASONRY • MORTAR • SOIL STABILISATION

1.3 Details of the supplier of the product

Supplier name BORAL AUSTRALIA
Address Level 18, 15 Blue Street, North Sydney, NSW, 2060, AUSTRALIA
Telephone (02) 9220 6300
Website http://www.boral.com.au

1.4 Emergency telephone numbers

Emergency 1800 555 477 (8am – 5pm WST)
Emergency (A/H) 13 11 26 (Poisons Information Centre)

2. HAZARDS IDENTIFICATION

2.1 Classification of the substance or mixture

CLASSIFIED AS HAZARDOUS ACCORDING TO SAFE WORK AUSTRALIA CRITERIA

Physical Hazards
Not classified as a Physical Hazard

Health Hazards
Skin Corrosion/Irritation: Category 2
Serious Eye Damage / Eye Irritation: Category 1
Specific Target Organ Toxicity (Single Exposure): Category 3 (Respiratory Irritation)
Carcinogenicity: Category 1
Specific Target Organ Toxicity (Repeated Exposure): Category 2

Environmental Hazards
Not classified as an Environmental Hazard

2.2 GHS Label elements

Signal word DANGER

Pictograms

Page 1 of 9

SDS Date: 03 Jan 2020
Version No: 3

Apply safe work practices

Safe work practices means working in a way that minimises risk to yourself, other people, equipment, materials, the environment and work processes.

Carry out tasks in a manner that is safe for operators, other personnel and the general community

Safe work practices need to be established and implemented in the workplace. Too often work practices focus only on getting the job done as quickly as possible. This type of behaviour and culture in the workplace places everybody at risk.

The first step to establish and maintain safe work practices is to have regular toolbox talks so all workers have a clear understanding of what the goals are and how to achieve them. If there are work instructions, it is essential that you follow them. If there are aspects of the work instruction that you can offer feedback on for improvement, the practices and culture of your workplace will improve. The improvements may not only improve aspects of WHS but also result in higher quality work.

Basic safe work practices on construction sites include:
- keeping a clear and clean workplace
- ensuring that sufficient consideration is given to others in the workplace when conducting a high-risk activity
- using appropriate PPE
- ensuring no practical jokes are practised
- not accepting bullying or any form of harassment
- ensuring first aid equipment is available
- ensuring facilities such as lunch rooms, change rooms and bathrooms are available on site and correctly maintained by everybody who uses the facilities.

Use plant and equipment in accordance with specifications, regulations and Australian Standards

The use of plant and equipment is a common activity on a construction site and some workers will modify equipment to suit their personal preferences. This type of practice should not be encouraged or supported in the workplace. The alteration of safety equipment will increase the chance that someone will be injured. The types of injuries that can occurred include:
- amputation of limbs
- crushing injuries
- broken bones
- burns from electric shock
- death.

There is a wide range of Australian Standards that relate to plant and the workplace (see Table 5.2). These Standards cover several aspects relating to plant and equipment, including design, manufacture and use.

TABLE 5.2 Table indicating the Australian Standards related to a particular piece of equipment

Plant description	Reference number	Standard title	Design	Make	Use
Cranes, including hoists and winches	AS 1418 (Series)	Cranes including hoists and winches	•	•	
	AS 4991 – 2004	Lifting devices	•	•	•
	AS 2550 (Series)	Cranes – Safe use			•
Earth-moving machinery	AS 2294.1	Earth-moving machinery – Protective structures – General	•	•	
	AS 2958.1	Earth-moving machinery – Safety, Part 1: Wheeled machines – Brakes	•	•	•
	ISO 6165	Earth-moving machinery – Basic types – Identification and terms and definitions	•		
	ISO 6746-1	Earth-moving machinery – Definitions of dimensions and codes – Part 1: Base machine	•		
	ISO 6746-2	Earth-moving machinery – Definitions of dimensions and codes – Part 2: Equipment and attachments	•		
	ISO 7133	Earth-moving machinery – Tractor scrapers – Terminology and commercial specifications	•		
Explosive powered tools	AS/NZS 1873 (Series)	Power-actuated (PA) hand-held fastening tools	•	•	•
Hand-held electric tools	AS/NZS 60745	Hand-held motor operated electric tools – Safety – General requirements	•	•	•

Plant description	Reference number	Standard title	Design	Make	Use
Fall arrest	AS/NZS 1891.1	Industrial fall-arrest systems and devices – Harnesses and ancillary equipment	•	•	
	AS/NZS 1891.4	Industrial fall-arrest systems and devices – Selection, use and maintenance			•
	BS EN 1263-1:2002	Safety nets – Safety requirements, test methods	•		

Follow procedures and report hazards, incidents and injuries

Reporting a hazard is a step in the right direction to prevent injury from occurring. Many organisations will have a reporting process for staff to follow.

The act of following procedures when an incident has occurred can be difficult, as the first priority is to provide first aid. However, it is important that when the immediate danger has passed, appropriate authorities are contacted. This may include the police and your local Safe Work authority. Many insurance companies will rely on such reports to determine their involvement in supporting the recovery of any injured workers.

Do not use tools and equipment in areas containing identified asbestos

Any activity to do with asbestos material should be left for people who are appropriately trained and licensed for such activity. The risk to your health and that of your co-workers is of utmost importance. There are no circumstances where you should be using any tools or equipment on material that has been identified as potentially containing asbestos. Information relating to asbestos on a construction site is covered in detail in Chapter 10.

Identify and follow requirements of worksite safety signs and symbols

The ability to follow the requirements of safety signs and symbols on a construction site requires you to have a full understanding of the different types of signs and the categories that the signs represent (see Chapter 4 for a full discussion of signs and symbols). Following the requirements of site signage also relies on you using approved equipment and understanding the safety requirements of a task or site before you commence working. It is common practice for workers to have their own kit, which in the most part will consist of PPE.

Maintain the worksite area to prevent incidents and accidents, and meet environmental requirements

Housekeeping on a construction site should be a high priority. There is a whole range of reasons for such practices to be in place. The quality of the finished product can be affected by the work environment. If there is too much debris and equipment in the way, there is the potential for workers not being able to undertake the most appropriate approach to a particular task. A messy site will also increase the risks of a workplace injury occurring. In some instances builders have been known to hold a bond from subcontractors until they have satisfactorily cleaned the work area.

The use of a checklist can be useful (see Figure 5.7).

LEARNING TASK 5.3

TOOLBOX TALK PARTICIPATION

Watch the video of a toolbox talk on 'Safety rules for using power tools' at https://www.youtube.com/watch?v=JYnD_nMcnJ8.

Indicate the main safety tips that you have picked up from watching this video. What are the key safety rules? Discuss the safety rules within your class group. Questions for discussion include the following:
- Have you been trained to use the tool?
- Is the machine or appliance in perfect working order?
- Are the guards in place?
- Does it have a current safety tag?
- Has it been serviced to the manufacturer's specifications?
- Are you drug and alcohol free?
- Do you have the right PPE?
- Are you fully focused and concentrating on the task?
- Have you planned your workload to eliminate risks?
- Is your work area tidy?
- Are you keeping your workmates safe?

 COMPLETE WORKSHEET 3

Site Inspection Checklist – Construction site housekeeping			
This checklist is to be used on construction sites to ensure that the site is maintained to a standard that will allow for all workers to undertake their work in a safe manner with reduced hazards.			
Site address:		Date of Inspection:	
Name of inspector/s:			
	YES	NO	ACTION REQUIRED
FLOOR AREAS			
Loose free objects not in any walkways limiting access	☐	☐	☐
Objects protruding out of flooring	☐	☐	☐
Clean of debris / waste (saw dust) etc	☐	☐	☐
STORAGE			
Is there adequate rubbish bins and recycling bins provided?	☐	☐	☐
Has any waste material been placed in the appropriate bins?	☐	☐	☐
Are there areas to safely store materials or plant for the construction work?	☐	☐	☐
Has the material delivered to site is stored in a safe manner and will not fall?	☐	☐	☐
Are combustible and flammable substances and other hazardous chemicals safely stored and clearly identified?	☐	☐	☐
ENVIRONMENT			
Is adequate lighting available to perform work?	☐	☐	☐
Is there sufficient ventilation to confined areas?	☐	☐	☐
Is firefighting equipment available?	☐	☐	☐
HEIGHT WORK			
Are there fall prevention systems in place e.g., scaffolding?	☐	☐	☐
Is the fall prevention system in proper working order?	☐	☐	☐
SERVICES			
Is there a water supply on the site?	☐	☐	☐
Is there sufficient drainage for the water run off?	☐	☐	☐
Are all power leads utilised in a safe manner?	☐	☐	☐
NOTES:			

FIGURE 5.7 Site housekeeping inspection checklist

Follow emergency procedures

Following emergency procedures includes identifying the correct personnel, following safe workplace procedures in the event of an emergency, and effectively carrying out emergency response and evacuation procedures, when required.

Identify designated personnel

Any site induction should include information about the personnel to contact in the event of an emergency situation. This information will be in the emergency plan for the site. Table 5.3 is a typical list of contacts in an emergency plan. The name and direct contact details of the particulars for the specific site will be filled in.

TABLE 5.3 Typical list of contacts in an emergency plan

Contact	Name	Phone number
Emergency services – triple zero	Fire/police/ambulance	000
Police		
Fire warden		
First aid officer		
WHS officer		
Security office		
Reception		
State Emergency Services (SES)		
Nearby businesses		
Poison information line		
Utilities		
COVID-19 information line		

Given that most people have mobile phones, it is generally to contact the appropriate person. Some sites may have an internal communication system, such as a two-way radio system. This assists to keep communication to site-specific interactions.

Having a list of emergency contacts also gives guidance about who to contact depending on the type of emergency. In many instances, the first aid officer will check to ensure that any further contacts have been notified if required. One of the biggest mistakes made in emergency situations is to assume that someone else has made contact with the relevant officer or emergency service.

Follow safe workplace procedures for dealing with accidents, fire and other emergencies

It is important to understand an organisation's policies and procedures regarding accidents and emergencies. The use of equipment, such as firefighting gear, will also require a level of basic training in order to be able to operate it effectively. The need to have a basic level of training is also applicable to the area of first aid. Regardless of the requirements of having a first aid-trained person on site, depending on the number of people, it is in the best interest of yourself and other workers if you have had appropriate first aid training. Such training will provide you with sound current techniques that will assist in addressing any situation that you may be exposed to in the workplace. Revisit Chapter 4 to review information about safe workplace procedures and responses.

Describe, practise and effectively carry out emergency response and evacuation procedures

Safe Work Australia's *Model Code of Practice: Managing the work environment and facilities* stipulates that evacuation practice drills should be held at least every 12 months. Depending on the construction site and the organisation's requirements, practice drills may occur on a more regular basis. The organisation may have several drills that are required at the different stages of construction. Evacuation practices required when the project is at excavation stage will be different from those required once the structure is in place. Drills will test the effectiveness of the emergency plan and the supporting systems and procedures.

 COMPLETE WORKSHEET 3

SUMMARY

In Chapter 5 you have learnt about the application of work health and safety requirements, policies and procedures. In addition, the chapter has:
- explored different aspects of identifying and assessing risks through reporting procedures and following safe work procedures
- identified hazardous materials and other hazards on site
- discussed the different types of work practices on a construction site and how they are applied
- identified how to follow basic emergency procedures.

REFERENCES AND FURTHER READING

NSW Government, *Work Health and Safety Act 2011 No 10*.

Safe Work Australia, *Emergency Plan Template*, https://www.safeworkaustralia.gov.au/doc/emergency-plan-template.

Safe Work Australia, *Guide for Major Hazard Facilities – Information, training and instruction for workers and other people at the facility*, https://www.safeworkaustralia.gov.au/doc/guide-major-hazard-facilities-information-training-and-instruction-workers-and-other-persons.

Safe Work Australia, *Model Code of Practice: Construction work*, https://www.safeworkaustralia.gov.au/search/site?search=Code+of+Practice+Construction+Work.

Safe Work Australia, *Model Code of Practice: How to manage work health and safety risks*, https://www.safeworkaustralia.gov.au/search/site?search=Code+of+Practice%3A+How+to+manage+work+health+and+safety+risks.

Safe Work Australia, *Model Code of Practice: Managing risk of plant in the workplace*, https://www.safeworkaustralia.gov.au/doc/model-code-practice-managing-risks-plant-workplace.

Safe Work Australia, *Model Code of Practice: Managing the work environment and facilities*, August 2019, https://www.safeworkaustralia.gov.au/doc/model-code-practice-managing-work-environment-and-facilities.

Relevant Australian Standards
AS 1418 Cranes including hoists and winches
AS 2294.1 Earth-moving machinery – Protective structures – General
AS 2550 Cranes – Safe use
AS 2958.1 Earth-moving machinery – Safety, Part 1: Wheeled machines – Brakes
AS 3745 Planning for emergencies in facilities
AS 4687 Temporary fencing and hoardings
AS 4991 Lifting devices
AS/NZS 1873 Power-actuated (PA) hand-held fastening tools
AS/NZS 1891.1 Industrial fall-arrest systems and devices – Harnesses and ancillary equipment
AS/NZS 1891.4 Industrial fall-arrest systems and devices – Selection, use and maintenance
AS/NZS 60745 Hand-held motor operated electric tools – Safety – General requirements
BS EN 1263 – 1:2002 Safety nets – Safety requirements, test methods
ISO 6165 Earth-moving machinery – Basic types – Identification and terms and definitions
ISO 6746-1 Earth-moving machinery – Definitions of dimensions and codes – Part 1: Base machine
ISO 6746-2 Earth-moving machinery – Definitions of dimensions and codes – Part 2: Equipment and attachments
ISO 7133 Earth-moving machinery – Tractor scrapers – Terminology and commercial specifications

GET IT RIGHT

The photo below shows a person using the wet saw unsafely. Identify the unsafe practices and provide reasoning for your answer.

WORKSHEET 1

To be completed by teachers
Student competent ☐
Student not yet competent ☐

Student name: _____

Enrolment year: _____

Class code: _____

Competency name/Number: _____

Task

Read through the sections from the start of the chapter up to 'Plan and prepare for safe work practices', then complete the following questions.

1 Identify two different hazards and provide an example of each.

_____
_____

2 What is the purpose of a hazard report form?

_____
_____

3 What two possible areas are compromised if proper work practices are not followed?

_____
_____

4 Name two important benefits from participating in any process that identifies hazards.

_____
_____

5 What is the purpose of a hazardous material management plan?

_____
_____

6 List three common hazards that may be found in existing structures.

_____
_____
_____

7 When identifying a material such as asbestos, what other factors can be taken into consideration when determining what the material may be?

_____
_____

APPLY WHS REQUIREMENTS, POLICIES AND PROCEDURES IN THE CONSTRUCTION INDUSTRY

WORKSHEET 2

To be completed by teachers
Student competent ☐
Student not yet competent ☐

Student name: _____

Enrolment year: _____

Class code: _____

Competency name/Number: _____

Task

Read through the sections from the start of 'Plan and prepare for safe work practices' up to 'Apply safe work practices', then complete the following questions.

1 How could you determine the likely PPE requirements before you turn up on the construction site?

   _____
   _____

2 Having the correct PPE is essential, but what other aspect of PPE is required in its use?

   _____
   _____

3 What aspect of a task will be improved by the selection of correct tools?

   _____
   _____

4 Identify two outcomes that may occur if the incorrect tool is used to perform a task.

   _____
   _____

5 At what stage of the project do you determine the extent of the fencing required for a construction site?

   _____
   _____

6 List six different categories of information that are supplied in an SDS.

   _____
   _____
   _____
   _____
   _____
   _____

WORKSHEET 3

Student name: _____

Enrolment year: _____

Class code: _____

Competency name/Number: _____

To be completed by teachers
Student competent ☐
Student not yet competent ☐

Task
Read through the sections from the start of 'Apply safe work practices' to the end of the chapter then complete the following questions.

1 What is the meaning of the phrase 'safe work practices'?

   _____

   _____

2 What is the first step to establishing and maintaining safe work practices?

   _____

   _____

3 List five basic safe work practices that should be followed on every construction site.

   _____

   _____

   _____

   _____

   _____

4 Identify three injuries that may occur due to plant and equipment being modified.

   _____

   _____

   _____

5 What do you need to have in order to follow the requirements of safety signs?

   _____

   _____

6 What will assist you to follow safe workplace procedures when dealing with accidents?

7 Why is it important to have regular emergency drills on a construction site?

 **EMPLOYABILITY SKILLS**

Using the following table, describe the activities you have undertaken that demonstrate how you developed employability skills as you worked through this unit. Keep copies of material you have prepared as further evidence of your skills.

Employability skills	The activities undertaken to develop the employability skill
Communication	
Teamwork	
Planning and organising	
Initiative and enterprise	
Problem-solving	
Self-management	
Technology	
Learning	

UNDERTAKE BASIC ESTIMATION AND COSTING

This chapter covers the outcomes required by the unit of competency 'Undertake basic estimation and costing'. These outcomes are:
- Gather information.
- Estimate materials, time and labour.
- Calculate costs.
- Document details and verify where necessary.

Overview

The task of estimating and costing a project in the construction industry is an essential skill.

This has been a major factor for any worker in the construction industry as long ago as the exchange for services has been in existence. When quoting a customer the cost of a job, you need to have confidence that you have allowed for all expenses that will be incurred in completing the work. To be able to continue in the industry you need to maintain financial viability as a business owner.

The estimating process will also allow you to ensure that you have sufficient material to undertake tasks and complete the job without having to stop partway through to acquire further materials. This will assist in efficiency being maintained in the process of completing the work.

A basic requirement to be successful at estimating and costing is to have a sound understanding of building material and construction terminology. In addition to this you will be required to have a strong understanding of basic mathematics. When it comes to material, the main calculations that need to be able to be performed relate to length, area and volume (see Chapter 3 for a full discussion of these calculations). Other factors such as labour, time and associated overheads are usually calculated as a percentage.

Employers value people who fit into their workplace, solve day-to-day problems, manage their time and are keen to continue learning. These types of skills are known as employability skills. The employability skills you develop while working through this unit will be assessed at the same time as the skills and knowledge.

It is important to show evidence that you have developed these skills. Your trainer or assessor will discuss with you how to record your employability skills. The following table provides some examples of things you might do to develop employability skills in this unit.

EMPLOYABILITY SKILLS

Employability skill	What this skill means	How you can develop this skill
Communication	Speaking clearly, listening, understanding, asking questions, reading, writing and using body language.	• Liaise with others when organising work activities. • Interpret plans and working drawings.
Teamwork	Working well with other people and helping them.	• Work as part of a team to prioritise tasks. • Offer advice and assistance to team members and also learn from them.
Planning and organising	Planning what you have to do. Planning how you will do it. Doing things on time.	• List tasks in a logical order of action. • Select appropriate tools and materials.
Initiative and enterprise	Thinking of new ways to do something. Making suggestions to improve work.	• Initiate improvements in using resources. • Respond constructively to workplace challenges.
Problem-solving	Working out how to fix a problem.	• Participate in job safety analyses. • Check and rectify faults in tools and equipment if appropriate.
Self-management	Looking at work you do and seeing how well you are going. Making goals for yourself at work.	• Maintain performance to workplace standards. • Set daily performance targets.
Technology	Having a range of computer skills. Using equipment correctly and safely.	• Apply computer skills to report on project work. • Follow manufacturers' instructions.
Learning	Learning new things and improving how you work.	• Use opportunities for self-improvement. • Trial new ideas and concepts.

Gather information

The gathering of information is a process that relies on several different sources. The different sources may consist of:
- plans and specifications
- site visits
- manufacturers' specifications
- material rates
- production rates
- historical cost analyses.

The information needs to be gathered in a manner that is easily accessible and without any ambiguity. If there is any missing information, it is important to get clarification from the person who supplied the original information. In many cases, this will be the architect and engineer who produced the plans and specifications of the project.

Plans and specifications

Plans for construction work are essential. The plan for the construction of a new structure or alteration to an existing structure needs to be reviewed by the local council and approved for the proposed construction. The plans are a diagrammatical representation of the proposed construction. The drawings are a scaled version of the proposed structure. The plans have several drawings that demonstrate the floor plan of the building and the different floor levels, if applicable. The floor plan has dimensions of the components of the proposed structure (see Figure 6.1).

The other types of drawings are called elevations (see Figure 6.2). The elevations will provide vertical dimensions of the structure. By having drawings that have different perspectives of the same item, additional information can be provided.

To complement the floor plans and elevations there are drawings called sections, which are a slice through the building where the internal elements of the building are shown. In some instances there are detailed sections that demonstrate specific structural junctions and the appropriate construction method.

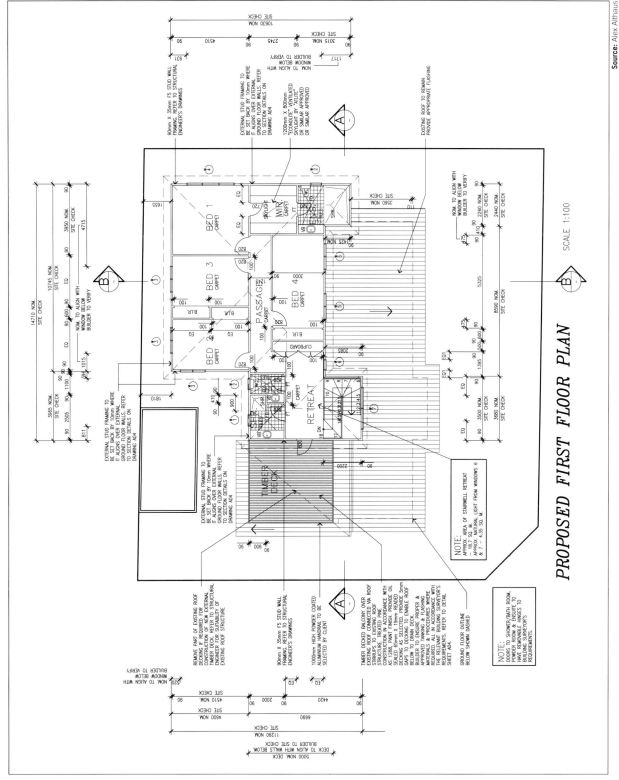

FIGURE 6.1 Typical floor plan

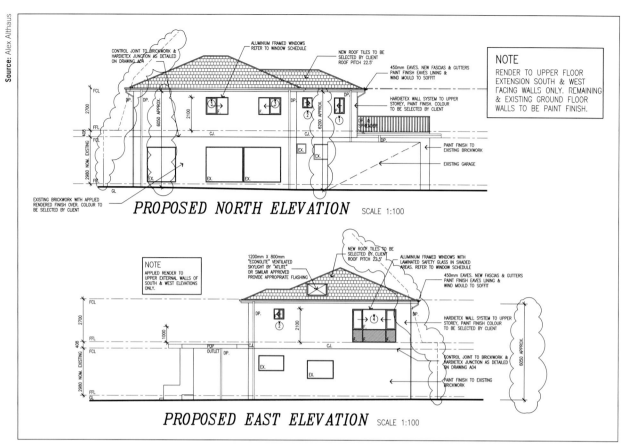

FIGURE 6.2 Typical drawing of elevations

The specifications for the proposed construction will list all of the elements for the structure. The specifications will nominate the appropriate sizes of material so that each element is structurally sound and complies with the relevant building regulations and Australian Standards.

The order of information in the specification will normally follow the sequence of construction to ensure it addresses all elements of the proposed structure. Natspec is one organisation that has established a construction specification system that provides a consistent approach to managing information in the construction industry (see Figure 6.3). Specification contents will commence with general requirements and site preparation and continue through to the final stages of the project such as landscaping.

The specification will also include additional information in the form of specific schedules. An example of this is a window schedule. As illustrated in Figure 6.4, a window schedule will specify window sizes, types and finishes (material and colour). The schedule will also refer to the location of the window in the plans for the proposed construction.

Site visits

Before finalising the costs and a quote being submitted to a customer, it is necessary to make a visit to the actual job site. Such an activity will provide the opportunity to make observations of the environment in which the proposed works are to take place, including ease of access to the site and the characteristics of the site itself. This may be relevant if the site is on, for example, a sloping site as this may extend the time for on-the-job activity.

The site visit will also give the opportunity to take any measurements that may vary from any of the plans that have been drawn up. The plans are usually focused on the new proposed structure and not on some of the on-site conditions.

Manufacturers' specifications

It is recommended that reference is made to manufacturers' specifications and other documents such as safety data sheets (SDSs). A manufacturer's specification may have specific advice on the installation of the material and any other applications that may be related to the material. This may include cutting requirements and the type of paint that is most suitable to the product (see Figure 6.5).

The manufacturer's specification will give details for the different methods of application of the product. In the case of paint, the specification will provide the amount of coverage in square metres that can be achieved with a certain amount of the product. The specification will also indicate the number of coats of

NATIONAL CLASSIFICATION SYSTEM – APRIL 2021

Refer to the National Worksection Matrix for the worksections included in each NATSPEC/AUS-SPEC package: full, basic/rural cut-down version and branded worksections. See www.natspec.com.au. [Square brackets] indicate that no generic worksection is available. A NATSPEC branded worksection may be available.

00 PLANNING AND DESIGN (AUS-SPEC)

001 General
- 0001 Design reference and checklist
- 0010 Quality requirements for design
- 0012 Waterfront development
- 0013 Bushfire protection (Design)

002 Open space
- 0021 Site regrading
- 0022 Control of erosion and sedimentation (Design)

004 Road reserve
- 0041 Geometric road design
- 0042 Pavement design
- 0043 Subsurface drainage (Design)
- 0044 Pathways and cycleways (Design)

005 Road reserve - rural
- 0051 Geometric rural road design - sealed
- 0052 Geometric rural road design - unsealed
- 0053 Rural pavement design - sealed
- 0054 Rural pavement design - unsealed

006 Bridges
- 0061 Bridges and related structures

007 Public utilities
- 0071 Water supply - reticulation (Design)
- 0072 Water supply - pump stations (Design)
- 0074 Stormwater drainage (Design)
- 0076 Sewerage systems - reticulation (Design)
- 0077 Sewerage systems - pump stations (Design)

01 GENERAL

011 Documentation

012 Tendering
- 0120 Pre-tendering contract preparation (AUS-SPEC)
- 0121 Tendering
- 0122 Information for tenderers (AUS-SPEC)
- 0123 Conditions of tendering (AUS-SPEC)
- 0124 Tender submission documents (AUS-SPEC)
- 0125 Standard contract checklists (AUS-SPEC)
- 0126 Period supply and service checklists (AUS-SPEC)
- 0127 Commissioning - information

013 Generic preliminaries
- 0131 Preliminaries (Generic)
- 0133 Preliminaries (Interior and alterations)
- 0134 General requirements (Supply) (AUS-SPEC)
- 0135 General requirements (Services) (AUS-SPEC)
- 0136 General requirements (Construction) (AUS-SPEC)
- 0138 Multiple contracts

014 Contract preliminaries
- 0140 Preliminaries - ABIC BW-2018C
- 0141 Preliminaries - ABIC MW-2018
- 0142 Preliminaries - ABIC SW-2018
- 0143 Preliminaries - AS 2124
- 0144 Preliminaries - AS 4000
- 0145 Preliminaries - AS 4905
- 0146 Preliminaries - AS 4902
- 0147 Conditions of contract (AUS-SPEC)
- 0148 Preliminaries - ABIC EW-1
- 0149 Preliminaries - NCW4

015 Schedule of rates (AUS-SPEC)
- 0152 Schedule of rates (Construction)
- 0153 Schedules - period supply and service

016 Quality assurance
- 0160 Quality
- 0161 Quality management (Construction) (AUS-SPEC)
- 0162 Quality (Supply) (AUS-SPEC)
- 0163 Quality (Delivery) (AUS-SPEC)
- 0164 Commissioning
- 0167 Integrated management (AUS-SPEC)

017 General requirements
- 0171 General requirements
- 0172 Environmental management
- 0173 Environmental management (AUS-SPEC)

018 Common requirements
- 0181 Adhesives, sealants and fasteners
- 0182 Fire-stopping
- 0183 Metals and prefinishes
- 0184 Termite management
- 0185 Timber products, finishes and treatment

019 Sundry installations
- 0191 Sundry items
- 0192 [Structural components]
- 0193 Building access safety systems
- 0194 [Door seals and window seals]
- 0195 [Tactile indicators and stair edgings]

02 SITE, URBAN AND OPEN SPACES

020 Demolition
- 0201 Demolition
- 0202 Demolition (Interior and alterations)

022 Preparation and groundwork
- 0221 Site preparation
- 0222 Earthwork
- 0223 Service trenching
- 0224 Stormwater - site

024 Landscape structures
- 0241 Landscape - walling and edging
- 0242 Landscape - fences and barriers
- 0243 Landscape - water features

025 Landscape cultivation
- 0250 Landscape - combined
- 0251 Landscape - soils
- 0252 Landscape - natural grass surfaces
- 0253 Landscape - planting
- 0254 Irrigation
- 0255 Landscape - plant procurement
- 0256 Landscape - establishment
- 0257 Landscape - road reserve and street trees (AUS-SPEC)
- 0259 Landscape - maintenance

026 Landscape finishes
- 0261 Landscape - furniture and fixtures
- 0262 External sports and playground surfacing

027 Pavements
- 0271 Pavement base and subbase
- 0272 Asphalt
- 0273 Sprayed bituminous surfacing
- 0274 Concrete pavement
- 0275 Paving - mortar and adhesive bed
- 0276 Paving - sand bed
- 0277 Pavement ancillaries
- 0278 Granular surfaces
- 0279 Paving - on pedestals

028 Pathways (AUS-SPEC)
- 0281 Fire access and fire trails
- 0282 Pathways and cycleways (Construction)

029 Retaining walls (AUS-SPEC)
- 0292 Masonry walls
- 0293 Crib retaining walls
- 0294 Gabion walls and rock filled mattresses

03 STRUCTURE

030 Foundations
- 0301 Piling
- 0305 [Foundation isolation systems]

031 Concrete - in situ
- 0310 Concrete - combined
- 0311 Concrete formwork
- 0312 Concrete reinforcement
- 0313 Concrete post-tensioned
- 0314 Concrete in situ
- 0315 Concrete finishes
- 0318 Shotcrete
- 0319 Auxiliary concrete works (AUS-SPEC)

032 Concrete - systems
- 0321 Precast concrete
- 0322 Tilt-up concrete
- 0325 [Concrete protection]

033 Masonry
- 0331 Brick and block construction
- 0332 Stone masonry
- 0333 Stone repair
- 0334 Block construction
- 0335 Brick construction

034 Steel
- 0341 Structural steelwork
- 0342 Light steel framing
- 0343 Tensioned membrane structures
- 0344 Steel - hot-dip galvanized coatings
- 0345 Steel - protective paint coatings
- 0346 Structural fire protection systems

036 Earth
- 0361 Monolithic stabilised earth walls
- 0362 Mud brick walls
- 0363 Straw bale walls

038 Timber
- 0381 Structural timber
- 0382 Light timber framing
- 0383 Sheet flooring and decking
- 0385 Cross-laminated timber (CLT)

04 ENCLOSURE

041 Tanking and damp-proofing
- 0411 Waterproofing - external and tanking

042 Roofing
- 0421 Roofing - combined
- 0423 Roofing - profiled sheet metal
- 0424 Roofing - seamed sheet metal
- 0425 Roofing - shingles and shakes
- 0426 Roofing - slate
- 0427 Roofing - tiles
- 0428 Roofing - insulated panel systems
- 0429 Roofing - glazed

043 Cladding
- 0431 Cladding - combined
- 0432 Curtain walls
- 0433 Stone cladding
- 0434 Cladding - flat sheets and panels
- 0435 Cladding - planks and weatherboards
- 0436 Cladding - profiled and seamed sheet metal
- 0437 Cladding - insulated panel systems
- 0438 [Cladding - cement board]
- 0439 [Cladding - systems]

045 Doors and windows
- 0451 Windows and glazed doors
- 0453 Doors and access panels
- 0454 Overhead doors
- 0455 Door hardware
- 0456 Louvre windows
- 0457 External screens
- 0458 [Automatic doors]

046 Glass
- 0461 Glazing
- 0462 Structural silicone glazing
- 0463 Glass blockwork
- 0466 Structural glass assemblies
- 0467 Glass components

047 Insulation
- 0471 Thermal insulation and pliable membranes
- 0472 Acoustic insulation
- 0473 [Acoustic floor underlays]

05 INTERIOR

051 Linings
- 0511 Lining

052 Partitions
- 0520 Partitions - combined
- 0521 Partitions - demountable
- 0522 Partitions - framed and lined
- 0523 Partitions - brick and block
- 0524 Partitions - glazed
- 0525 Cubicle systems
- 0526 Terrazzo precast
- 0527 Room dividers
- 0528 [Partitions - composite systems]

FIGURE 6.3 Sample of National Classification system

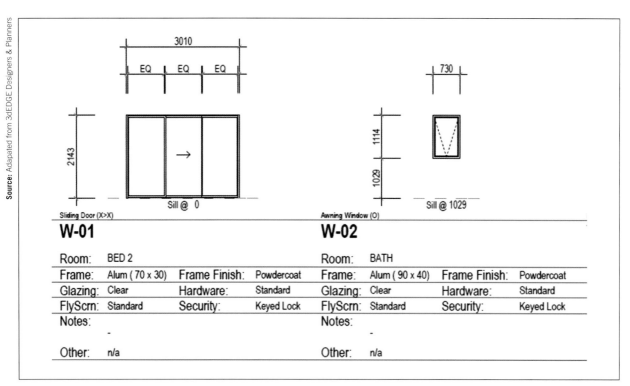

FIGURE 6.4 Example of typical window schedule

paint recommended for that particular product. The information will vary depending on the specific product that you will be using.

Material rates

The cost of each product must also be available so you can start allocating dollar amounts to materials that have been quantified. The cost of materials can be usually obtained from the supplier of the material – most suppliers will have a list that they will make available to you. Some suppliers also may have their material rates available on their website (see Figure 6.6 and Figure 6.7). There are also companies that specialise in supplying construction cost data that cover a range of materials and a wide variety of construction-related costings.

Production rates

When developing the full cost of a project, the cost of labour needs to be considered. Even though a tradesperson's hourly amount can be easily obtained, it is harder to estimate the productivity of a particular individual. In many circumstances, especially for small projects, this will be a guesstimate. This can be avoided if you engage contractors on a cost-per-job basis rather than an hourly rate. However, the ability to estimate time is required when undertaking smaller projects that you will be working on yourself.

Historical cost analyses

Another source of information is via the cost analysis completed at the end of each project. This does require close tracking and recording of information while the project is being completed, which can be very complex. If it is identified that sufficient allowances were not made in the first place, then the original information will need to be adjusted.

LEARNING TASK 6.1 MATERIAL RATES RESEARCH

Find three different online suppliers and source the cost for the materials in the table below. Based on the overall cost, which supplier would be the most cost-effective? Ensure you supply the details of each of the suppliers.

Material	Supplier 1	Supplier 2	Supplier 3
20 kg bag of concrete			
90 × 45 mm MGP12 H2F Termite Treated Pine Blue Timber			
Cladding spotted gum 128 × 19 mm L/m Shiplap			
50 × 2.8 mm 2 kg flathead galvanised nails			

COMPLETE WORKSHEET 1

wattyl Solagard®

DATA SHEET D4.14

ULTRA PREMIUM LOW SHEEN

RESOURCE CODE 11345-Line
OCTOBER 2020
PAGE 1 OF 4

EXTERIOR   WASH-UP (WATER)   TINTABLE

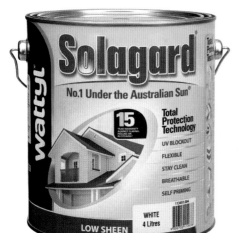

Description & Uses

Wattyl Solagard® Low Sheen is a tough and durable exterior paint which offers protection for your home in any weather or climate. Solagard® Low Sheen Total Protection Technology is designed to keep your home looking better for longer. Solagard® Low Sheen's gloss level provides the best balance between appearance, dirt resistance and flexibility for greater protection. Wattyl Solagard® Low Sheen is backed by a 15 year warranty against flaking, peeling and blistering*.

* For full details on the warranty refer to: www.wattyl.com.au

Features & Benefits

- UV Blockout
- Flexible – Resists Cracking & Peeling
- Self Priming
- Stay Clean – UV Cross Link
- Breathable – Resists Blistering

Colours

White, Light, Mid and Strong bases; selected coloured bases and factory colours. Tint using Ecotint® to Wattyl Colour Designer® range of colours or other competitor colours.	Low Sheen (Gloss Level 10-20 at 60 degrees)

Environmental Data & Certifications

Approved to Australia Paint Approval Scheme Specification APAS 0280/3&5.
Australian Standard AS3730.8

WATTYL.COM.AU   SW-COMPANY USE-OTHER

FIGURE 6.5 Paint SDS demonstrating manufacturer's specifications for coverage

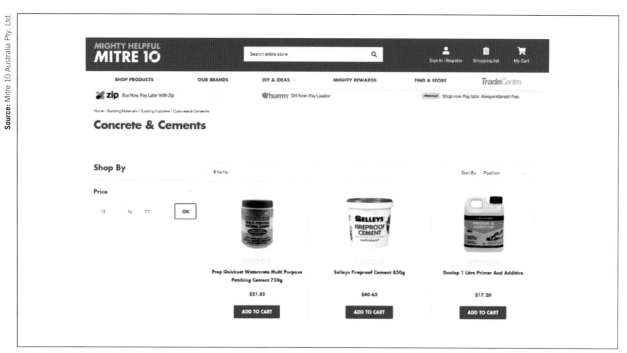

FIGURE 6.6 Online search results for individual item costs

FIGURE 6.7 Online search results for items that are categorised for building and construction

Estimate materials, time and labour

The estimation of materials, time and labour is an important part of the building process. If errors are made in this process, the business may not survive.

Of materials, time and labour, the most difficult to predict is time. There are several factors beyond anybody's control that may delay the project; for example, during bad weather some work cannot proceed.

Materials

The selection of materials will vary depending on the characteristics of materials, how they will be used, the desired look required for the project and the structural requirements of the proposed structure.

Estimates of the quantities of materials required for the project are usually done in a structured manner. The estimate of material will follow the same sequential structure of the specification. It is important to have a consistent approach so that no elements of the project are missed.

It is also important that the recording of material follows a standard approach. This may vary depending on the organisational practices. The calculations used for the Garage Project example in this chapter follow the standard format presented in Figure 6.8.

Item	Trade:		Reference:		
	Particulars	Unit	Quantity	Rate	$

FIGURE 6.8 Common format used to calculate material required for a project

Figure 6.9 shows the dimensions of the garage floor. Figure 6.10 shows how the sheet should be filled out. As concrete is the first material, it is listed as item number 1. In the Particulars column you write a description of the material and show all of your calculation workings.

Time and labour

Labour allowances are based on an hourly rate per unit of measure for the material being used. It is important to allow for all aspects of the particular task. Figure 6.11 indicates the type of reference data that you will need to build up over time. The common phrase for this type of data is known as a labour constant.

The allowance of sufficient time for a particular task is important to determine the costs for a particular project. It is also important that the time periods are also determined for an overall project time estimation. This type of information will inform further project planning and scheduling of activities.

Calculate costs

Calculating the costs of a project entails several steps. Determining material quantities is the first step. The next step is to calculate the amount of labour required, as this is not possible without knowing the amount of material involved. The application of the costs is the next logical step for the cost of the material and labour. There are several other factors that need to be included when the collating of the different material and labour costs have been completed.

Steps involved in calculating costs:
1. quantify material
2. quantify the labour
3. cost the material
4. cost the labour
5. add on overhead
6. add on profit margin.

EXAMPLE 6.1

GARAGE PROJECT, PART I

Specification
- Concrete slab 120 mm thick
- 25 Mpa
- Allow 2.5 % for wastage
- Order in increments of 0.02 m^3

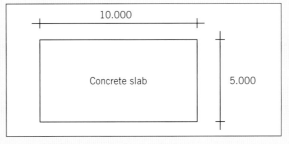

FIGURE 6.9 Floor plan for Garage Project

	Trade: Concreter		Reference: Garage project		
Item	Particulars	Unit	Quantity	Rate	$
1	Concrete 120 mm thick, 25 Mpa				
	10.000 length × 5.000 width × 0.120 thickness = 6.000 m^3				
	Allowance for wastage 2.5%				
	6.000 m^3 × 1.025 = 6.150 m^3				
	Round up to 6.200 m^3	m^3	6.2		

FIGURE 6.10 Calculation of concrete required for Garage Project

Details	Unit	Tradesperson	Labourer
Unreinforced concrete			
Strip footings, pads, piers			
Poured directly from truck	m³	0.6 hours	0.6 hours
15 m average wheel	m³	1 hour	1 hour
Reinforced concrete			
Slab on ground and paving			
Poured directly from truck	m³	0.8 hours	0.8 hours
15 m average wheel	m³	1.2 hours	1.2 hours

FIGURE 6.11 Labour constant of pre-mixed concrete for Garage Project

EXAMPLE 6.2

GARAGE PROJECT, PART II

This example is a continuation of the Garage project. Based on the details in Figure 6.11, calculate the hours required to perform the following task.

	Trade: Concreter			Reference: Garage Project		
Item	Particulars		Unit	Quantity	Rate	$
1	Concrete 120 mm thick, 25 Mpa					
	10.000 length × 5.000 width × 0.120 thickness = 6.000 m³					
	Allowance for wastage 2.5%					
	6.000 m³ × 1.025 = 6.150 m³ Round up to 6.200 m³		m³	6.2		
	Labour to place and finish concrete with a 15 m average wheel					
	Tradesperson 6.2 m³ × 1.2 hours = 7.44 hours 7.44 hours rounded up to 8 hours		hr	8		
	Labourer 6.2 m³ × 1.2 hours = 7.44 hours 7.44 hours rounded up to 8 hours		hr	8		

FIGURE 6.12 Calculation of labour requirements for Garage Project

Oncosts

The rate paid to workers needs to include several items other than the hourly amount paid directly to the worker. If they are a direct employee and not a contractor you will need to allow for several direct costs that you will incur as an employer. This may be referred to as a charge-out fee. The items that need to be included as oncosts include:
- public holidays
- sick days
- annual leave, including a 17.5% leave loading
- lost time due to travel or wet weather
- workers' compensation insurance
- payroll tax allowance
- superannuation allowance.

These allowances in some circumstances equate to 80% on top of the actual amount the worker is paid. This may vary depending on the different rates that are required to be paid under items such as workers' compensation. It is important that the hourly rate for a worker includes oncost before the rate is applied to working out the cost of the work.

Overheads

Overheads are costs not directly related to the employment of a worker that need to be factored in on every project. Overheads include items such as:
- office expenses such as phones, rent of any premises, electricity and furniture
- other staff salaries, such as the employment of a foreperson
- professional services, such as accountancy and legal services
- business fees such as licences
- insurances that are required for the business, such as public liability insurance and property insurance.

As the items are different for different organisations, this figure is established as an additional percentage that is added to every project costing. The percentage amount is usually calculated based on the costs of the overheads for a one-year period and turnover for the organisation for one year. For example:

Turnover for 1 year = $ 500 000.00
Overhead costs for the same period = $45 000.00
$45 000.00 ÷ $500 000.00 = 0.09 × 100 = 9%.

Profit

The objective of being in business is to make a profit. The profit is added at the end of the process of determining the final amount that will be quoted for the proposed project. The profit margin will allow for any differences that may occur between what was estimated to be the estimated cost to the builder compared to the amount that was allowed for in the estimating process. The accuracy of the original estimation will have a high impact on the final amount of profit that is realised against the project. If the original estimate is too low then the profit will be lower also. Conversely, if the estimation is too high then the application of the profit may push your quote well above what the client is willing to pay.

The percentage allowed for profit will be determined by the owner of the business. Some factors that will come into the decision-making process are the level of risk that is taken, and how much the current marketplace will bear for the overall cost of the project.

COMPLETE WORKSHEET 2

Recording and calculating costs

The recording and calculation of costs is broken down into different trade areas. The calculations will be in different measures as the nature of the material will determine how the cost of material will be calculated. This chapter's Garage Project includes concrete, brickwork and carpentry, and each of these trade areas measures the required materials by different units.

Each of the items that will be costed require a description of the material that is to be used and, in some circumstances, how it will be used. The examples below do not cover all aspects of the proposed structure and omit some building elements.

It is essential that there is a total amount of each trade area, which is collated on a final sheet where allowances for overheads and profit can be calculated.

EXAMPLE 6.3

GARAGE PROJECT, PART III

This example is a continuation of the Garage Project. The example works through the:

- plans and diagrams
- specifications of material (including costs)
- labour constant for three trade areas: concreter, bricklayer and carpenter.

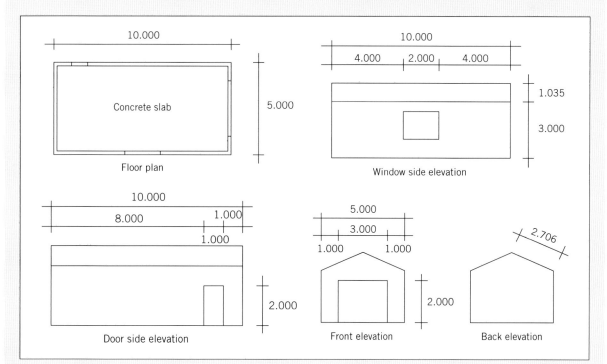

FIGURE 6.13 Floor plans and diagrams for Garage Project

Specification

Concrete
- Concrete slab 120 mm thick, 25 Mpa: $285.00 per m^2

Brickwork
- 110 brickwork – face brick laid in stretcher bond: $1200.00 per 1000 bricks
  $1200.00 ÷ 1000 bricks = $1.20 per brick × 50 bricks per m^2 = $60.00 per m^2

Carpenter – timber roof
- Rafters – @ 0.900 c/c cut and fixed into position 140 × 45 mm MGP10 H2F Termite Treated Pine: $13.85/m
- Ridge – 190 × 45 mm MGP10 H2F Termite Treated Pine: $14.11/m
- Ceiling joists – 90 × 45 mm MGP10 H2F Termite Treated @ 1.800 c/c: $8.60/m
- Roofing battens – 70 × 35 mm MGP10 H2F Termite Treated: $5.20/m

Roofer – metal roof
- Lysaght 0.42 mm Colorbond Spandek Metal Sheeting: $32.00/m^2
- Ridge capping: $10.70/m
- Barge capping: $12.10/m
- Earthwool Space Blanket R-1.8 under metal roof insulation: $80.00 per roll – 12 m^2

Details	Unit	Tradesperson	Labourer
Pre-mixed concrete			
Charge-out rate		$52.00	$46.00
Reinforced concrete – slab on ground and paving			
15 m average wheel	m^3	1.2 hours	1.2 hours
110 brickwork			
Charge-out rate		$58.00	$46.00
110 brickwork – face brick			
110 brickwork laid from slab level	m^2	1 hour	1 hour
Timber roof			
Charge-out rate		$68.00	
Gable roof			
Construct gable roof – rafters	100 lm	15 hours	
Construct gable roof – ridge	100 lm	15 hours	
Construct gable roof – ceiling joists	100 lm	8 hours	
Construct gable roof – roof battens	100 lm	10 hours	
Metal roofing			
Charge out rate		$58.00	
Metal roofing			
Fix metal roofing sheets roof	m^2	0.5 hours	
Fix metal ridge capping	m	0.2 hours	
Fix metal barge capping	m	0.3 hours	

FIGURE 6.14 Total labour constants for Garage Project

Item	Particulars	Unit	Quantity	Rate	$
	Concreter				
1	Concrete 120 mm thick, 25 Mpa				
	10.000 length × 5.000 width × .120 thickness = 6.000 m³ Allowance for wastage 2.5% 6.000 m³ × 1.025 = 6.150 m³ Round up to 6.200 m³	m³	6.2	$285.00	$1767.00
2	Labour to place and finish concrete with a 15 m average wheel				
	Tradesperson 6.2 m³ × 1.2 hours = 7.44 hours 7.44 hours rounded up to 8 hours	hr	8	$52.00	$416.00
	Labourer 6.2 m³ × 1.2 hours = 7.44 hours 7.44 hours rounded up to 8 hours	hr	8	$46.00	$368.00
	Total amount concrete				**$2551.00**
	Brickwork				
1	110 thick face bricks in stretcher bond having joints raked on one face				
	→ 10.000 − 0.110 = 9.890 m ↓ 5.000 − 0.110 = 4.890 m ← 10.000 − 0.110 = 9.890 m ↑ 5.000 − 0.110 = 4.890 m Total length = 29.560 m Area = 29.560 × 3.000 = 88.680 m²				
	Gable = 1.035 × 2.500 = 2.588 m² × 2 = 5.176 m² Total area = 88.680 + 5.176 = 93.856 m²				
	Deduct openings				
	Door 1 = 2.000 × 1.000 = 2.000 m² Door 2 = 2.000 × 3.000 = 6.000 m² Window 1 = 1.500 × 2.000 = 3.000 m² Total amount = 11.000 m² Total Area = 93.856 m² × 11.000 m² = 82.855 m²	m²	83	$60.00	$4980.00
2	Labour to lay 110 brickwork face brick with a raked finish on one face, single storey				
	Tradesperson 83 m² × 1 hour = 83 hours	hr	83	$58.00	$4814.00
	Labourer 83 m² × 1 hour = 83 hours	hr	83	$46.00	$3818.00
	Total amount brickwork				**$13 612.00**

FIGURE 6.15 Total labour and material calculations for Garage Project

Item	Particulars	Unit	Quantity	Rate	$
	Carpenter				
1	Rafters @ 0.900 c/c cut and fixed into position				
	140 × 45 mm MGP10 H2F Termite Treated Pine				
	10.000 ÷ 0.900 + 1 = 12 of 12 × 2 sides of roof = 24 of 24 × 3.000 = 72 lineal metres	lm	72	$13.85	$997.20
2	Ridge 190 × 35 mm MGP10 H2F Termite Treated Pine				
	10.000	lm	10	$14.11	$141.10
3	Ceiling Joists 90 × 45 mm MGP10 H2F Termite Treated @ 1.800 c/c				
	10.000 ÷ 1.800 = 6 of 6 × 5.4 = 32.400 m	lm	33	$8.60	$283.80
4	Roofing battens 70 × 35 mm MGP10 H2F termite treated				
	2.706 ÷ 0.900 +1 = 4 of 4 × 2 sides of roof = 8 of 8 × 10.000 = 80 lineal metres	lm	80	$5.20	$416.00
5	Carpenter to cut and fix into position gable roof – rafters				
	(15 hours ÷ 100 m) × 72 m = 10.8 hours	hr	11	$68.00	$748.00
6	Carpenter to cut and fix into position gable roof – ridge				
	(15 hours ÷ 100 m) × 10 m = 1.5 hours	hr	2	$68.00	$136.00
7	Carpenter to cut and fix into position gable roof – ceiling joists				
	(8 hours ÷ 100 m) × 33 m = 2.64 hours	hr	3	$68.00	$204.00
8	Carpenter to cut and fix into position gable roof – roof battens				
	(10 hours ÷ 100 m) × 80 m = 8 hours	hr	8	$68.00	$544.00
	Total amount carpentry				**$3470.10**

FIGURE 6.15 Continued

COMPLETE WORKSHEET 3

Document details and verify where necessary

The final step of the estimating and costing process is the summary of the different trade areas. The final sheet should clearly indicate the amount for each particular trade area and then the subtotal of the project.

It is at this point that you apply the percentage amount that has been determined for your organisation in the way of overheads. The final calculation is the application of the profit margin that is set as the standard for the organisation.

The documents that are produced from the process of calculating a quote for a proposed project need to be systematically stored in a secure location. The use of software is the most efficient method of calculating costs and storing project records for future reference. There are several software applications available to assist in this administrative process.

When first starting out, many contractors establish their own system of managing documentation. No matter which system is used, records management needs to support business processes including costing, submitting quotations and recording expenditure, so an analysis can be conducted at the end of a project to inform future projects.

EXAMPLE 6.4

GARAGE PROJECT, PART IV

This part of the project is where the total amount for each trade area is collected on to one page.
This allows for the overheads and profit to be applied across the whole project.

	Trade: Summary		Reference: Garage Project		
Item	Particulars	Unit	Quantity	Rate	$
1	Total amount concrete				$2551.00
2	Total amount brickwork				$13 612.00
3	Total amount carpentry				$3470.10
4	Subtotal				$19 633.10
5	Overheads 50% $19 633.10 × 1.5 =				$29 449.65
6	Profit 20% $29 449.65 × 1.2 = $35 339.58				$35 340.00

FIGURE 6.16 Overhead and profit calculations for Garage Project

SUMMARY

In Chapter 6 you have learnt how to undertake basic estimation and costing. In addition, the chapter has:
- identified the different sources of information needed to cost a basic project
- explored the different types of information that are used to calculate the material and labour required to undertake a basic project
- discussed a systematic approach to estimating the material, time and labour for several different trade areas
- identified additional costs of a project to address business operational elements such as overhead and profit.

REFERENCES AND FURTHER READING

Buildsoft, **https://www.buildsoft.com.au**
Buildxact, **https://www.buildxact.com.au**
Constructor, **https://constructor.com.au**
CoreLogic, **https://www.corelogic.com.au**
Natspec Construction Information, **https://www.natspec.com.au**
Rawlinsons Publishing, **https://www.rawlhouse.com.au**

GET IT RIGHT

Mary has completed a project that she quoted on from the plans and specifications only. Going through a cost analysis of the project Mary noticed that the expected profit margin has not been as high as estimated. When looking into each trade area, Mary noticed that across all trade areas the costs for labour were higher than estimated, even though the material costs were within an acceptable level.

Which aspect of information gathering did Mary miss from the estimating process? Identify possible aspects that could have impacted on miscalculation of the overall project.

WORKSHEET 1

To be completed by teachers
Student competent ☐
Student not yet competent ☐

Student name: _____

Enrolment year: _____

Class code: _____

Competency name/Number: _____

Task

Read through the sections from the start of the chapter up to 'Estimate materials, time and labour', then complete the following questions.

1 What information can be gained from a floor plan that will make it possible to calculate the materials required?

_____
_____

2 What other types of drawings are required other than floor plans?

_____
_____

3 What other documents need to be read in conjunction with the plans for the proposed project?

_____
_____

4 Why is it important to undertake a site inspection when quoting on a project?

_____
_____

5 What two pieces of information can you gain from a manufacturer's specifications?

_____
_____

6 Material rates are an important part of costing a project. Where do you obtain material rates from?

_____
_____

7 How does a historical cost analysis affect the production rates that you would use?

_____
_____

WORKSHEET 2

Student name: _____

Enrolment year: _____

Class code: _____

Competency name/Number: _____

To be completed by teachers
Student competent ☐
Student not yet competent ☐

Task

Read through the sections from the start of 'Estimate materials, time and labour' to 'Recording and calculating costs', then answer the following questions.

1. Identify two reasons why different materials may be used for different buildings.

   _____
   _____

2. What type of sequence should costing sheets follow?

   _____
   _____

3. What is a labour constant?

   _____
   _____

4. Where would you find the information to make up your labour constant?

   _____
   _____

5. What other purposes can the estimation of time be used for when planning a new project?

   _____
   _____

6. List three factors that are included to determine the total oncost.

   _____
   _____
   _____

7. Why could the overhead allowance be different for different organisations?

   _____
   _____

UNDERTAKE BASIC ESTIMATION AND COSTING

8 What is the objective of being in business?

9 Name two factors that can have an effect on the profit margin.

WORKSHEET 3

To be completed by teachers
Student competent ☐
Student not yet competent ☐

Student name: _____

Enrolment year: _____

Class code: _____

Competency name/Number: _____

Task
Complete the costing of the garage project to the specifications. This will include the metal roofing works and an overall summary of the costing for the project.

Roofer – Metal roof

	Trade: Metal roofing		Reference: Garage Project		
Item	Particulars	Unit	Quantity	Rate	$
1	Lysaght 0.42 mm Colorbond Spandek Metal Sheeting				
	?				
	?	m²	?	?	
2	Earthwool Space Blanket R-1.8 under Metal Roof Insulation				
	?	ea	?	?	?
3	0.42 mm Colorbond Ridge capping	lm	?	?	?
4	0.42 mm Colorbond Barge capping				
	?	lm	?	?	?
5	Roofer to cut and fix roof sheeting into position				
	?	hr	?	?	?
6	Roofer to cut and fix roof sheeting into position				
	?	hr	?	?	?
7	Roofer to cut and fix roof sheeting into position				
		hr	?	?	?
	Total amount metal roofing				?

Summary of project costing

	Trade: Summary		Reference: Garage Project		
Item	Particulars	Unit	Quantity	Rate	$
1	Total amount concrete				$2551.00
2	Total amount brickwork				$13 612.00
3	Total amount carpentry				$3470.10
4	Total amount metal roofing				?
5	Subtotal				?
6	Overheads 50%				
	?				?
7	Profit 20%				
	?				?

UNDERTAKE BASIC ESTIMATION AND COSTING

MY SKILLS

EMPLOYABILITY SKILLS

Using the following table, describe the activities you have undertaken that demonstrate how you developed employability skills as you worked through this unit. Keep copies of material you have prepared as further evidence of your skills.

Employability skills	The activities undertaken to develop the employability skill
Communication	
Teamwork	
Planning and organising	
Initiative and enterprise	
Problem-solving	
Self-management	
Technology	
Learning	

UNDERTAKE A BASIC CONSTRUCTION PROJECT

This chapter covers the outcomes required by the unit of competence 'Undertake a basic construction project'. These outcomes are:
- Review and prepare to undertake basic construction project.
- Manufacture components for basic construction project.
- Assemble project components.
- Clean up.

Overview

Undertaking a construction project involves the application of skills and knowledge to achieve the desired outcome. It includes the safe and proper use of basic tools, equipment and materials that are relevant to your project. The focus is on developing generic or basic skills with a broader building and construction industry context, and developing the foundation for more specialist trade qualifications.

This chapter requires the completion of a basic project using basic tools, equipment and materials. The chapter is the opportunity for you to apply the concepts, skill and knowledge you have gained from previous chapters. The application of the concepts is not limited to the examples provided in this chapter. There may be alternative practical activities that will give you the opportunity to demonstrate your understanding and abilities. In this chapter we will be focusing on a pergola as a basic construction project.

Completing the tasks successfully requires a good understanding of the project, careful planning and preparation, an appropriate work health and safety (WHS) strategy and interpretation of plans and instructions. It also involves selection and use of appropriate tools and materials for all stages from setting out to final finish, skills in calculation and record keeping, and the ability to work effectively in a team environment.

Employers value people who fit into their workplace, solve day-to-day problems, manage their time and are keen to continue learning. These types of skills are known as employability skills. The employability skills you develop while working through this unit will be assessed at the same time as the skills and knowledge.

It is important to show evidence that you have developed these skills. Your trainer or assessor will discuss with you how to record your employability skills. The following table provides some examples of things you might do to develop employability skills in this unit.

EMPLOYABILITY SKILLS

Employability skill	What this skill means	How you can develop this skill
Communication	Speaking clearly, listening, understanding, asking questions, reading, writing and using body language.	• Liaise with others when organising work activities. • Interpret plans and working drawings.
Teamwork	Working well with other people and helping them.	• Work as part of a team to prioritise tasks. • Offer advice and assistance to team members and also learn from them.
Planning and organising	Planning what you have to do. Planning how you will do it. Doing things on time.	• List tasks in a logical order of action. • Select appropriate tools and materials.
Initiative and enterprise	Thinking of new ways to do something. Making suggestions to improve work.	• Initiate improvements in using resources. • Respond constructively to workplace challenges.
Problem-solving	Working out how to fix a problem.	• Participate in job safety analyses. • Check and rectify faults in tools and equipment if appropriate.
Self-management	Looking at work you do and seeing how well you are going. Making goals for yourself at work.	• Maintain performance to workplace standards. • Set daily performance targets.
Technology	Having a range of computer skills. Using equipment correctly and safely.	• Apply computer skills to report on project work. • Follow manufacturers' instructions.
Learning	Learning new things and improving how you work.	• Use opportunities for self-improvement. • Trial new ideas and concepts.

Review and prepare to undertake basic construction project

The process of planning will establish the time frames, finances, labour and physical resources required for the project. The desired outcome is to carry out the work in a safe and efficient manner, and to avoid delays, waste and any accidents. Consideration must always be given to your WHS obligations to anybody who may be affected, from the workers on the site to the general public.

Review plans and specifications to undertake a basic construction project

Having a full understanding of the tasks required in a project will place you in a position to complete the project as it was intended. When reviewing the plans and associated documents, first check that the plan and documents are for the correct job. It is worth spending some time reading over the documents and identifying if there is any conflicting information. A common mistake that occurs with diagrams is the mislabelling of measurements. At times, the dimensions on a plan may have been miscalculated or written down incorrectly.

Problems that can occur with measurement errors include ordering the wrong types or quantities of materials, which may only be discovered partway through the construction process. There are also examples of buildings being constructed on the wrong location on a site as the measurements on the plan were incorrect. The planning process should include inspecting the worksite, assessing work conditions and requirements and identifying faulty or defective equipment or other workplace hazards.

For every task that is to be performed it is essential that all staff have the same understanding. This includes you, your team, your supervisor and any other tradesperson who may be affected by the activities related to completing a task. It is essential for there to be an understanding of the quality aspects as well as the processes of performing the task.

The information may come from different sources and from different people, so a clear, common understanding is essential. This information will usually be available from the person responsible for supervising the work and kept on the worksite for quick reference.

It may include diagrams and work drawings, specifications, project plans and schedules, notes and emails, and work diaries or reports. Information may also come from equipment or materials suppliers or manufacturers and from government or safety authorities. These sources of information may refer to the:

- task itself, such as how to construct a pergola according to the standard specifications or work drawings
- material to be used, such as pre-treated timber
- standard of finish required, such as concreting driveways or paths
- equipment or plant to be used, such as an electric circular saw.

Other sources of information also need to be considered and some of this information may be more difficult to obtain directly. For example, every workplace is subject to legislation to ensure compliance with safety requirements such as personal protective equipment (PPE) and the use of safety barricades and signage. This information should also be included in your workplace safe work procedures, safety data sheets (SDSs) or job safety analyses (JSAs) for the project.

Regulations may also refer to permitted hours of work in residential areas, limits on noise and dust levels, requirements for particular licences and approvals, and compliance with Australian Standards and industry-specific codes of practice. The planning and preparation stage provides the opportunity to research these issues and to determine how they apply to the particular workplace and each task to be carried out. The following example describes the planning that goes into constructing a pergola.

The construction of a pergola may involve:

- formal approval or authorisation to perform the work, which are usually required by local governments and environmental authorities
- approvals to use nominated materials and building methods, including the correct disposal of any waste materials
- confirmation of work (for example, electrical work) to be carried out and signed off by licensed people; the planning stage will need to identify when these people will be available to do the work, and to build this into the sequence of events
- requirements to prevent unauthorised entry and to protect all people on the worksite; for example, by using barricades and warning signs, and wearing specific safety protection
- access to instructions on operating particular plant or equipment.

Prepare all work to comply with project and regulatory requirements

When diagrams and specifications are prepared, they are done in line with all building regulations. Typically, when plans and specifications are submitted to the local council for approval, such checks will be undertaken.

Work health and safety (WHS) legislation in Australia requires employers to provide a safe workplace for employees and visitors. The legislation also requires employees to comply with the relevant WHS requirements such as wearing supplied PPE and following workplace procedures. Penalties may be imposed where there is non-compliance and an injury or illness occurs.

Your organisation's WHS policy and procedures will specify how hazards will be identified and the risk of occurrence eliminated or minimised for each workplace. This is usually undertaken through a risk assessment, which will include the development of a safe work method statement.

Instructions on the safety requirements for tools and equipment are available from the advice of competent tradespeople, manufacturers' specifications and industry codes of practice. Resources such as guidance notes, safety checklists, wall charts and fact sheets are also available from your state or territory WorkSafe authority.

When a hazard has been identified, you should apply the process of hierarchy of control. As discussed in Chapter 4 (see Figure 4.5), this is the best way to control hazards and manage or minimise risk. When a hazard is identified, you start at the top of the hierarchy and try to eliminate the hazard. If it can't be eliminated, you move to the next control in the hierarchy, and so on, until you have controlled the hazard.

It is wise to use a combination of methods from the hierarchy of control to ensure the safety of all people on the worksite. For example, use a fence (see Figure 7.1) to isolate a construction site (isolation), follow specific procedures for tasks being undertaken (administrative control), wear safety glasses (PPE) and use a hammer instead of a nail gun (substitution).

Every organisation will have safety plans and programs that must be understood and used. These may include:

- safe operating procedures, including operational risk assessment and management of hazards associated with:
  - earth leakage boxes
  - lighting
  - power cables, including overhead cables and conduits
  - restricted access barriers
  - surrounding structures
  - traffic control
  - trip hazards

authorities such as local councils, state, territory and federal governments and the Environmental Protection Authority (EPA). Workplaces are required to have environmental plans or strategies to minimise their environmental impact. Substantial fines and cancellation of permits may result where activities continue to exceed the environmental limits.

Table 7.1 lists strategies to comply with environmental requirements.

TABLE 7.1 Strategies to comply with environmental requirements

Environmental issue	Strategies to meet the requirements
Chemical spills	Have appropriate spill kits in place, and use chemicals in a designated controlled location.
Excessive dust	Wet down and keep damp the material that is generating the dust.
Excessive noise	Investigate alternative construction methods to reduce the noise. Limit the hours of operation to avoid the public being affected. Place hoardings around the site.
Hazardous goods	Have appropriate storage in place. Ensure staff have full knowledge of how to handle such material.
Vibration	Seek alternative methods and equipment that will reduce impact. Search for methods that include padding to absorb noise and vibration.
Waste management	Reduce, recycle and reuse materials.

Select and use personal protective equipment (PPE) for each part of the task

The selection of the appropriate PPE wholly depends on the task that you are to perform.

There are, however, standard levels of PPE on a construction site. Footwear and protective clothing are standard pieces of PPE that all construction workers must wear at all times. Where there is any overhead work being conducted, hard hats are required.

The use of any equipment typically requires the use of eye and ear protection. The nature of the machines can produce loud noise and when the machine is cutting the noise can become louder. Eye protection from flying debris is also required.

The type of PPE required for particular pieces of equipment is identified in documentation provided by the manufacturer of the equipment. You still need to assess every task that is to be performed and identify the required PPE.

Chapter 4 has detailed information about different types of PPE and their purpose. Figure 7.2 shows some of the PPE that you might use when undertaking a basic construction project.

FIGURE 7.1 Safety fence

- worksite visitors and the public
- working at heights
- working in confined spaces
- working close to others
■ appropriate manual-handling activities including the use of load shifting equipment
■ use of firefighting equipment
■ first aid and evacuation procedures
■ safe working with tools and equipment
■ correct methods for handling and storing dangerous or hazardous goods
■ use of safety data sheets and job safety analyses
■ correct use of appropriate PPE.

Compliance with environmental regulations is also considered when the application for construction is being assessed by the council. This is to ensure construction tasks are carried out to a standard quality, with minimum risk to people and property, and with minimal negative impact on the environment.

You should be aware of the different laws, regulations, guidelines and codes of practice that may affect your work and workplace from an environmental perspective. For example, you may need to have strategies in place to avoid unacceptable levels of noise, dust, fumes or vibration, to ensure waste is correctly managed, and to ensure the site and affected surrounds are adequately cleaned up on completion of the work.

Limits on acceptable levels have been set by legislation and regulations developed by various

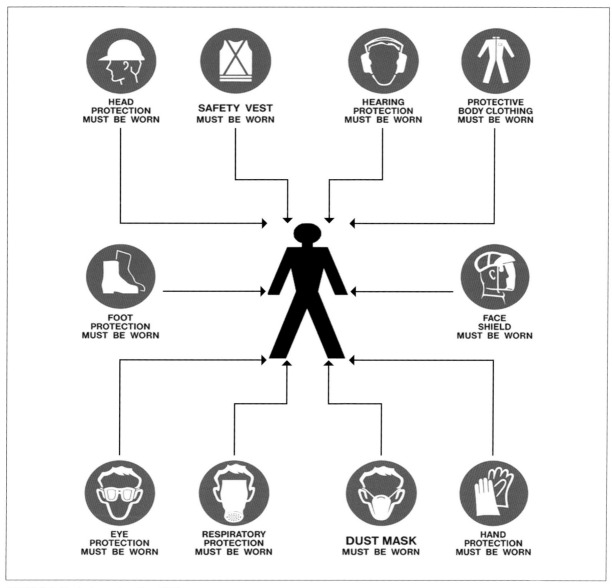

FIGURE 7.2 Typical construction PPE

Select tools and equipment, check for serviceability and report any faults

The selection of the correct tools and equipment will depend on your understanding of the particular task. While there are standard tools and equipment that are required with most tasks, there are also specialised tools and equipment for different types of work.

To select the correct tools and equipment, you need to have a clear understanding of what the tools and equipment are intended for. You also need to be able to identify if the tools and equipment are in correct operating order. Common tools and equipment are listed in Table 7.2.

The initial assessment is to ensure that the tool or piece of equipment that is selected is going to be used for its intended use. Once the correct tool has been selected, there should be a thorough check of each item to ensure that it is in good operating condition. If the equipment has been regularly and properly maintained, this should not be a problem. While many hand tools are maintained by the tradesperson who uses them, powered equipment should be maintained by an authorised repairer.

The main issues when checking hand tools are:
- handles are not loose
- no parts are missing
- cutting teeth or blades are sharp.

The main issues to inspect when checking power tools are:
- power cords are intact and not frayed or wires exposed
- air hoses are not leaking
- all guarding is in place
- all moving parts move freely without binding or hitting other parts of the machine.

TABLE 7.2 Common tools and equipment

Hand tools	Use
Chisel	To remove wood (or concrete for hardened cold chisels); available in a wide range of types and sizes
Claw hammer	To drive and remove nails
Hand saw (trim, rip, crosscut, keyhole, coping, hacksaw)	To cut appropriate materials accurately
Rule and tape	To take measurements of material and building elements
Shovel	To dig and clean up the site
Sledgehammer	To break up concrete and drive wedges
Square	To help marking out material to be cut and to set out or check building components once installed
Wheelbarrow	To transport material across the site
Power or pneumatic tools	**Use**
Compressor (pneumatic)	To provide a high-pressure air supply
Electric plane (rotary)	To shave off small amounts of timber
Impact power drill	To drill brittle materials such as brickwork and concrete
Nail gun (pneumatic)	To drive nails
Power saw (circular, table, mitre/radial arm)	To cut materials

Determine project components, prepare a component list and calculate quantities of materials

To determine the quantity of material for a project, you should use a systematic approach. Chapter 3 provides a guide to the types of calculations required plus an approach to work through each trade area of a project.

Being able to quantify the material required will become the basis for determining the costs of the project. Depending on the size and complexity of the project, parcels of work may be contracted out. From a business perspective, it is at times easier to have a subcontractor provide a quote for them to supply the material and labour to perform the work required. This will especially be the case when there are works that fall under a specialised or licensed trade area.

The details of the required material for any job will be specified by the appropriate documentation. It is essential that you have a systematic approach to working through the supplied documentation to ensure that you clearly identify all of the material required for the project. This is usually done by following the sequence of construction (typically from the ground up).

When calculating material quantity requirements, it is important to have an understanding of the manner in which different materials are supplied. Materials may be supplied by length, area, volume or weight. Refer to Chapter 6 for a detailed discussion on estimation and costing approaches.

COMPLETE WORKSHEET 1

LEARNING TASK 7.1 PROJECT SELECTION

Select a project (or use the pergola project identified in Learning task 7.2) to complete as part of this unit of competence. Based on the project you have chosen, you are required to keep a journal of the project that you undertake as a part of this chapter. The journal will be a daily entry that will record information such as:
- stage of the project/task
- how many hours working on the different stages, including the number of people
- any new skills or knowledge that you gained from the activity during the day
- what safety practices occurred during the day's activity.

Manufacture components for basic construction project

Construction projects are made up of several different components, and the combination of these components ends up making the final product; for example, wall frames, roof trusses and roof tiles. It is essential that these components are assembled with accuracy as the next component is reliant on the quality of the previous component.

At the planning phase, you will have identified which components can be assembled independently of each other and which components need to be assembled in a sequence. Planning will determine the allocation of the time and resources to develop a logical sequence of assembly that can assist in determining the development of an assembly line, if possible.

It may be necessary for other trade areas to become involved in the sequence of work that you have in place. You will need to ensure that you have consulted with other trade areas to ensure that the progress of work does not prevent the other trade areas from being able to complete their works. This will minimise any potential delays and conflict on the job site.

Select processes to manufacture components

Construction work is usually a combination of work conducted on site and prefabricated components that are prepared off site, usually at a factory and to standard formats, and then transported and assembled on site. Consideration needs to be given to which components are assembled on site compared to off site.

Obtaining the required components may involve:
- selecting from a series of standard components such as windows, cabinets or doors, which are provided fully assembled
- modifying a standard component such as adapting a standard kitchen unit to accommodate a larger refrigerator
- commissioning the off-site manufacture of components (for example, roof trusses or stairs) to a subcontractor
- assembling a component on site; for example, from pieces provided in kit-form
- constructing the component on site; for example, wall frames, brick piers, footings or concrete slabs.

There are several factors that come into the decision-making process when determining whether to assemble components on site. Some of these factors are the expertise of workers on site, the amount of time available, the required quality or standard, and the availability of components during the project. Consideration also needs to be given to costs and any additional plant and equipment that may be required. All of these will have an impact on costs and the timeline of construction.

It is expected that construction projects will be completed within the estimated budget and time frames. It is also expected that the project will meet the client's needs as well as expected quality standard.

Determining the construction process requires that:
- there is a clear understanding of the contents of the plans and specifications
- there is a reasonable and achievable project time frame and budget
- the various work tasks follow a logical sequence; these tasks may be basic or very complex and be carried out concurrently or consecutively as each stage is completed

- worker skills and technology support the construction process and suit the level of project scope and complexity
- opportunities are taken to use specialist skills and resources; for example, by outsourcing the supply of components to competent manufacturers.

Construction methods must comply with several requirements, including:
- building regulations and Australian Standards
- conditions by authorities issuing building and environmental approvals
- work health and safety legislation
- environmental protection requirements.

As each project is different, construction methods and processes also may vary. For example, today slab construction is used in many domestic houses whereas in the past bearers and joists with timber flooring were used (see Figure 7.3 and Figure 7.4). The most appropriate methods need to be identified in the planning stage of the project.

FIGURE 7.3 Domestic dwelling construction using a slab on the ground

FIGURE 7.4 Domestic dwelling construction using bearers and joists

Manufacture components and check for accuracy, quality and suitability

The final product needs to meet a specified standard and quality level for all construction projects. There will be several Australian Standards that the final product will need to meet. For example, components that are used in a concrete slab – such as the steel reinforcement and the cement that is used in the concrete – are specified in accordance with the relevant Australian Standard.

There are other quality standards that need to be checked and complied with, including industry standards, manufacturer standards, approved supplier accreditations and accepted workplace procedures. Reports provided by consumer protection groups and safety authorities provide examples of problems that occur when substandard or inappropriate component parts have been used. This is important when imported products are being used and may not have been manufactured to the relevant Australian Standards. Many manufacturers pride themselves and their products as being above standard, which will make them more competitive with other manufacturers of the same type of product.

It is important that the component parts are checked for accuracy, quality and suitability before the construction project is started, regardless of the scope or complexity of the project. These checks should be done according to the plans, drawings, specifications and procedures. From a client's perspective, if you select a product then you take on the responsibility for any failures that may occur.

Where detailed inspection of each part is not possible, check for certification that shows the manufacturer has complied with required procedures and standards. For example, certification may be required when prefabricated components have been supplied (see Figure 7.6).

The installation of the component is a big factor in the end quality of the product. Even if a product is not faulty, if it is installed incorrectly the quality may fall below standard. Experienced or correctly trained tradespeople are needed to undertake the installation of specialised components.

 COMPLETE WORKSHEET 2

LEARNING TASK 7.2 PROJECT PLANNING

Based on the project that you have chosen (or the pergola project identified below):
- calculate the material required to complete the project as per the plans and specification (use the templates and processes identified in Chapter 6)
- calculate the material required to complete the project as per the plans and specification
- identify the tools and equipment required for each step of the project.

Once you have completed these tasks, you will be in a position to start the construction of your chosen project.

PERGOLA PROJECT

Below are typical plans and specifications for a pergola project.

		Specification	
Item	Type	Size	
Footings	Concrete	300 mm × 300 mm × 450 mm	One per post
Post bracket	Galvanised		One per post
Posts	H3 treated pine	90 mm × 90 mm	
Beams	H3 treated pine	240 mm × 45 mm	
Rafters	H3 treated pine	190 mm × 45 mm	600 c/c
Battens	H3 treated pine	70 mm × 35 mm	1.500 c/c
Bracing	H3 treated pine	70 mm × 35 mm	
Bolts for posts	Galvanised cup head	10 mm diameter	
Bolts for beams	Galvanised cup head	10 mm diameter	
Pergola angle bracket	Galvanised	88 mm × 63 mm × 36 mm	One bracket each side of each rafter
Timber connector nails for brackets	Galvanised	35 mm × 3.15 mm	
Griplock nails to fix battens	Galvanised	100 mm × 4.5 mm	Two nails per junction of batten and rafter

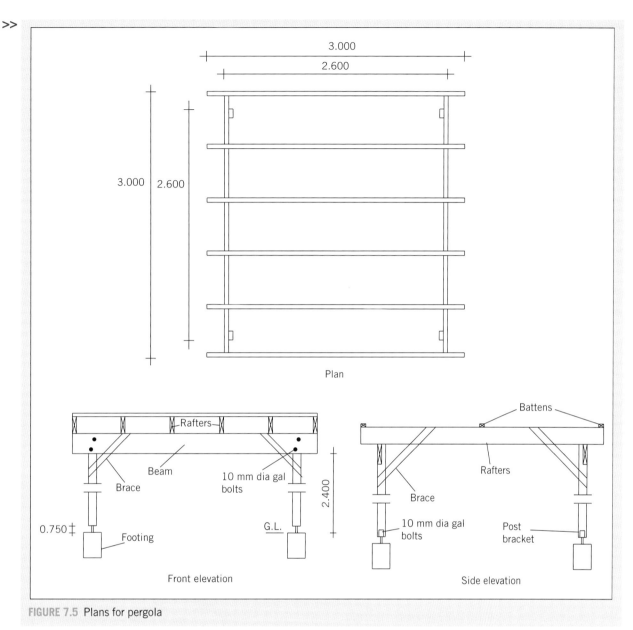

FIGURE 7.5 Plans for pergola

FIGURE 7.6 Industry certification for building components such as windows

Assemble project components

Once all the materials have been ordered and delivered, and the appropriate tools and equipment have been set up and found to be in good operating order, it is time to construct the project. It is important that throughout the assembly of the project you check that the progress of the work is in line with the intended project. Depending on the project and construction methods, some variation of the assembly process may be required.

Select assembly process

Before you commence the construction of a project, you need to confirm that all relevant permits have been approved and received. In the case of a pergola (the example project discussed throughout this section), this will be the local council. The checks will relate to the new structure as it will be attached to an existing building and potentially extend to boundary lines where setbacks may need to be enforced. Consideration will need to be given to any easements, height restrictions and environmental concerns.

The next step is to organise the required material and goods to be supplied so the works can commence.

The work area will involve several preparation activities, which may include:

- signage to designate the area as a construction site, provision for access to site, authorisation of entry, prohibitions on site, provision for warning notices and barricades for excavations
- location of permanent mark or datum for floor levels
- access to power supplies or switchboard, water and worker facilities
- WHS policy, provision of PPE and first aid facilities
- provision for storage, reuse, recycle or disposal of waste materials
- adequate clearance of site to allow set-out for construction and sufficient working space
- provision for storage of materials, chemicals and equipment.

The site preparation for a project such as a pergola will require that any excavation works that are required do not clash with any existing services. As these situations are mostly on existing structures, consideration needs to be given at the planning phase for such potential issues.

In instances of a pergola, the structure is typically built in situ. This process will follow a logical sequence of construction working its way up from the ground. As the process of construction is typically one piece at a time, it is important to have as many of the components as possible ready to be assembled.

Set out, level and erect/install project in line, level and plumb

The process set out below is a standard approach to commencing a project such as a pergola. The context will change depending on the nature of the project that you are undertaking, but the setting out of the project is typically the first step.

The construction process has three main steps:
- Stage 1 – set-out and levelling
- Stage 2 – excavation as required
- Stage 3 – construction/erection.

Some of the steps may not be required; for example, if you are installing a pergola on to a concrete slab, there may be no need to excavate.

Stage 1 – Set-out and levelling

The following actions are required for the first stage of erecting a free-standing pergola:

1. Drive a peg into the ground or mark on the slab of the first post.
2. Measure out to the location of the second post.
3. Set up profiles for the first string line, ensure that the height is consistent and level.
4. Using the 3-4-5 method mark out a line at 90 degrees.
5. Set up profiles for the second line.
6. Measure and mark out the two remaining lines parallel to the first two lines established.
7. Set up profiles for the remaining two lines in position and at the correct height.
8. Check the height of each post location.
9. Check for square by measuring diagonals.

Figure 7.7 is a diagrammatical representation of how the setting out of a site will look from a plan view and demonstrates the check of square by measuring and comparing the diagonal lengths from external corner to external corner. Figure 7.8 demonstrates the location and position of the pegs in relation to the intersection of the string lines.

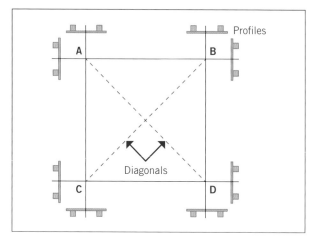

FIGURE 7.7 Plan view of setting out with string lines and profiles

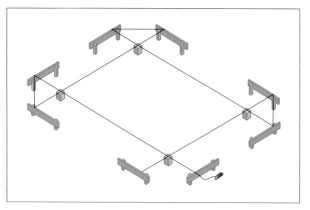

FIGURE 7.8 Position of pegs when setting out structure

Stage 2 – Excavation as required

Undertaking the excavation process needs to be done with great caution. The location and size of the excavation needs to be accurate as this will be the beginning of the structure, and if done incorrectly will put a stop to the construction process. Excavating the area is important

because it provides a level, stable foundation for the construction project. For this stage, you need to:
- excavate post holes to the required depth and width
- fill the holes with concrete mix
- insert galvanised post supports while the concrete is wet
- recheck levels are identical with stringline/profile boards
- allow four to five days for the concrete to cure or harden.

The desired outcome is to have a stable, secure footing for the posts (see Figure 7.9).

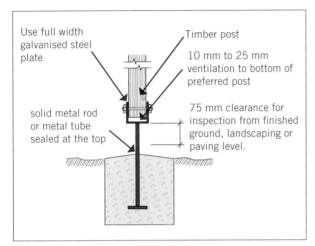

FIGURE 7.9 Section view of post secured to footing

Stage 3 – Construction/erection

Once the footings are ready, you can begin constructing and erecting the timber framework. For this stage of the project, you will need to:
- check the timber quality on delivery, store timber away from damp ground and pre-treat timber if required
- cut the posts to the required length, pre-drill for galvanised bolts into supports and beams, erect the posts into position, check for true vertical and fix securely with temporary supports
- cut the timber beams to length, attach them to posts, cut braces to shape and fix them to the beams or posts, check for plumb, cut rafters to length and attach them to the beams
- ensure all items are securely attached and finished to the required standards.

Depending on the configuration of the project, there will be different steps and components. Remember you should always read and understand the plans and specifications for any project.

Assemble components to specification and quality, and check conformity to plans and specifications

All construction projects have a quality standard that they must meet. This quality standard appears in the specifications and the associated plans and diagrams. On large projects there will be local government inspections required to certify that the construction meets the specifications and Australian Standards. There are instances where structural engineers will certify the work being performed.

Many construction companies have their own quality assurance processes that consist of policies and procedures that are established with the project specifications and relevant codes of practice and building regulations. The benefit of having such processes in place is to ensure that the final product is of high quality, allowing the company to build a strong reputation of high quality that becomes a selling point for future business.

When subcontractors are being used, builders will use checklists as a quality assurance measure before the work is signed off for final payment.

It is also essential to have a clear understanding of the client's expectations in relation to quality. Even though there is at times a lot of documentation, it is advisable to have the client work through it and provide feedback.

COMPLETE WORKSHEET 3

Clean up

A project does not finish at the end of construction. All projects, large or small, need to be left in a clean and tidy condition.

The task of cleaning should be a part of your regular daily routine. You should constantly clear any debris from where you are working to ensure that the working area is kept in a manner that is safe for all workers. This includes the use of any tools and equipment throughout the process of completing a task.

On all construction sites there are legal requirements to manage waste, including a strong emphasis on the reuse and recycling of materials. There are other environmental aspects that need to be managed during the process of construction, such as managing airborne dust and water run-off. Minimising on-site waste can save the company money previously spent on transport and landfill charges. These strategies also present a positive image of the company as a responsible contributor to the community.

Meet legislative and workplace requirements for safety, waste disposal and materials handling

Cleaning up a construction site is not something that is left to the end of the job. Cleaning up is a key part of the project from the beginning and is a continuous activity throughout the life of the project.

Construction site clean-up strategies may include:

- putting away all power leads, tools and equipment to prevent unauthorised use and to prevent slips, trips and falls
- controlling waste water so that run-off does not affect waterways and stormwater systems
- managing waste including resource recovery and recycling
- controlling and managing hazardous materials
- avoiding site contamination such as that from damage to underground tanks or pipelines; leakage of chemicals or waste fluids by inappropriate storage, handling or disposal; importing contaminated fill
- managing soil erosion and sediment
- restoring the site to its original condition, where appropriate, via revegetation or environmental repair, backfilling and levelling off of excavations
- ensuring hazardous items such as solvents and fuel are safely locked away or removed from the work area in accordance with safety regulations.

The act of cleaning needs to be considered in the same way as any other task on the job site and having a job safety analysis (JSA) carried out. When hazards such as chemicals and high-pressure washers are present, appropriate measures should be in place, including access to safety data sheets (SDSs), appropriate personal protective equipment and training, and adequate ventilation.

Appropriate measures need to be put into place to manage the disposal or reuse of any material. If any material is to be reused then it must be stored and secured in a manner that does not put anybody at risk and avoids damage to the material. A common material that is reused in the construction process is the formwork used when pouring the concrete members of a building.

The process of cleaning up is a process that is more than just sweeping up and putting offcuts into a bin. There are environmental considerations that need to occur in the cleaning process. On many construction sites chemical-based materials have been used and, depending on the material, there may need to be appropriate and specialised disposal protocols in place.

There are codes of practice that will provide a guide to managing waste. In some situations, such as material that contains asbestos, licensed people with specialised skills might need to be engaged.

Regulations by councils and environmental authorities apply to the disposal of items such as solvents, chemicals and plastics, paints and cleaning fluids, sealants and fixing compounds, electrical wiring, and the removal of asbestos. Even sweeping can raise particles of silica, treated timber dust, paint dust and other chemical residues. Other regulations relate to materials that may be suitable for recycling such as metal, bricks, concrete and timber.

Check, maintain, store and secure tools and equipment and report any faults

The use of tools is the main activity that occurs on the job site, so maintaining tools and equipment should be a high priority. Most tradespeople develop a practice of cleaning down their tools and equipment either each day or after each use. This assists in maintaining both the longevity and the functionality of the tools and equipment. In some circumstances, it is recommended to refer to the manufacturer's recommendations for cleaning and equipment maintenance.

Many large organisations have tool and equipment registers for the scheduling of regular off-site maintenance of equipment. The register will also provide a life expectancy of the equipment so that the organisation can plan for replacement costs. The life cycle of tools and equipment needs to be systematically monitored to ensure that time is not lost and that the equipment is in good condition to perform the required tasks.

Storage is an important part of the maintenance of tools and equipment. Many manufacturers will recommend methods for storage of their product, particularly for medium to large pieces of equipment. Many tradespeople store their hand tools and equipment in lockable toolboxes, such as those on the back of a ute. When it comes to vans, a shelving storage system is commonly used. Many of the tools are kept in smaller toolboxes that can be carried and also contain any supporting hardware for the functionality of the tools. For example, it is not uncommon to see hand saws with the teeth protected by a removable plastic sleeve.

LEARNING TASK 7.3 FEEDBACK ON PROJECT

Once you have completed your project, have your teacher and other class members inspect the final product.

Each person who is inspecting the project should list items under two categories:
1 areas for improvement
2 areas that have met or exceeded expectations.

Based on the feedback, consider what or how you would perform the project differently if you were to do it again. Discuss with your class the common areas of improvement for all projects and what skills need to be developed.

 COMPLETE WORKSHEET 4

SUMMARY

In Chapter 7 you have learnt about undertaking a basic construction project. In addition, the chapter has:
- explored the different aspects of the planning process for a project
- identified different types of PPE that are required in different situations
- discussed the selection of tools and equipment
- identified different phases used in the assembly process of a basic construction process
- looked into the benefits of regular cleaning and maintenance of tools.

REFERENCES AND FURTHER READING

Jaybro, *Tools & Equipment*, **https://www.jaybro.com.au/tools-hardware-chemicals/building-tools-equipment.html**.

National Safety Signs, *Site Combination Signs*, **https://nationalsafetysigns.com.au/product-category/building-construction-signs/site-combination-signs**.

NSW Government, *The Pocket Guide to Construction Safety*, http://www.safework.nsw.gov.au/__data/assets/pdf_file/0004/386446/Pocket-Guide-to-Construction-Safety.pdf.

Staines, A. 2014, *The Australian decks and pergolas construction manual*, 7th edn, Pinedale Press, Caloundra, Qld.

Timber Promotion Council, *Carports & Pergolas Design & Construction Manual*, Melbourne, 1996.

Total Tools, *Construction Equipment*, **https://www.totaltools.com.au/construction-tools/construction-equipment**.

GET IT RIGHT

Steven has undertaken a project to build a deck for a client. While he has done this type of work in his previous employment, this is his first solo project.

On the morning of the first day of the job, Steven calls into the hardware store, only to discover that the store does not have all the materials he requires. In fact, it will take a week to secure sufficient quantities to complete the project.

Steven arrives at the job site late and without all the required materials. The client has been waiting as they need to leave for their own job.

As the days pass, the client becomes dissatisfied with the progress of the job and realises that the estimated date of completion will not be met.

The contractual agreement between Steven and the client states that Steven will charge an hourly rate for the time spent on the project. On completion of the project, well past the discussed date, the client is unwilling to pay the amount Steven has charged based on the hours he has worked.

What steps could Steven have taken before the start of the job to avoid this situation?

 **WORKSHEET 1**

To be completed by teachers
Student competent ☐
Student not yet competent ☐

Student name: _____

Enrolment year: _____

Class code: _____

Competency name/Number: _____

Task

Read through the sections from the start of chapter up to 'Manufacture components for basic construction project', then complete the following questions.

1 List two examples of the types of documents that are used to gain information about a construction project.

   _____
   _____

2 Why should you spend time reviewing the project documents?

   _____
   _____

3 Identify two environmental issues and list what strategies could be put into place.

   _____
   _____
   _____
   _____

4 Other than the presence of safety signs, how would you know which PPE should be used?

   _____
   _____

5 What information do you need to be able to select the correct tools and equipment for a particular task?

   _____
   _____

6 What faults should you be looking for when selecting a hand tool?

   _____
   _____

7 What faults should you be looking for when selecting a power tool?

　

8 What information do you need to be able to calculate the quantity of material required?

WORKSHEET 2

To be completed by teachers
Student competent ☐
Student not yet competent ☐

Student name: _____

Enrolment year: _____

Class code: _____

Competency name/Number: _____

Task

Read through the sections from the start of 'Manufacture components for basic construction project' up to 'Assemble project components', then complete the following questions.

1 Construction work is usually a combination of two types of work. Name and provide a brief explanation of each type.

_____

_____

2 What are three factors to consider when determining whether to assemble components on site?

_____

_____

_____

3 Name two areas where information about quality standards may be found.

_____

_____

4 What should you look for if you are not able to check the quality of a manufactured product?

_____

_____

WORKSHEET 3

To be completed by teachers
Student competent ☐
Student not yet competent ☐

Student name: _____

Enrolment year: _____

Class code: _____

Competency name/Number: _____

Task

Read through the sections from the start of 'Assemble project components' up to 'Clean up', then complete the following questions.

1 What considerations need to be taken into account before excavation takes place?

   _____
   _____

2 What are the main steps when undertaking a basic project?

   _____
   _____
   _____
   _____
   _____

3 How do you check for square of the job at the setting out stage?

   _____
   _____

4 What would building companies base their quality checklists on?

   _____
   _____

5 What standards should all goods be manufactured to?

   _____
   _____

WORKSHEET 4

Student name: _____

Enrolment year: _____

Class code: _____

Competency name/Number: _____

To be completed by teachers

Student competent ☐

Student not yet competent ☐

Task

Read through the sections from the start of 'Clean up' up to the end of the chapter, then complete the following questions.

1. Why is it good practice to keep your immediate work area clear and clean?

   _____

   _____

2. List four construction site clean-up strategies.

   _____

   _____

   _____

   _____

3. What are some of the benefits of cleaning and maintaining your tools and equipment?

   _____

   _____

4. List three benefits of having a tool/equipment register in place.

   _____

   _____

   _____

5. Identify two methods of storage of tools and equipment.

   _____

   _____

MY SKILLS 7

 **EMPLOYABILITY SKILLS**

Using the following table, describe the activities you have undertaken that demonstrate how you developed employability skills as you worked through this unit. Keep copies of material you have prepared as further evidence of your skills.

Employability skills	The activities undertaken to develop the employability skill
Communication	
Teamwork	
Planning and organising	
Initiative and enterprise	
Problem-solving	
Self-management	
Technology	
Learning	

HANDLE AND PREPARE BRICKLAYING AND BLOCKLAYING MATERIALS

This chapter covers the outcomes required by the unit of competency 'Handle and prepare bricklaying and blocklaying materials'. These outcomes are:
- Prepare for work.
- Handle, sort and stack materials.
- Mix bricklaying and blocklaying mortar.
- Clean up.

Overview

Bricks are one of the oldest and most durable forms of building material, dating back to early Egyptian and Middle Eastern civilisations of between 5000 and 4000 BCE. These bricks were handmade from clay and grass or straw, then pressed into bottomless moulds and left to dry in the sun before use. When protected from the weather, these 'adobe'-type bricks would last for many hundreds of years. During the time of the Roman Empire, to produce a more durable product, the bricks were burnt in primitive wood-fired kilns. This high heat enabled the clay particles to fuse together, or become vitrified.

Brick making has improved over the centuries from the adobe-type to the handmade sandstocks produced by convicts after 1788 in Australia, up to the present day, where a wide variety of brick types, shapes, colours and textures are available with a high level of durability.

Employers value people who fit into their workplace, solve day-to-day problems, manage their time and are keen to continue learning. These types of skills are known as employability skills. The employability skills you develop while working through this unit will be assessed at the same time as the skills and knowledge.

It is important to show evidence that you have developed these skills. Your trainer or assessor will discuss with you how to record your employability skills. The following table provides some examples of things you might do to develop employability skills in this unit.

EMPLOYABILITY SKILLS

Employability skill	What this skill means	How you can develop this skill
Communication	Speaking clearly, listening, understanding, asking questions, reading, writing and using body language.	• Listen carefully to work instructions. • Ask questions to clarify task requirements.
Teamwork	Working well with other people and helping them.	• Work individually and as a team to complete tasks in a timely manner. • Liaise with others to avoid conflicting work tasks.
Planning and organising	Planning what you have to do. Planning how you will do it. Doing things on time.	• Plan resources required for each task. • Plan sequence of activities in order to successfully complete work task.
Initiative and enterprise	Thinking of new ways to do something. Making suggestions to improve work.	• Prepare information before you discuss issues. • Suggest improvements to processes.
Problem-solving	Working out how to fix a problem.	• Manage task sequencing to cater for changes to the work environment. • Adapt to cultural differences in communications.
Self-management	Looking at work you do and seeing how well you are going. Making goals for yourself at work.	• Evaluate and improve own methods of communication. • Aim for quality results within specified task time frames.
Technology	Having a range of computer skills. Using equipment correctly and safely.	• Implement safe work procedures for machinery. • Use a range of communication tools such as two-way radios, telephone and email.
Learning	Learning new things and improving how you work.	• Research online product and work practice information. • Observe more experienced construction workers and adopt their practices to improve your work processes.

Prepare for work

The planning and preparation for brick or blocklaying work is specific to the material that is to be used. There are several high-risk materials that workers need to be aware of. This awareness will allow you to correctly identify the most appropriate PPE that is required when the material is in use.

It is important that environmental factors are also considered, as working outdoors is common for bricklayers. The other aspect of the environment that needs to be considered is the location of the work. As brickwork is an envelope around a building, the material will have to be transported to all locations on the site.

The constant flow of material on a job site for bricklaying is a key factor. It is important that first there is sufficient material available on the job site for the required amount of activity. The second factor is to keep a steady supply of the material up to the bricklayer so there is no pause in the laying of bricks. This requires a sequence of coordinated activities to have bricks in place and mortar mixed and in place, while progressively moving around the worksite both horizontally and vertically.

> ### LEARNING TASK 8.1
> #### PREPARING FOR THE USE OF CEMENT
> Research and download a copy of an SDS for cement and clay bricks.
> 1. Identify the potential hazards from a toxicological perspective.
> 2. Identify the PPE precautions that need to be in place before handling the identified material.

Bricks and blocks

Bricks and blocks come in a range of different finishes. Brickwork is considered to be universal as a construction material. Brickwork can be used as a structural component, which is commonly known as load bearing, but it can also be considered as a visual finish to a structure, which is referred to as non-load bearing.

Bricks can be identified by their finish, which is determined by the production process. Common types (see Figure 8.1) include the following:
- clinkers – overburned bricks that are very hard and distorted in shape
- callows – underburned bricks that are light in colour and very absorbent
- sandstocks – handmade by convicts and early settlers (and still produced using machine methods)
- commons – regular in shape but can have flaws or uneven colouring
- face – regular in shape, colour and texture.

The broad classifications of bricks are listed in Table 8.1 in accordance with AS 3700 – 2011 Masonry structures. Figure 8.2 shows images of these categories.

TABLE 8.1 Categories of bricks

Category of brick	Description
Solid unit	These bricks are made via a pressing process, and are dried through mechanical means such as large ovens. These bricks are easily identified as they have an indent in the top of the brick referred to as a frog.
Cored unit	These bricks are manufactured in a process that is known as extruded, so many bricks manufactured in this manner are referred to as extruded bricks. These bricks have holes that go through the entire brick from the bottom to the top. The holes allow for the mortar to extend into the holes, which helps to lock the bricks together.
Hollow unit	These masonry units are made via a pressing process. They are larger than standard size bricks and are commonly used in reinforced masonry walls.
Horizontally cored unit	These units are typically extruded and are made of clay. The production method is the same as the standard process for extrusion bricks. These units are usually used for internal walls and are lighter in weight due to the hollow nature of the unit.
Special purpose unit	These bricks take on several different shapes as they are for specific purposes and are mainly used as decorative bricks.

Types of bricks

Calcium silicate

Calcium silicate masonry bricks are manufactured from calcium silicate, sand, water and pigment. Sand is the dominant material that is used for this type of brick (see Figure 8.3).

Clay masonry

Clay masonry bricks are manufactured from burnt clay or shale. The quality of the clay can determine the final colour of the manufactured brick. This is the brick most commonly used in domestic construction in Australia. The bricks come in a broad range of colours and combinations of colour (see Figure 8.4). The range of textured finishes allows for different visual design appearances. The textured finishes can range from a smooth face to impressions made in the face to give more of a grainy or deep pattern (see Figure 8.5).

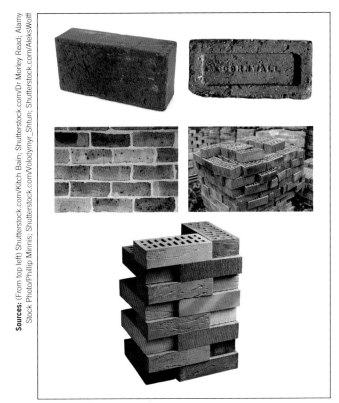

FIGURE 8.1 Brick types (from top left): clinkers, callows, sandstocks, commons and face

FIGURE 8.2 Brick classifications (from top left): solid unit, cored unit, hollow unit, horizontally cored unit and special purpose unit

FIGURE 8.3 Calcium silicate bricks

FIGURE 8.4 Range of brick colours

FIGURE 8.5 Range of brick textures available

Concrete masonry

Concrete masonry blocks are manufactured from concrete, though the mixture of material is a lot finer than what is used for general purpose concrete. The concrete blocks come with a range of different textured finishes. There is also a wide range of colours available, which are produced during the manufacture of the concrete blocks.

Blocks, or concrete masonry, are made from a mixture of coarse and fine aggregates with cement and water, manufactured in units of rectangular prismatic shape and then hollowed. They are intended for use in bonded masonry construction (see Figure 8.6).

FIGURE 8.6 Concrete blocks stacked ready for transporting

Retaining walls

Concrete blocks are commonly used for the construction of retaining walls as the hollow structure of the blocks allows for reinforced bars to be built into the wall. Once the wall has been constructed, the hollow blocks are filled with a concrete mix to fill any voids. The wall becomes a reinforced concrete wall with the ability to take horizontal pressure to the face of the wall (see Figure 8.7).

There is a wide range of different sizes manufactured, which allows concrete blocks to be used in several different construction situations.

Breeze blocks

Breeze blocks are another form of concrete block. They can be moulded into different designs to create a

FIGURE 8.7 Retaining wall with concrete-filled hollow cement blocks

decorative wall that allows for air to flow freely through it. Such walls are common in tropical environments to allow breezes to cool down enclosed areas. The walls also allow strong winds to pass through without adding stress to the wall (see Figure 8.8).

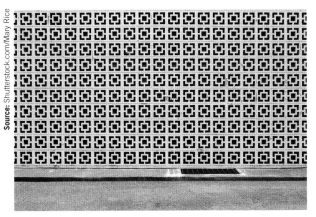

FIGURE 8.8 Wall constructed of breeze blocks

Autoclaved aerated concrete

Autoclaved aerated concrete (ACC) masonry is made from fine concrete and is manufactured with hydrogen gas to make it light in weight (see Figure 8.9). The trapped air also helps the product to maintain a high thermal rating.

AAC can be formed into various sizes and shapes during the manufacturing process, including into panels used for walls, floors and roofing. When AAC is used as

FIGURE 8.9 ACC blocks and large panels

panels, it is supported by various engineered systems for fixing purposes. However, when ACC products are used as blocks in the traditional sense, the blocks are bonded with specialised adhesive. The AAC blocks are laid on a bed of mortar to achieve a level course and the following blocks are bound by an adhesive that is usually only 2 mm to 3 mm in thickness.

ACC blocks are commonly known as Hebel blocks, the name of the company that manufactures them.

LEARNING TASK 8.2

MATERIAL RESEARCH ASSIGNMENT I

Prepare a folder with a cover sheet containing the following details:
- your name
- your class/group name or code
- name of the assignment ('Material Research Assignment I')
- your teacher's name
- the due date.

Part 1

Refer to the chapter text so far, then provide the following details:
1. a brief description of each material discussed
2. a brief description of the manufacturing process
3. at least two uses for the material
4. how the material should be safely handled during use or preparation for use.

   The materials to be identified and described are:
   - calcium silicate
   - clay masonry
   - concrete masonry
   - autoclaved aerated concrete (AAC)

   Aim to write approximately half a page for each of these materials.

Part 2

For each of the materials listed above, search the internet for a manufacturer's brochure and/or details. Attach these details to the assignment at the end of the description of each material.

Stone masonry

Natural stone has been used in walls and structures for thousands of years. Today, granite, sandstone, slate, limestone and marble are the favoured types of natural stone used for masonry construction. The style is classified according to the shape and surface finish of the stone; for example, rubble, ashlar, cut stone or dimension stone. Within each of the classifications, variations can be used for interest or to bring out the characteristics of a particular type of stone. Surface finishes range from split face (a rough natural face of stone used for masonry walls with small, uneven, narrow stones) to polished face (a mirror-like, glossy finish produced by computer-assisted grinding techniques). These can all be achieved using hard, durable, natural stone (see Figure 8.10).

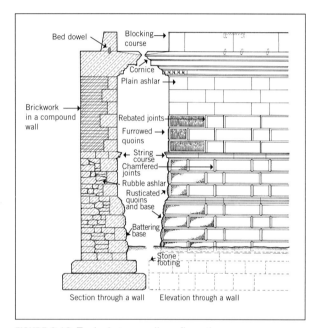

FIGURE 8.10 Typical stone wall configuration

FIGURE 8.11 Natural stone wall with different coloured stone

Several different natural stones are used in contemporary construction including granite, quartzite, basalt, marble, travertine, limestone, onyx and semi-precious stone. The use of stone is not limited to walls – it is common for stone, both natural and artificial, to be used for kitchen bench tops, splashbacks and ornamental decorations.

As well as natural stone, there is also reconstituted stone, which is crushed fragments or stone powder that are bonded together with polyester resins or mortar. This material can be shaped into blocks, pavers, brick or even sheets. The wall shown in Figure 8.12 is not made from real stone but sheets of crushed stone with resin made to look like real stone set in mortar. These are sometimes referred to as faux stone wall panels. The advantages of this system are cost (it is much cheaper than real stone) and lighter weight.

FIGURE 8.12 Faux stone wall construction

Size and quantity

Different masonry units have different sizes and require different consideration when calculating the number of bricks required for a job. First, you need to confirm the size of an individual brick. If you are using a brick that is larger than the standard brick then you will have to order fewer bricks. Standard-sized bricks measure 230 mm (L) × 110 mm (W) × 76 mm (H).

There are two basic methods for calculating the number of bricks required:

- using a coursing chart (see Figure 8.13)
- determining the area of the face of the wall and multiplying by the number of bricks or blocks per square metre.

Many brick manufacturers also supply an online brick calculator that will assist in quantifying the number of bricks or blocks required.

Source: Midland Brick

BRICKWORK DIMENSIONS
NOTE: All dimensions in mm

STANDARD BRICK (Bricks per m² in wall = 48.5 approx.)
FORMAT SIZE: 240 × 120 × 86mm
MANUFACTURING SIZE: 230 × 110 × 76mm
VERTICAL GAUGE: 7 Courses to 600mm

No. of Bricks	Length	Opening	Height	No. of Bricks	Length	Opening	Height
1	230	250	86	26	6230	6250	2229
1½	350	370		26½	6350	6370	
2	470	490	172	27	6470	6490	2314
2½	590	610		27½	6590	6610	
3	710	730	257	28	6710	6730	2400
3½	830	850		28½	6830	6850	
4	950	970	343	29	6950	6970	2486
4½	1070	1090		29½	7070	7090	
5	1190	1210	429	30	7190	7210	2572
5½	1310	1330		30½	7310	7330	
6	1430	1450	514	31	7430	7450	2657
6½	1550	1570		31½	7550	7570	
7	1670	1690	600	32	7670	7690	2743
7½	1790	1810		32½	7790	7810	
8	1910	1930	686	33	7910	7930	2829
8½	2030	2050		33½	8030	8050	
9	2150	2170	772	34	8150	8170	2914
9½	2270	2290		34½	8270	8290	
10	2390	2410	857	35	8390	8400	3000
10½	2510	2530		35½	8510		
11	2630	2650	943	36	8630		3086
11½	2750	2770		36½	8750		
12	2870	2890	1029	37	8870		3172
12½	2990	3010		37½	8990		
13	3110	3130	1114	38	9110		3257

FIGURE 8.13 Typical course chart

EXAMPLE 8.1

USING A COURSING CHART

If you had a wall that measured 3.000 m in length × 0.900 m high, how many bricks are required?

The red outlined area indicates that 12.5 bricks are required for the length and the blue area indicates that 11 bricks are required for the height.

Therefore: 12.5 × 11 = 137.5 bricks are required.

Usually a percentage is allowed for wastage and can vary depending on the make of the bricks to be used.

Durability

The durability of masonry products in general is fairly high. Other construction materials – such as timber if used as cladding for a building – require a high frequency of maintenance and protection from the atmospheric elements.

Natural stone is very durable, but there are environmental considerations to consider, particularly if using limestone. Some stones, such as limestone, are dissolved and eroded by the action of acid. In cities, where rain catches atmospheric contaminants that change the pH level of the rain water to a slight acid, this 'acid rain' can have an impact on stone structures.

Efflorescence in brickwork is another major consideration. Efflorescence is the presence of salt coming through the brickwork and usually sitting on the face of the wall. The amount of salt in any part of the process of constructing a brick wall needs to be carefully considered. Salt may naturally be present in many of the natural products that are used, such as sand and water. The presence of salt may be due to moisture from an external source after the wall has been constructed.

Fire rating

Bricks and masonry material all have a good fire rating. They have been and still are used to line the interiors of chimneys, kilns and most refractory units as they are able to withstand high temperatures. Bricks and blocks also have high thermal ratings and are used to provide dwellings with a higher energy efficiency rating.

Cutting

Brick, concrete and masonry materials can be cut using similar types of tools. Manual tools include lump hammers and bolsters, brick hammers and scutch hammers or chisels (see Figure 8.14). Figure 8.15 shows how a scutch hammer should be used and the type of support needed under bricks when they're being shaped.

The most appropriate mechanical tool that is used for cutting is a Diamond blade wet-cutting brick saw (see Figure 8.16).

During the cutting process, water and small pieces of masonry may be thrown up into the operator's face; therefore, it is essential that wrap-around safety goggles or glasses be worn to prevent possible permanent eye injury.

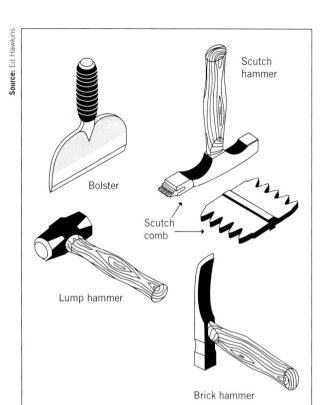

FIGURE 8.14 Common hand tools used to cut bricks and masonry

FIGURE 8.16 Diamond-dust blade wet-cutting brick saw

Handle, sort and stack materials

Handling

Due to the weight involved in masonry products, handling is usually carried out by mechanical means. Generally, the masonry units are transported to the building site on a truck with a hiab. Another way that masonry may be moved from the truck is by a truck-mounted forklift. Larger pieces of stone can be moved around using a small front-end loader known as a bobcat (see Figure 8.17), or by using a small crane.

FIGURE 8.17 Front-end loader – bobcat

FIGURE 8.15 A scutch hammer being used to shape a brick

The final procedures are usually done by hand. Bricks, blocks and masonry can be moved around with the aid of brick carriers (see Figure 8.18). Leather gloves and rubber or leather cots may be used to protect the

FIGURE 8.18 Typical brick barrow and brick carrier

hands and especially the tips of the fingers, which could end up with 'bird's eyes' (round areas of skin worn away until they bleed). More information about PPE, including gloves, can be found in Chapter 4.

Stacking

Bricks, blocks and masonry should be stacked on timber pallets for delivery and site storage. The packaging of bricks varies from manufacturer to manufacturer. It is common to have brick or block loads delivered that are completely wrapped in plastic (see Figure 8.19). The stacking of bricks by the manufacturer also may mean that a pallet is not required. It is essential that when bricks are delivered on site that sufficient space is allocated to accommodate them.

FIGURE 8.19 Blocks on a pallet and wrapped in plastic

Just prior to being laid, bricks can be stacked onto plywood sheets or directly onto the ground in a formation that ensures the stack will not collapse. Remember, if masonry is stacked onto wet clay, the clay will need to be cleaned off before bricks are laid into the wall as the clay will weaken the bond between brick and mortar.

Storage

As most bricks, blocks and masonry products are porous, they need to be protected from exposure to heavy rain and from lying in water. If these units soak up too much water they will tend to 'float' when laid. This means water in the mortar is not taken up and the bricks tend to float on top of the mortar bed rather than bonding to it. This may result in the bricks or blocks moving out of alignment. Plastic covers should be placed over stacks outside until they are ready to be laid. This applies also to 'green' or freshly laid brickwork during wet weather. Some covering may be necessary to allow the work to set before excessive moisture is absorbed.

LEARNING TASK 8.3

MATERIAL RESEARCH ASSIGNMENT II

Prepare a folder with a cover sheet containing the following details:
- your name
- your class/group name or code
- name of the assignment ('Material research assignment II')
- your teacher's name
- the due date.

Identify two types of natural stone that are used in construction today. Obtain photos from the internet or photos you have taken yourself. Then write where the stone is typically used and why you would use this type of stone in the location (approximately one page).

Mix bricklaying and blocklaying mortar

Mortars are mixed and used to bond brickwork, blockwork and all types of masonry to form a homogeneous unit that is capable of carrying great loads and can also be decorative in form. Mortars are made by mixing cement, hydrated lime or lime putty, and clean, sharp sand – normally proportioned by volume – with clean, fresh water (these materials are discussed later in this chapter).

Under AS 3700 – 2011 Masonry structures, mortars are classified as M1, M2, M3 or M4. The Standard also stipulates mix proportions for each mortar type, plus the units for which the mortar is suitable.

Commonly used mortars include the following:
- Lime mortar consists of one part hydrated lime or lime putty to three parts sand. It is usually a fairly weak mix, depending on the lime used – hydraulic lime putty tends to give the most strength. This mix may be used on its own for internal walls as it tends to weather badly when used externally, or it may be used as a base mix for composition (compo) mortar.
- Cement mortar consists of one part cement to one part hydrated lime or lime putty to six parts sand. It is a good-strength general-purpose mortar that is very workable.
- Compo mortar consists of one part cement to five or six parts sand, plus a plasticiser such as Bycol added to the water. It usually contains type GP cement with a 'larry-mix' or 'brickies sand', which is normally a mixture of fine and coarse sand particles, plus an aerating admixture such as the detergent-like Bycol (also called liquid ball bearings, due to its action in the mix). It is a general-purpose mix commonly used for both internal and external work, especially on brick veneer construction.
- Pre-mixed mortar is bagged and ready for use without having to add any other components apart from water. Pre-mixed mortar is not commonly used on construction sites as bricklayers prefer to use their own proportions of the different components.

Many variations are possible with these mixes, which will be determined by the climate and type of work to be performed. All of these mixes can be coloured using powdered oxides to allow variations in joint colour or to match brickwork. Red and black are the most common, but yellow, green and brown are also available. The colour of the cement can also affect the colour of the mortar.

Quantity requirements

Determining the total amount of mortar required is based on approximately 0.6 m³ (1.2 tonnes) of damp sand, which will lay approximately 1000 standard bricks (see Example 8.2). Note that cement and lime come in

EXAMPLE 8.2 MORTAR RATIOS AND QUANTITIES

Figure 8.20 reproduces a mortar guide used to calculate the material required for mortar, depending on the number of bricks to be used on a project.

Table 8.2 demonstrates how the mortar guide can be used to calculate the amount of raw material required to make sufficient cement mortar for 3500 bricks at M3 classification with a mix of 1:1:6.

AS 3700 code	Mortar composition (C:L:S)	No. of 20 kg bags of cement	No. of 20 kg bags of lime	Cubic metres of sand	Tonnes of damp sand
M4	1:0.5:4.5	17.3	3.2	0.55	1.04
	1:0.25:3	25.9	2.4		
M3	1:1:6	12.9	4.8		
M2	1:2:9	8.6	6.4		
M1	1:3:12	6.5	7.2		
M1	0:1:3	–	9.7		

FIGURE 8.20 Mortar guide based on intended use of mortar

TABLE 8.2 Calculating mortar material quantities

Material	Ratio	Per 1000	Bricks per 1000	Final amount
Cement	1	6 bags	3.5	45.15 bags = 46 bags
Lime	1	4.9 bags	3.5	17.15 bags = 18 bags
Sand	6	1.04 tonnes	3.5	3.64 tonnes

Source: Think Brick Australia, Manual 19 - Industry Reference Guide. First published 1994. Latest edition revised and republished June 2019.

20 kg bags, although some pre-blended products – such as Lime 'n' Lite and Lime 'n' Grey, which blend cement and lime at 1:1 by volume – are 17.85 kg net weight.

Durability

The durability of mortar is significantly impacted by the different classes of mortar that are used for the construction of the brick wall. Other factors that play a role are the different types of components used and the finish to the joints between the brickwork.

Fire rating

The fire rating of mortar is required to be consistent with the bricks or blocks that it is used with. The main requirement is that the masonry wall must be able to maintain its structural integrity in the event that it is exposed to fire.

Cutting

The cutting of mortar once it has dried follows the same principles that are used for any masonry material. The most effective method is mechanical, through a wet diamond-based cutting disc or a drill using a masonry drill bit. However, the use of plugging chisels is effective for clearing out existing mortar.

Handling

Mortar can be handled by mechanical means such as a crane and kibble for large loads on sites that cover a large area. However, in most situations mortar is mixed close to the location of use and is poured from the mixer straight into a wheelbarrow. The mortar is then transferred and placed onto mortar boards with a shovel. The bricklayer then uses a trowel to place the mortar into position on the line of bricks or blocks.

> The weight of mortar means that manually loading up mortar boards with a shovel can place high levels of strain on the body. It is essential that appropriate techniques are used to minimise wear and tear on the body, particularly bending the knees when lifting in order to protect the back.

Mortar is not stored for long periods as it hardens quickly so is mixed for a specific use at a specific time. Pre-mixed mortars also need to be stored in a dry location away from any moisture. Once pre-mixed mortars come into contact with moisture, the chemical reaction of the cement drying and going hard begins, rendering it unworkable if not used within a specified time.

LEARNING TASK 8.4 SET UP TO LAY BRICKS

This activity can be a standalone activity or the first steps of the construction of a brick wall (see Learning task 9.2 in Chapter 9). You are required to:
- set up masonry units in preparation for constructing a brick wall. This will include the correct positioning of the bricks and mortar board in relation to the proposed location of the wall.
- cut the bricks in half with a hammer and bolster.
- mix mortar as specified by your teacher and transport the mix to the proposed location of the wall.

COMPLETE WORKSHEET 2

Mortar components

Mortar is made up of different materials that when mixed in different combinations result in a product that will bind bricks in different situations and with different finishes. All components need to be of high quality in their own right to produce a quality product. The components that make up mortar are cement, sand, lime, oxides and plasticisers.

Cement

Cement typically comes in 20 kg bags. This weight allows for manual handling and takes into consideration an acceptable safe lifting load. When cement is being measured for mixing mortar, it can be done in half-bag batches.

Cement comes in an off-white colour in addition to the standard grey colour (see Figure 8.21). Off-white cement is used to allow a different colour finish to the mortar through the use of oxides (discussed below).

FIGURE 8.21 General purpose and off-white 20 kg bags of cement

Several different cements are used in construction, with the most common cement used for the production of mortar classified as type GP (General Purpose). AS 3972 – General purpose and blended cements defines the different types of cement that are available and in common use, including the following:

1. General purpose cement (Type GP). This hydraulic cement contains Portland cement plus various minerals, depending on the particular manufacturer.
2. General purpose limestone cement (Type GL). This hydraulic cement contains Portland cement and limestone.
3. Blended cement (Type GB). This hydraulic cement contains general purpose cement plus fly ash or ground granulated iron blast-furnace slag (or both) and/or amorphous silica.
4. Special purpose cements. These cements are based on general purpose cement or a blended cement mixed with various additives to obtain particular qualities. They include high early-strength cement (Type HE), low-heat cement (Type HE), sulfate-resisting cement (Type SR) and shrinkage limited cement (Type SL). (Adapted from Standards Australia)

Health implications of handling cement

There are several possible health aspects that need to be considered and managed when using any cement. Safe Work Australia has published national guidance material that provides information on working with silica and silica-containing products.

Cement contains silica that can be harmful and adversely affect your health. Exposure to silica dust can result in breathing problems and diseases such as chronic bronchitis, emphysema, lung cancer, sarcoidosis, silicosis and progressive massive fibrosis. Common injuries include eye irritation and eye damage. Safe Work Australia states:

Silicosis and progressive massive fibrosis are irreversible and often fatal. Symptoms of these diseases may not appear for many years after exposure. Workers may be diagnosed with these diseases and not present with any symptoms, even at the point of initial diagnosis, which is why prevention and health monitoring are critical.

Source: Safe Work Australia © Commonwealth of Australia. CC BY 4.0 International Licence. (https://creativecommons.org/licenses/by/4.0/)

 When using cement, appropriate PPE should be worn. The impacts from working with cement can have long-lasting health implications.

Safety data sheets for cement recommend the use of safety measures, including PPE. Cement Australia's recommendations for general purpose cement are as follows:

- Eye/face. Wear safety glasses or dust-proof goggles when handling material to avoid contact with eyes.
- Hands. Wear PVC, rubber or cotton gloves when handling material to prevent skin contact.
- Body. Wear long-sleeved shirt and full-length trousers.
- Respiratory. Where an inhalation risk exists wear a Class P1 (Particulate) respirator, dependent on a site-specific risk assessment. (Source: Cement Australia)

Sand

Sand is an integral part of the make-up of mortar. Sand is considered as an aggregate. The colour of the sand that is used can influence the final colour of the mortar.

Any sand that is used for mortar should be kept free of impurities, such as chips or splinters of brick, as they will deteriorate over time and affect the workability of the mortar. The sand also should be free of salts, organic materials and clay.

Sand is available in 20 kg bags, but on building sites it is usually delivered by the tonne either from a tip truck or in one-tonne bags (see Figure 8.22). The one-tonne bags are commonplace on large construction sites as the bag can be craned in to different locations on the site, whether it is across the floor area of the site or to different levels of construction on a multi-story site.

FIGURE 8.22 Tonne-sized bags of sand

It is recommended that sand be stored in a location close to where it will be used to create mortar.

Lime

Lime is used in cement-based mortar mixes to enhance plasticity, binding qualities and durability, and to help slow down the drying process, which provides more time for the mortar to be used. Lime also assists in minimising cracks in the mortar as it dries (see Figure 8.23).

Oxides

Oxide is an additive that is added to a mortar mix to create a colour that is aesthetically desirable. The inclusion of an oxide does not change the structural strength workability of the mortar – just the final colour. If colour coordination is considered at the planning stage, the final finish of the brickwork will complement

FIGURE 8.23 20 kg bags of hydrated lime

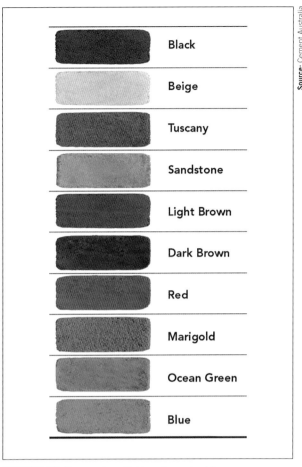

FIGURE 8.24 Range of available coloured oxides

the overall design of the structure. There is a wide range of colours available (see Figure 8.24).

Plasticisers

Plasticisers are liquid additives that help to give mortar more adhesive qualities and to more evenly spread the mix of the components within the mortar. Such additives also introduce small air bubbles that slow down the drying process of the mortar. Bycol is a commonly used plasticiser.

Jointing

Jointing is the term given to the horizontal (bed) and vertical (perpend) connections between bricks, blocks and masonry. These joints, normally 10 mm thick, can be finished in many ways, depending on the required appearance or weathering properties (see Figure 8.25).

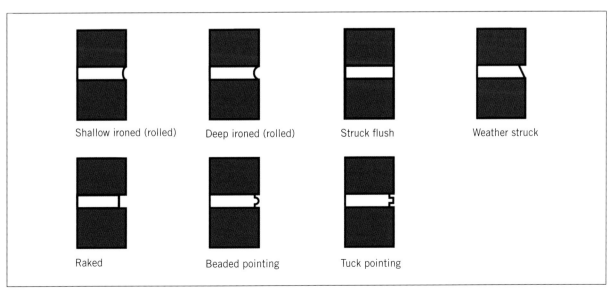

FIGURE 8.25 Jointing finishes

The jointing finish also adds to the decorative aspects of the brickwork.

The absence of a perpend is common to see in brick walls. This is usually in cavity wall construction as the opening allows for any water that builds up in the cavity too escape. Weep holes are constructed from plastic that is inserted between the bricks to allow water to escape and to prevent vermin from entering the building.

COMPLETE WORKSHEET 3

Clean up

The cleaning up of work after laying bricks can be a large task. It is best to progressively clear the work area of any excess material that has been created during the day.

If there are brick offcuts or other solid material, it is straightforward to collect it and deposit it in the appropriate area on the site. This may be a in a skip bin or a designated fenced-off area. Wet material also needs to be cleaned up – ideally before it dries.

The extent of the cleaning up will depend on how tidy and clean the bricklayer is when laying the bricks. Some brickies are messy, while others clean off excess mortar as they lay the bricks. The usual practice is to hose down any wet material at the end of the day before the mortar or mortar droppings go hard. If they are left, then a scraper will usually be required.

SUMMARY

In Chapter 8 you have learnt about masonry materials and components, and their storage and handling. In addition, the chapter has:
- considered the physical and aesthetic considerations that are important when considering brick and blocklaying materials
- identified the uses, durability, handling and manufacturing considerations of masonry products
- covered the different types of mortar and the components that make up mortar
- identified storage, handling and stacking requirements, and discussed the durability and fire rating considerations for construction trade brick and blocklaying materials
- considered various jointing techniques for laying bricks and blocks.

REFERENCES AND FURTHER READING

Cement Concrete & Aggregates Australia, Durability of Masonry Mortar, **https://www.ccaa.com.au/imis_prod/documents/Library%20Documents/CCAA%20Datasheets/TN67-2007Durability%20masonry%20mortarTBR.pdf**.

Commission for Occupational Safety and Health, *Concrete and masonry cutting and drilling – Code of practice*, WA Department of Mines, Industry Regulation and Safety, 2019.

Hebel, **https://hebel.com.au**.

Safe Work Australia, *Working with Silica and Silica Containing Products*, **https://www.safeworkaustralia.gov.au/doc/working-silica-and-silica-containing-products-pdf-doc**.

Relevant Australian Standards
AS 3700– 2011 Masonry structures
AS 3972 General purpose and blended cements

GET IT RIGHT

The picture below shows the finished use of blocks creating a retaining wall. Identify the error that has occurred during its construction.

_____
_____
_____
_____
_____

WORKSHEET 1

To be completed by teachers
Student competent ☐
Student not yet competent ☐

Student name: _____

Enrolment year: _____

Class code: _____

Competency name/Number: _____

Task

Read through the sections from the start of the chapter up to 'Mix bricklaying and blocklaying mortar', then complete the following questions.

1 Complete the following statement: Bricks are one of the _____ and most _____ forms of building material.

2 Brickwork can be both load bearing and non-load bearing. (Circle the correct answer.)

   TRUE   FALSE

   _____

3 List the four categories of masonry bricks.

   _____
   _____
   _____
   _____

4 What is the dominant material used in the construction of calcium silicate bricks?

   _____

5 What type of brick is most commonly used in domestic construction?

   _____

6 Identify two aspects of finishes to bricks that contribute to the overall visual appearance of a wall.

   _____
   _____

7 What other construction materials are used with concrete blocks to construct a retaining wall?

   _____
   _____

8 What manufacturing process provides autoclaved aerated concrete (AAC) with its desired qualities of being lightweight with high thermal qualities?

_____

_____

9 Identify three different types of natural stone that are used in the construction industry.

_____

_____

_____

WORKSHEET 2

To be completed by teachers

Student competent ☐

Student not yet competent ☐

Student name: _____

Enrolment year: _____

Class code: _____

Competency name/Number: _____

Task

Read through the sections from the start of 'Mix bricklaying and blocklaying mortar' up to 'Mortar components', then complete the following questions.

1. Name the Australian Standard that specifies the mixing proportions of material to make different classes of mortar.

   _____

2. Identify at least two materials that can affect the final colour of mortar.

   _____

   _____

3. How many cubic metres (m^3) of sand are required to make mortar for 1000 standard bricks?

   _____

WORKSHEET 3

To be completed by teachers
Student competent ☐
Student not yet competent ☐

Student name: _____

Enrolment year: _____

Class code: _____

Competency name/Number: _____

Task

Read through the sections from the start of 'Mortar components' up to the end of the chapter, then complete the following questions.

1 Name two diseases that can result from the handling of cement.
   _____
   _____

2 Identify two items of PPE that should be used when working with cement.
   _____
   _____

3 Name two impurities that sand needs to be free of.
   _____
   _____

4 Identify two uses for special purpose cement.
   _____
   _____

5 Which type of additive is used to colour mortar?
   _____

6 What is the purpose of a plasticiser in mortar?
   _____
   _____

HANDLE AND PREPARE BRICKLAYING AND BLOCKLAYING MATERIALS

7 Complete the diagrams below to the specified finish of the joint.

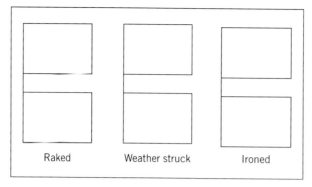

EMPLOYABILITY SKILLS

Using the following table, describe the activities you have undertaken that demonstrate how you developed employability skills as you worked through this unit. Keep copies of material you have prepared as further evidence of your skills.

Employability skills	The activities undertaken to develop the employability skill
Communication	
Teamwork	
Planning and organising	
Initiative and enterprise	
Problem-solving	
Self-management	
Technology	
Learning	

USE BRICKLAYING AND BLOCKLAYING TOOLS AND EQUIPMENT

This chapter covers the outcomes required by the unit of competency 'Use bricklaying and blocklaying tools and equipment'. These outcomes are:
- Plan the work.
- Prepare for work.
- Use plant, equipment and tools.
- Clean up.

Overview

In the bricklaying and blocklaying trade area you will be called on to use and operate a wide range of tools and equipment. In order to work efficiently, you must be able to identify, use and maintain these tools correctly. Selection of the right tool will assist in producing a higher quality of work with maximum efficiency.

The construction of masonry structures has been a means of survival for many thousands of years. Many of the methods and processes for the construction of masonry structures require manual labour, just as in years gone by, although there have been several technological enhancements that have improved output and safety in the industry today.

Safe work method statements (SWMSs) and job safety analyses (JSAs) have also become crucial factors in the planning and execution of works in the construction industry. When undertaking any bricklaying tasks on a construction site, an SWMS will identify the correct use of hand tools and power tools. The SWMS will also identify the appropriate personal protective equipment (PPE) to be used during the performance of work-related tasks.

Employers value people who fit into their workplace, solve day-to-day problems, manage their time and are keen to continue learning. These types of skills are known as employability skills. The employability skills you develop while working through this unit will be assessed at the same time as the skills and knowledge.

It is important to show evidence that you have developed these skills. Your trainer or assessor will discuss with you how to record your employability skills. The following table provides some examples of things you might do to develop employability skills in this unit.

◎ EMPLOYABILITY SKILLS

Employability skill	What this skill means	How you can develop this skill
Communication	Speaking clearly, listening, understanding, asking questions, reading, writing and using body language.	• Listen carefully to work instructions. • Ask questions to clarify task requirements.
Teamwork	Working well with other people and helping them.	• Work individually and as a team to complete tasks in a timely manner. • Liaise with others to avoid conflicting work tasks.
Planning and organising	Planning what you have to do. Planning how you will do it. Doing things on time.	• Plan resources required for each task. • Plan sequence of activities in order to successfully complete work task.
Initiative and enterprise	Thinking of new ways to do something. Making suggestions to improve work.	• Prepare information before you discuss issues. • Suggest improvements to processes.
Problem-solving	Working out how to fix a problem.	• Manage task sequencing to cater for changes to the work environment. • Adapt to cultural differences in communications.
Self-management	Looking at work you do and seeing how well you are going. Making goals for yourself at work.	• Evaluate and improve own methods of communication. • Aim for quality results within specified task time frames.
Technology	Having a range of computer skills. Using equipment correctly and safely.	• Implement safe work procedures for machinery. • Use a range of communication tools such as two-way radios, telephone and email.
Learning	Learning new things and improving how you work.	• Research online product and work practice information. • Observe more experienced construction workers and adopt their practices to improve your work processes.

Plan and prepare for work

Prior to using any tool on the job – whether it be a hand tool or power tool – it is important to prepare and organise the task to be done.

There are five things to consider before starting to use the machine or tool:
1. Is the work area safe, clean and tidy, with adequate lighting and ventilation?
2. Has the most appropriate tool been selected to do the task?
3. Is the tool ready for use, with the appropriate cutter or blade?
4. Is the tool safe to use?
5. Is the person using the tool familiar with the tool and able to operate it safely?

Upon identifying a tool, you will then need to learn what it is designed for, its correct use and any important safety aspects. The following examples outline some common tools that you are likely to use in the construction industry in relation to bricklaying and blocklaying.

Identifying and selecting hand tools

The hand tools in this chapter have been placed into five categories: hammers and cutting tools, setting-out tools, bricklaying tools, supplementary tools and levelling tools. These categories are grouped based on the usage undertaken by bricklayers. The explanation of each tool will highlight the task it performs and the material it is commonly used on.

> Always wear eye protection when using hand tools to prevent loose debris from striking your eyes. Even when hand tools are maintained in good operating order, debris can be very dangerous.

Hammers and cutting tools

Hammers and cutting hand tools that are used for masonry work are very robust and heavy duty. They are designed and manufactured to withstand high levels of repetitive impacts as the intended use is for the tools to strike each other or the material. Tools must be used for their intended purpose for the best results and for the longevity of the tools.

Lump hammer

A lump hammer, also known as a club or mash hammer, is used with a bolster to cut bricks and stone (see Figure 9.1). The handle, which may be made of timber, fibreglass or steel, is forged with the head.

FIGURE 9.1 Lump hammers with timber handle and fibreglass handle

Bricklayers' (brickies) hammer

A brickies hammer is used mostly for rough cutting bricks in common brickwork. It can also be used to drive nails and line pins as required (see Figure 9.2).

Scutch hammer

A scutch hammer is a hammer that has a comb inserted in the end of the head of the hammer that can be replaced when it wears down. The scutch hammer is used to chip off dry mortar from bricks, blocks and any other masonry construction materials. There are two types of scutch hammers – single-ended (see Figure 9.3) and double-ended – with the name indicating whether combs appear on one or both ends of the head.

FIGURE 9.2 Typical brickies hammer

FIGURE 9.3 Single-ended scutch hammer with replaceable comb

Brick bolster

A bolster is a steel-bodied tool with a broad flat face that forms a thin cutting edge (see Figure 9.4). It is used with a lump hammer by bricklayers and stone masons to provide a shear cut through masonry materials; for

FIGURE 9.4 Brick bolsters

example, cutting dry-pressed bricks to length. Like a scutch hammer, it might also feature a comb.

Bolsters come in a variety of sizes, from 50 mm blades up to 115 mm. Many bolsters come with a plastic hand grip that also has a surrounding flange to protect the user's hand from accidentally being hit by the striking hammer.

Cold chisel

Cold chisel is a single solid piece of metal that has an end that is shaped to resemble a pointed wedge (see Figure 9.5). It is used in many applications to chip or knock away at concrete as well as brickwork; for example, knocking out brick channels when chasing in a brick wall to run pipes or electrical cable. A cold chisel can also be used to cut bricks when there is no powered tool available and only a small number of bricks to be cut. The pointed end of the chisel will need to be ground down to keep a sharp edge.

FIGURE 9.6 Scutch chisels

FIGURE 9.5 Medium-sized cold chisels

Scutch chisel

A scutch chisel is similar to a cold chisel, except the scutch chisel has an interchangeable end where a comb can be placed (see Figure 9.6). These chisels are used where a scutch hammer is not able to reach. A scutch chisel can also allow for more accurate location of a required cut.

Plugging chisel

A plugging chisel has a long thin cutting edge and is used to remove mortar from between two bricks; for example, to insert timber plugs into existing brickwork to fix items into place. Plugging chisels are also very effective when a single brick is to be removed from an existing wall. The cutting blade on the chisel is thinned down to fit between bricks without damaging the bricks themselves (see Figure 9.7).

FIGURE 9.7 Plug chisels

Setting-out tools

Setting-out tools come in a range of different forms as they all have specialised roles to play. Setting out is a highly important aspect of brickwork that needs to be thoroughly conducted throughout the whole process of construction of a brick wall. When used correctly, these tools are highly effective in ensuring that brick walls maintain their structural integrity as well as the aesthetic requirements of brickwork.

String

A string line is a length of string that is stretched from one fixed point to another to create a straight line for bricks to be lined up with when constructing a brick wall (see Figure 9.8). String lines are made from nylon, which is rot resistant, and can come in different strengths depending on the number of strands used to construct the string (see Figure 9.9). String lines come in a variety of bright colours to make the line more visible.

FIGURE 9.8 Bricks being placed into position guided by the line of the string

FIGURE 9.9 Nylon string line

Block set

A block set (see Figure 9.10) is used to hold a string line from one corner of a brick wall to the other to provide a

FIGURE 9.10 Block set

line to guide the laying of bricks in between (see Figure 9.11). Used in pairs, corner blocks have traditionally been constructed of timber. Today there are several different patterned brands of corner blocks that are mass produced from hard-wearing plastic.

FIGURE 9.11 Block set holding a string line in position

Pin set

A pin set is used for the same purposes as a block set, but in situations where using a corner block set is not possible. This may be due to the corner being too far away, or where there is an internal corner. It must be noted that the use of pins in mortar that has not dried or has recently been laid may cause the pin to come loose or move. The pins are manufactured out of metal so they can be driven into the mortar without damaging the pins (see Figure 9.12).

Chalk line

A chalk line is a roll of string that sits in a container that also holds powdered chalk (see Figure 9.13). As the string is pulled out of the container, the string is loaded with chalk. The line is then stretched between two points, pulled away from the centre and flicked against the surface that the line is required to be marked on.

FIGURE 9.12 Metal pins and string

FIGURE 9.14 A profile pole being used to guide the position of the bricks

FIGURE 9.13 Chalk line

This provides a straight line along which bricks or blocks can be laid.

Brick tape measure

This tape not only has measurements set at millimetre and metre increments – as is the case for standard tape measures – but also has measurements that indicate gauge heights for bricks and blocks, including the height of brickwork and blockwork. This tape measure is effective when setting out the length for a brick or block wall.

Profile pole

A profile pole is a piece of straight metal tubing that is placed into a plumb position (see Figure 9.14) to allow bricks to be laid up to the metal profile to create a plumb finish to the brick wall. The profile pole also allows for the gauge of the brickwork to be set out, which gives overall guidance for the height of the brickwork and each layer of bricks.

There are patented profile poles available that are fitted with clamps so the pole can be secured to the brickwork. These profiles also have the gauge of standard brickwork marked out.

Profile clamps and braces

Profile clamps are specifically designed to clamp the profile pole firmly into position when being held against a brick wall under construction. These clamps come in different combinations, though the common clamp is designed to hold the profile in one direction only.

Braces are designed to assist in setting the profile up in a plumb position and to maintain the position while the brick wall is being constructed.

Gauge rod

A gauge rod is a straight piece of metal that has the standard height of brickwork marked on its face (see Figure 9.15). This allows for a consistent distance to be maintained between each brick that is laid. In many circumstances, the functions of the profile pole and the gauge rod are combined. The use of a gauge rod provides a consistent finish for the brickwork across the brick wall and the overall job.

 COMPLETE WORKSHEET 1

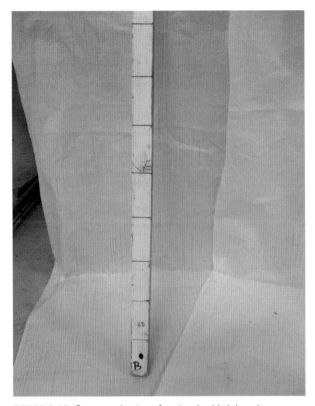

FIGURE 9.15 Gauge rod set up for standard brickwork

FIGURE 9.16 Standard brickies trowel

FIGURE 9.17 Trowel being used to create a furrow in the mortar

Bricklaying tools

Bricklaying tools are directly used to place the bricks and mortar in place. These tools are used with such high frequency that they wear out in a shorter amount of time than other tools in the bricklayer's toolbox.

Trowels

Several different types and sizes of trowels are used by bricklayers, from the basic brickies trowel to more specialised tools, such as those used to patch existing walls or rebuild partial walls.

Brickies trowel

The brickies trowel is used to lay bricks and so is the most used tool in the bricklayer's tool kit. It is approximately 250 mm in length and can have some variation in shape, depending on the manufacture of the trowel (see Figure 9.16). The brickies trowel is used to place the mortar on top of bricks, to prepare the mortar for the next layer of bricks to be placed on top, and to place mortar on the ends of each brick to butt up against the previous brick. The trowel is also used to scoop off any excess mortar before and after bricks have been laid, and to tap the brick into the correct position and ensure that the brick is sitting level.

Once the mortar has been placed onto the face of a brick, the point of the trowel is then used to create a furrow in the mortar to allow for bricks to be pushed down and levelled up (see Figure 9.17).

Bucket trowel

The front edge of a bucket trowel has a flat straight edge (see Figure 9.18) to allow the bricklayer to scoop the contents out of the bottom of a bucket. Its usefulness for placing of mortar onto a brick wall is limited due to its small size.

Brick jointer trowel

A brick jointer trowel is used to create a finish in the mortar between bricks once the bricks have been laid. The task of finishing off the joint between the bricks is done after the bricks have been laid and the mortar allowed to semi-harden.

Tuck pointer

A tuck pointer is a specialised trowel that is used to fill any areas between bricks that do not have enough mortar between them; for example, when existing brickwork is being refurbished. The trowel's blade has a narrow width (see Figure 9.19), so when the mortar is being pushed into position the concentration of effort is directly on the required area.

FIGURE 9.18 Typical bucket trowel

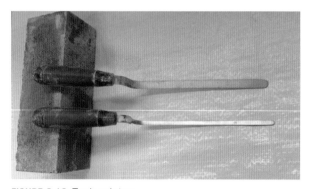

FIGURE 9.19 Tuck pointers

FIGURE 9.20 Joint raker with wheels

FIGURE 9.21 Alternative method to achieve a square finish to the mortar

FIGURE 9.22 Brick jointer

Mortar joint raker

A mortar joint raker has a handle with two wheels at one end and a nail protruding beyond the wheels (see Figure 9.20), which becomes the depth that the mortar is cleaned out from the front face of any brickwork. The jointer is run over the brickwork when the mortar is semi-dry. The finish that this particular jointer produces is commonly known as a square finish.

There are two other common methods of finishing off the mortar between bricks that create the same finish. One method is to drive a nail into a piece of timber with the head of the nail protruding by the depth of the joint to be raked out. The nail head is then run through the mortar joints to scoop out the mortar that is not required. The other method is to cut a profile into a piece of metal the depth of the mortar (see Figure 9.21). Both methods are effective in producing the desired finish.

Brick jointer

The brick jointer (see Figure 9.22) is also used to create a finish to the mortar between the bricks, and is also carried out when the mortar is semi-dry. Because of the rounded end of the jointer, the finish that this tool creates is known as a 'round ironed' joint and is a very common finish (see Figure 9.23). The diameter of the rod

FIGURE 9.23 Brick jointer being used to finish off mortar between brickwork

that is used will to some extent determine the depth of the rounded joint.

Supplementary tools

The supplementary tools that are discussed below are all tools that support the process of laying bricks and the cleaning of the brickwork.

Bucket

Buckets are used to carry water and to measure the amount of water that is added to mortar in its dry state. In some circumstances buckets can be used to transport mortar or hand tools. Buckets are available in several different sizes and types (see Figure 9.24).

FIGURE 9.24 Heavy-duty plastic bucket with metal handle

Drum

Drums are used to temporarily store water close to a mixer to make mortar. The drums are typically 200 L steel or plastic drums (also referred to as 44 gallon drums) and are drums that have been used for other storage purposes and are no longer required. Such drums can be manufactured from different materials such as metal or plastics (see Figure 9.25).

Tin snips

Tin snips are generally used to cut any wire or metal ties that are used in the process of constructing a brick wall.

FIGURE 9.25 Plastic drum used to store water

There are several different types of this tool; see Chapter 11 for more information.

Shovels

There are two shovels that are specifically used by bricklayers: the long-handled round point mouth and the short-handled square mouth (see Figure 9.26). The shape of the shovels helps to measure the materials that are mixed to make mortar. The flat edge of the square-mouth

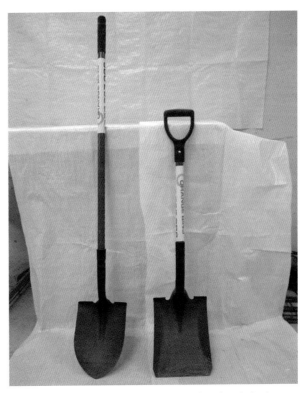

FIGURE 9.26 Long-handled round-point shovel and short-handled square-mouth shovel

USE BRICKLAYING AND BLOCKLAYING TOOLS AND EQUIPMENT

shovel also helps to pick up any mortar that has been dropped at the base of a brick wall.

Tool carriers

The typical tool carrier for a bricklayer is a heavy-duty bucket. There are several different firms that manufacture bucket buddies, which are specially designed pouches that hang off the rim of a bucket to allow for additional tools to be carried. There is also a wide range of tool carry bags that are constructed of heavy-duty material (see Figure 9.27).

FIGURE 9.27 Bag used by brickies to carry hand tools

Another commonly used tool carrier is the lockable metal toolbox. These toolboxes are constructed to be very robust and are able to carry the heavy load of tools required by bricklayers. There are also very large metal toolboxes that are fixed to the trays of utes and are used to transport tools and equipment from job site to job site.

Traditional timber toolboxes are now rarely used due to the weight of bricklayers' tools, which limits the life span of such timber toolboxes. Having said that, timber toolboxes still appear on construction sites on occasion.

Mortar boards

Mortar boards are typically made from sheets of plywood. They vary in size so they can be used in several work locations. The board is a temporary platform for the mortar to be stored just before use, its flat surface allowing the mortar to be scooped up by a bricklayer to load his or her trowel (see Figure 9.28).

Stands are available to hold mortar boards up off the ground for easier access and to save the bricklayer from constantly bending over to pick up mortar (see Figure 9.29).

Planks

Planks are often used when constructing a brick wall. They are used to construct a platform where the height of the wall is out of reach of bricklayers. The planks also carry material that needs to be stored and be available for the construction of the wall (see Figure 9.30). Refer to Chapter 11 for more information on planks.

Scaffolding

Bricklayers use a heavy-duty type of scaffolding so it can bear heavy loads. These loads are both live loads,

FIGURE 9.28 Mortar board

FIGURE 9.29 Raised mortar board stand

loads that move on a constant basis, and dead loads, which are loads that are stationary and so do not place movement strain on the scaffolding.

Scaffolding is used by bricklayers when the brickwork reaches a height that makes it difficult for bricks to be laid properly. This also is a work health and safety matter, as bending and lifting to height becomes a hazard for the bricklayer. The temporary work platform that is created also provides protection from any falling debris for anybody who may be working underneath or passing by the structure under construction.

The scaffolding is a temporary structure used only during the construction phase. Once the structure has been completed, the scaffolding can be removed. On

FIGURE 9.30 Single metal plank used by brickies

FIGURE 9.31 Scaffolding around a house

FIGURE 9.32 Lifting tongs used to carry multiple bricks or large blocks

FIGURE 9.33 Typical wheelbarrow used to transport mortar

many construction sites scaffolding is used by other tradespeople, such as carpenters, window fixers, renderers and roofers.

Brick and block lifting tongs

Lifting tongs are designed to lift and transport several standard-sized bricks in a safe and manageable manner. Lifting tongs also make it easier to handle larger blocks that need to be moved (see Figure 9.32).

Wheelbarrows

There are different types of wheelbarrows that are used for work related to constructing masonry walls. The standard wheelbarrow (see Figure 9.33) can be used to transport bricks around a construction site, but is mostly used to transport mortar from the mixer to the location where the bricklayer requires the mortar. There are occasions when mortar is mixed in a wheelbarrow, but usually only when small amounts of mortar are required.

The second type of wheelbarrow is the flat tray wheelbarrow, which is designed specifically to carry bricks (see Figure 8.18 in Chapter 8). The flat tray and high edge allow for bricks and blocks to be loaded and moved in a secure and safe manner. Refer to Chapter 11 for information on the range of wheelbarrows used in the construction industry.

Larry

A larry is a long-handled tool with a metal plate at the end that has two holes through it (see Figure 9.34). These holes allow the mortar to pass through the holes to assist in the mixing of the different materials. This way of mixing mortar is usually undertaken when only a small amount of mortar is required. This normally occurs in a wheelbarrow for ease of mixing and transport of the mortar.

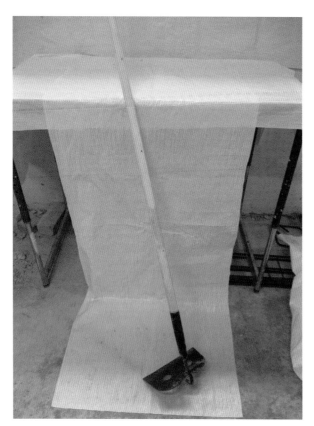

FIGURE 9.34 Larry used to mix mortar

Floor scraper

A floor scraper is used at the start and conclusion of the construction of a brick wall on a concrete slab. At the start of the job, the scraper will remove any foreign material from the floor to give a clean, clear surface for bricks to be laid. At the end of the job, the scraper is used to remove any pieces of dry mortar that have fallen onto the floor below the wall. The floor scraper consists of a handle with a holder at one end that allows for the blade of the scraper to be replaced when worn from repetitive use (see Figure 9.35).

Brush

Brushes are used to remove excess mortar from a brick wall (see Figure 9.36). If this cleaning is progressively undertaken during the construction of the wall, the work required for the final cleaning of the wall is minimised. If the mortar has dried too much, a heavier abrasive cleaning action will be required.

Brushes come with different strength bristles and are used for different applications. For example, wire brushes can be used to clean off dry mortar from tools. In some cases this is preferable to using water, as it reduces the opportunity for rust to develop on the tools.

COMPLETE WORKSHEET 2

FIGURE 9.35 Floor scraper used to clean floor slabs before and after the construction of a brick wall

FIGURE 9.36 Brush used to clean down brickwork

Levelling tools

Levelling tools have been used for thousands of years and the basic principles have not changed in all that time. Gravity and water are still used in the majority of tools used today, although technological advances have seen the introduction of laser levels, which have proven to be as reliable as traditional methods of determining level.

It is important to become familiar with some of the terms used in the levelling process. These terms include levelling, level line, plumb, datum and traverse. Refer to the glossary for definitions of these terms.

Plumb bob

A plumb bob is a weight attached to a piece of string to create a line that is plumb, preventing walls from being built with a lean. The weight of the plumb bob uses gravity to create a straight line from a fixed elevated position (see Figure 9.37). Plumb bobs are not effective on windy days.

FIGURE 9.37 Typical plumb bob

Spirit level

Spirit levels come in a range of different lengths. The typical size that is used by a bricklayer is a 1.2 m level (see Figure 9.38). This gives the opportunity for the level to sit across several bricks during the process of constructing a brick wall. The length assists in levelling and keeping a wall plumb, and ensures the consistent alignment of the bricks being laid (see Figure 9.39). Refer to Chapter 11 and Figure 11.56 for information on reading spirit levels.

FIGURE 9.38 Using a level to check for plumb and straight

Hydrostatic level

A hydrostatic level is a clear plastic tube that is filled with water. The water at each end of the tube will always end up at the same height because water will

FIGURE 9.39 Using a level to check for level and straight

always find its own level (see Figure 9.40). The hydrostatic level is more commonly known as a water level. The water level is used to transfer a level point across a greater distance than a spirit level is capable of. Due to the flexible nature of the water level, if operated by two people it can be used to gauge levels around corners.

FIGURE 9.40 Water level

Laser level

There are two different types of laser level. The first type, called a rotary laser level, has a laser light that spins around inside the body of the casing. The body of the casing is set at a level or plumb position and the spinning light creates a line indicating the desired position or location. This type of laser level can be used for a high point of reference and can also be tilted to project the laser light onto the floor to help set out wall locations on the floor. The rotary lasers rely on a receiver to pick up the position of the laser and will emit different noises to go either up or down until the receiver is in the correct position (see Figure 9.41). Rotary laser levels are best suited for outdoor work and over long distances.

FIGURE 9.41 Laser level with rotary laser mechanism

The second type of laser level is a static laser level that produces a static line that is easily visible and will give a line of reference. The laser is usually brighter and more accurate over short distances (see Figure 9.42).

FIGURE 9.42 Static laser level

Both types of laser levels can be used in conjunction with a receiver and can be adapted to both vertical and horizontal points of reference. It is important to keep in mind the type of work you will be performing before you select the type of laser level (see Figure 9.43).

FIGURE 9.43 Plumb laser and laser receiver

LEARNING TASK 9.1 HAND TOOL USE

In this activity you are required to construct a brick wall. Under the guidance of your teacher you are required to:
- construct a brick wall with a minimum height of 943 mm and minimum length of 1430 mm
- cut bricks using a hammer and bolster where half-bricks are required
- construct the brick wall in stretcher bond with a range of hand tools, including a trowel, level and profile poles
- apply a round-ironed joint finish to the mortar between the bricks.

The preparation of the mortar may be conducted in conjunction with Learning task 8.4 in Chapter 8.

Identifying and selecting power tools

Pre-operational checks need to be performed before any power tools are used. The nature of the checks will depend on how the tool is driven – that is, petrol powered or electrically powered – and will cover the cutting component of the tool. If any faults are identified, the tool must not be used and reported immediately to a supervisor. Faults include exposed wires in the cord, appropriate guards not in place, and blunt cutting edges to any blades.

 Power tools have moving parts and some can reach high speeds. Hold the material and power tool in position and always wear eye and ear protection.

Mortar mixer

A mortar mixer has a vertical drum with paddles inside that rotate and mix the contents (see Figure 9.44). Unlike concrete mixers, where the drum rotates to mix the contents, the drum of a mortar mixer is static. Mixers are commonly used by bricklayers as they are readily

FIGURE 9.44 Typical mortar mixer

transportable. Mixers are driven by either electrical or petrol-powered motors. Any of these mixers need to be constantly maintained by washing down any mortar to prevent the machine becoming too heavy or reducing its capacity to mix mortar. Refer to Chapter 11 for information on the range of mixers used in the construction industry, as well as maintenance and safety requirements.

Brick saw

Brick saws are used to cut bricks and masonry products (see Figure 9.45). The saws are typically used where the finish of the cut edge is going to be exposed. An example of this is where sill bricks come together from two different angles. The brick saw is a heavy piece of equipment and is designed with lifting points for two people. The brick saw is positioned on a specialised stand so the saw can be secured into the stand and the weight is distributed in such a manner that the saw will not overbalance.

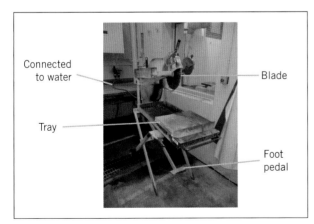

FIGURE 9.45 Brick saw

The saw has a handle at the top that allows the operator to pull the saw down and apply the required amount of pressure to cut through the masonry product. As there may be different types of materials to be cut, it is up to the operator to judge what the correct pressure is. The saw has a spring mechanism that pulls the motor and blade up. This allows material to be placed under the blade with sufficient clearance, and takes the weight of the saw when cuts are being made.

When using a brick saw, PPE requirements become obvious due to the noise that is emitted when the machine is turned on, which increases as it cuts into the material. While the noise can vary based on the material being cut, the high-pitched sound the saw emits without being under load warrants ear protection.

As with any cutting equipment, eye protection from flying debris is also required. It is also recommended that you wear appropriate protective clothing as the machine uses high amounts of water to operate effectively.

Foot pedal

The foot pedal assists to bring the cutting blade down onto the material to be cut. This is a means of applying additional pressure other than the handle at the top of the machine, spreading the physical effort required to use the brick saw.

Blades

A range of different blades can be fitted to the brick saw. These blades are made from different materials, including some that use industrial diamonds on the cutting edge, and provides the blade with the ability to cut through masonry products. Blades can also be solid or segmented.

Tray

The base of the brick saw has a tray that sits on the edges of the base. The tray has wheels that allow for the tray to slide on the base to give a broader reach of cutting distance for the blade. The tray usually has a rubber matting so that any material that is placed onto the tray will grip and be held in place with less chance of slipping.

Water connection

Brick saws are connected to a water supply that is used to continuously feed water onto the cutting blade to reduce the wear on the blade and minimise heat generated from the friction. The water is captured in the base of the saw, where there is a drainage point for the water to escape. Consideration needs to be given to where the excess water is flowing from the brick saw so it does not create a work health and safety hazard.

Setting up

When setting up the brick saw, the draining of excess water is not the only consideration. The saw also needs to be set up close to the area where the bricks are stored, and close to a water supply. There also needs to be sufficient space around the saw to allow for the cut bricks to be stored and so as any excess waste or offcuts can be appropriately stored for later removal.

Routine maintenance
- Check electrical connections for proper and secure attachment. No bare wire or coloured cables should be visible at any connection point.
- Make sure all electrical connections are waterproof.
- If petrol motors are used, check oil levels as required and clean, adjust and/or replace spark plugs as required.
- Ensure that the cutting blade has sufficient cutting edge remaining on the blade.

Safety
- Always wear ear and eye protection when using a brick saw.
- Ensure the material you are cutting is secured and will not move.
- Ensure that the extension lead and lead of the saw are positioned away from wet areas.

Brick elevator
A brick elevator is a conveyor belt system that carries bricks to a different location, typically at a higher elevation. The elevator is usually powered by a petrol motor. As with many types of machines that carry material, there is a safe working load (SWL). The SWL needs to be confirmed for the particular elevator that is being used.

A brick elevator requires two operators: one person at the bottom of the elevator loading the bricks, and another person at the top who takes the bricks off the elevator and stacks them in an appropriate location and manner (see Figure 9.46). As with all plant and equipment there needs to be constant maintenance and operational checks.

Routine maintenance
- Check electrical connections for proper and secure attachment. No bare wire or coloured cables should be visible at any connection point.
- If petrol motors are used, check oil levels as required and clean, adjust and/or replace spark plugs as required.
- Ensure that when the elevator is being transported or stored all locking in pins are in place.

Safety
- When setting up and packing away ensure that two people are available to lift the equipment.
- The setting up of a brick elevator requires an area at the base that is stable and has a clear surrounding area.
- Ensure that no loose garments can get caught in moving parts of the machine.
- Ensure that the loading and unloading points of the machine are at an appropriate height and there is sufficient space for the operators to manoeuvre around.

FIGURE 9.46 Brick elevator transporting bricks to a higher location

LEARNING TASK 9.2
POWER TOOL USE – BRICK SAW

Under the guidance of your teacher you are required to use a brick saw to cut the following:

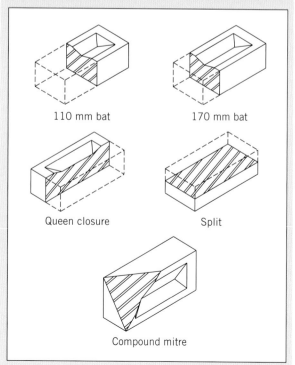

FIGURE 9.47 Saw cuts

COMPLETE WORKSHEET 3

Clean up

Due to the nature of the work, bricklaying and blocklaying tools and equipment require daily cleaning. As indicated in the chapter, if tools and equipment are not cleared of materials such as mortar then the daily build-up will reduce their effectiveness.

Special consideration is required when washing down the different machines used in the bricklaying trade. Cutting equipment, such as a brick saw, are subject to a build-up of fine slurry that can harden and reduce the efficiency of the machine. Consideration needs to be given to the maintenance of the motorised components of the equipment as well as the hoses and moving mechanisms.

SUMMARY

In Chapter 9 you have learnt how to use a selection of tools and equipment for brickwork and blockwork construction. In addition, the chapter has:
- identified a variety of tools and equipment used for brickwork and blockwork, including hand tools and power tools
- emphasised the importance of maintenance and safe use of power tools and equipment.

REFERENCES AND FURTHER READING

BT Engineering Australia, **https://www.btengpl.com.au**.

Commission for Occupational Safety and Health, *Concrete and Masonry Cutting and Drilling – Code of practice*, WA Department of Mines, Industry Regulation and Safety.

SafeWork NSW, *Cutting and Drilling Concrete and Other Masonry Products*, **https://www.safework.nsw.gov.au/__data/assets/pdf_file/0010/52867/Cutting-and-Drilling-Concrete-and-Other-Masonry-Products-Code-of-Practice.pdf**.

GET IT RIGHT

The photo below shows a mixer used to mix mortar. Identify the different safety issues of this situation. For each issue identified, indicate what should be the correct arrangements.

Source: Shayne Fagan

_____
_____
_____
_____
_____
_____

WORKSHEET 1

Student name: _____

Enrolment year: _____

Class code: _____

Competency name/Number: _____

To be completed by teachers	
Student competent	☐
Student not yet competent	☐

Task

Read through the sections from the start of the chapter up to 'Bricklaying tools', then complete the following questions.

1 Match the tool name with the picture letter and provide a brief summary of what the tool is used for.

Tool	Letter	What is the tool used for
Block set		
Brick bolster		
Caulk line		
Lump hammer		
Plug chisel		
Scutch chisel		

a

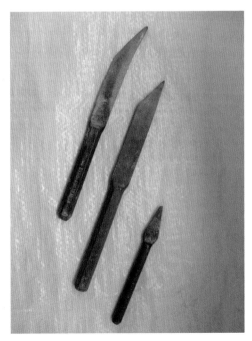

b

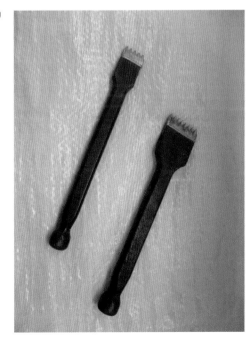

c

d

e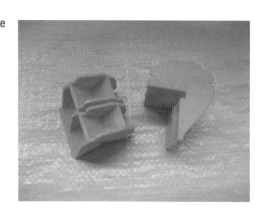

f

2. What two functions can a profile pole fulfil when used to construct a brick wall?

3. What is the purpose of corner blocks?

4. What factors contribute to the strength of a string line?

WORKSHEET 2

To be completed by teachers
Student competent ☐
Student not yet competent ☐

Student name: _____

Enrolment year: _____

Class code: _____

Competency name/Number: _____

Task

Read through the sections from 'Bricklaying tools' up to 'Levelling tools', then complete the following questions.

1 What is the purpose of a mortar board?

   _____

   _____

2 What are the different finishes to a mortar joint, and which tools are used to produce the different finishes?

   _____

   _____

3 Why are brushes used on a brick wall?

   _____

   _____

4 Identify two uses for a standard wheelbarrow.

   _____

   _____

5 Give an example of what a brickies trowel is used for.

   _____

   _____

6 What is scaffolding used for in relation to bricklaying tasks?

   _____

   _____

WORKSHEET 3

To be completed by teachers
Student competent ☐
Student not yet competent ☐

Student name: _____

Enrolment year: _____

Class code: _____

Competency name/Number: _____

Task

Read through the sections from 'Levelling tools' up to the end of the chapter, then complete the following questions.

1. What action should you take when a powered tool is found to be faulty during the pre-operational check?

   _____
   _____

2. How many people are recommended as a minimum to place a brick saw into position?

   _____
   _____

3. When setting up a brick saw, what are two environmental aspects that need to be considered?

   _____
   _____

4. What PPE should be used when operating a brick saw?

   _____
   _____

MY SKILLS 9

🎯 EMPLOYABILITY SKILLS

Using the following table, describe the activities you have undertaken that demonstrate how you developed employability skills as you worked through this unit. Keep copies of material you have prepared as further evidence of your skills.

Employability skills	The activities undertaken to develop the employability skill
Communication	
Teamwork	
Planning and organising	
Initiative and enterprise	
Problem-solving	
Self-management	
Technology	
Learning	

HANDLE CARPENTRY AND CONSTRUCTION MATERIALS

10

This chapter covers the outcomes required by the units of competency 'Handle construction materials' and 'Handle carpentry materials'. These outcomes are:
- Plan and prepare.
- Manually handle, sort, stack and store carpentry, construction materials and components.
- Prepare for mechanical handling of materials.
- Check and store tools and equipment.
- Clean up.

Overview

The selection and use of carpentry and construction materials is an important part of the construction process. This is not only for constructing the building but also for the longevity of the building. The construction method will influence which structural and finishing materials will be used. The range of building materials discussed in this chapter covers the most commonly used construction materials in contemporary domestic construction.

Throughout this chapter we will look at each material from several different perspectives such as use, durability, strength, fire rating, cutting, handling, stacking, storing and relevant environmental issues.

Employers value people who fit into their workplace, solve day-to-day problems, manage their time and are keen to continue learning. These types of skills are known as employability skills. The employability skills you develop while working through this unit will be assessed at the same time as the skills and knowledge.

It is important to show evidence that you have developed these skills. Your trainer or assessor will discuss with you how to record your employability skills. The following table provides some examples of things you might do to develop employability skills in this unit.

EMPLOYABILITY SKILLS

Employability skill	What this skill means	How you can develop this skill
Communication	Speaking clearly, listening, understanding, asking questions, reading, writing and using body language.	• Listen carefully to work instructions. • Ask questions to clarify task requirements.
Teamwork	Working well with other people and helping them.	• Work individually and as a team to complete tasks in a timely manner. • Liaise with others to avoid conflicting work tasks.
Planning and organising	Planning what you have to do. Planning how you will do it. Doing things on time.	• Plan resources required for each task. • Plan sequence of activities in order to successfully complete work task.
Initiative and enterprise	Thinking of new ways to do something. Making suggestions to improve work.	• Prepare information before you discuss issues. • Suggest improvements to processes.
Problem-solving	Working out how to fix a problem.	• Manage task sequencing to cater for changes to the work environment. • Adapt to cultural differences in communications.
Self-management	Looking at work you do and seeing how well you are going. Making goals for yourself at work.	• Evaluate and improve own methods of communication. • Aim for quality results within specified task time frames.
Technology	Having a range of computer skills. Using equipment correctly and safely.	• Implement safe work procedures for machinery. • Use a range of communication tools such as two-way radios, telephone and email.
Learning	Learning new things and improving how you work.	• Research online product and work practice information. • Observe more experienced construction workers and adopt their practices to improve your work processes.

Plan and prepare

As discussed in previous chapters, the planning and preparation of a construction project involves estimating the material required for the project, including its availability. This will provide a guide to the lead time required so that all material is on site for the construction process. Any lost time due to poor planning and organisation of material can become costly if it means that no other activity can occur on the construction site in the interim.

The safety of the workers using the materials also needs to be considered. A good guide for such matters will be the safety data sheet (SDS) for each of the products to be used. If there are potential impacts across the whole site, then specialised signage will also be required. An example of this is any building product that contains asbestos.

Other aspects to be considered include the delivery location on the job site and the safe and secure storage of the material. Weather conditions can have a detrimental effect on many different building materials, and the location of the store on the site will also need to take this into consideration. The limitations of the site are also a factor, as they may vary from large construction site with plenty of room for temporary storage to a domestic inner-city site where the building envelope covers the complete area of the land that is being used.

During the planning stage you will need to inspect the site and take note of the room that is available, as material delivery can take up a sizeable amount of space.

During the site inspection consideration needs to be given to any services that are located on or near the site. Many services are underground, though overhead services need to be considered at the same time. For all services, potential hazards need to be identified and control measures implemented. The control measures may be as straightforward as signage or all the way to hoardings around the whole job site.

> **LEARNING TASK 10.1**
> **PREPARING FOR SITE DELIVERY**
>
> As a class, identify an area close to your workshop where you need to prepare for a delivery of a sling of timber and two pallets of bricks.
> Brick pallet = L 1.200 m × W 1.200 m × H 0.700 m
> Timber pack = L 6.000 m × W 1.000 m × H 0.700 m
> Create a sketch of the location. In the sketch identify:
> - the location where the material is to be stored, considering the size of the material to be delivered
> - the access points for the delivery of the material
> - any services that may be in the area
> - what precautions are required for the services.

Selecting appropriate materials

For centuries, builders have used traditional materials like timber, brick, iron and stone for homes, and government and public buildings. Today, however, the materials being used in the construction industry are changing. Older traditional-type materials are now being combined with new technology or even newly developed materials to form prefabricated or pre-assembled products, which are also given new uses.

Building codes and standards are being modified to accept these new materials and combinations of materials and to allow for improved, cost-effective construction.

There is an ever-widening choice when it comes to selecting the best or most appropriate material for the job (see Figure 10.1).

Factors affecting the selection of appropriate materials for use in building include their physical characteristics and economic considerations.
- Physical characteristics:
  - the density or mass of the material (how light or heavy it is)
  - strength and ability to carry loads and resistance to being stretched or compressed
  - electrical resistance or conductivity
  - durability (how well it stands up to the elements)
  - fire resistance
  - insulation properties
  - moisture resistance
  - general appearance.

FIGURE 10.1 A contemporary residential building, using a variety of materials

- Economic considerations:
  - manufacture or production costs
  - retail costs
  - whether or not the material has a sustainable source
  - what effect the use of the material has on the environment.

Without knowing the physical characteristics or the economic considerations of the materials to be handled, stacked and stored, it could be easy to store a material in the wrong environment or location. An example of this is bags of cement left stacked outside in the rain. By being aware of these characteristics the material can be correctly handled, stacked and stored when it arrives on site, thus avoiding it being damaged on site and causing costs to the builder as s/he will need to replace or repair the damaged items before work can continue.

Construction materials and components

The following materials and components are discussed in this section:
- timber
- engineered timber products
- concrete components
- paints and sealants
- steel components
- insulating materials
- plasterboard sheeting
- asbestos
- glass
- adhesives.

Timber

Timber is an important building and construction material that actively competes with companion materials such as steel and masonry products. It is referred to as wood or lumber while it is in the form of a growing tree and as timber once it is felled and milled.

It consists of several chemical components that are given by approximate mass, such as: cellulose (45%) and hemicellulose (22%), which together make up the fibre; lignin (30%), which acts as a binding agent or glue for the fibres; and extractives (3%), which are minerals, gums and resins.

The main trunk of the tree is cone-shaped, which means that when the tree grows and produces new wood in the form of growth rings, it grows horizontally in girth or increases in diameter, rather than growing vertically. The increase in height occurs as a result of a new layer being added and the activity of special cells at the extreme tips of branches. A guide to the approximate age of the tree can be determined by counting the growth or annual rings found in the cross section of the tree (see Figure 10.2).

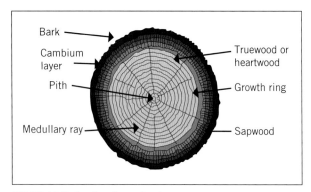

FIGURE 10.2 Section through a log

Identification

While timber is usually classified as either hardwood or softwood, this has little to do with its hardness or softness.

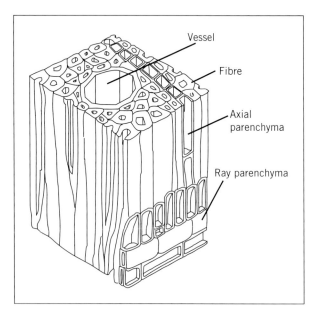

FIGURE 10.3 Diagram of a cube of hardwood (magnification × 250). The pits in the cell walls have been omitted.

Hardwood trees

These are broad-leafed trees (angiosperms) that flower and have either fruit or nuts for seed containment and generally have a slower growth cycle. The timber structure is made up of a series of hollow vertical pipes known as vessels, which are more easily identified when magnified. This can be checked by slicing the end grain of the timber with a very sharp knife or blade to reveal the pores or vessels more clearly (see Figure 10.3).

Some common types of hardwood in use are tallowwood (see Figure 10.4), spotted gum, brush box, meranti, river red gum, Tasmanian oak, ironbark, turpentine, messmate, stringy bark and meubau.

FIGURE 10.4 Tallowwood tree and timber

Balsa wood is technically a hardwood, but is actually so soft that it has limited uses in construction.

Softwood trees

Softwood trees are usually conifers or pines (gymnosperms), have needle-like leaves and generally a faster growth cycle. The timber structure is made up of a series of hollow vertical cells known as tracheids (see Figure 10.5). These cells allow the water and minerals to pass up the tree to the leaves where the food is produced. Some common types of softwood are western red cedar, Radiata pine (see Figure 10.6), Baltic pine, hoop pine, kauri pine, Oregon or Douglas fir and cypress pine.

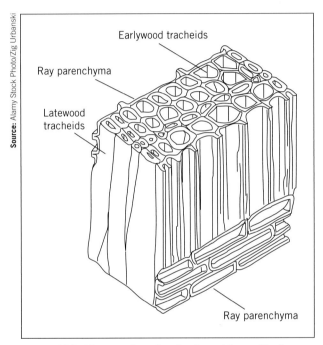

FIGURE 10.5 Diagram of a cube of softwood (magnification × 250). The pits in the cell walls have been omitted.

FIGURE 10.6 Radiata pine trees and timber

Uses

Timber and timber products may be used in a variety of situations and in the manufacture of many products. Some of these are extracted oils, fuel, paper, resins, manufactured boards, poles, posts, framing, furniture, boxes, toys, boats, weapons, sporting equipment, claddings and linings.

Structural uses within the construction industry include:
- sub-floor framing
- wall framing
- ceiling framing
- roof framing
- window and door frames
- formwork for concrete
- roof truss construction.
  Finishing uses include:
- architraves
- skirtings
- scotia and quad mouldings
- floor timber
- kitchen and built-in cupboards
- stairs and panelling.
  Other uses include:
- fencing
- pergolas and decking
- retaining walls and screens
- solid timber door construction.

Durability

The durability of a timber relates to its ability to withstand the destructive action of moisture and the elements, and of abrasion, plus its resistance to insect attack. When varieties of timber are exposed to hazards, such as insects, their durability will vary. The Hazard Level (H) is given a number, with H1 for low hazards and H6 for high hazards.

Some timbers, such as cypress pine and ironbark, are very durable and are grouped together with a Class 1 durability rating. Other poor-durability timbers such as Radiata pine and Oregon (otherwise known as Douglas fir) have a Class 4 durability rating (see Table 10.1).

TABLE 10.1 Natural durability classification of heartwood of some commonly used timbers

Class 1	Class 2	Class 3	Class 4	Hazard level of timber (treated timber)
White cypress pine	New England blackbutt	Brush box	Brownbarrel	H1: inside, above ground, fully protected
Ironbark	Blackbutt	Mixed open forest	Caribbean pine	H2: inside, above ground, protected from elements, no termite protection
Tallowwood	Kwila (Merbau)	Hardwoods from Nth NSW or Sth Qld	Douglas fir (Oregon)	H3: outside, above ground
Turpentine	Spotted gum	Rose/Flooded gum	Radiata pine	H4: in ground contact
Grey gum	Western red cedar	Sydney blue gum	Slash pine	H5: contact with ground or fresh water
	River red gum	Silver-topped stringybark	Mountain ash (Tasmanian oak)	H6: salt water contact
	Stringybark (yellow and white)		Alpine (Vic.) ash	
			Unidentified hardwood or softwood	

Untreated Class 3 and 4 timbers should not be used for weather-exposed structural members; that is, posts, joists and bearers of decks or unprotected beams protruding from the house.

These less durable timbers may require preservative treatment to make them more durable. These treatments may include CCA (copper chromium and arsenate), a pressure treatment that turns the timber a distinctive green colour; or LOSP (light organic solvent-borne preservative) – this preservative does not discolour the timber but may leave a waxy residue. Preservatives are available from your local hardware store (see Figure 10.7).

The heartwood or innermost section of a tree has inherent natural durability, whereas the sapwood or outermost section has poor durability qualities and is especially susceptible to insect attack.

Strength
Timber is strongest along the straight length of its grain. It will be weakened where short-, wavy- or cross-graining occurs and also where knots, caused by branches of the tree intersecting the trunk, occur.

Solid timber is generally not used for load-bearing spans greater than 4.8 m, as it deflects or bends excessively and becomes too expensive. Steel may be used as an alternative, or, if timber is preferred, engineered timber products such as laminated veneer lumber (LVLs) or Glulam beams may be used, allowing for greater spans to be bridged than if using natural timber.

Fire rating
All timber will eventually burn, but some timbers, such as teak, karri and jarrah, have a degree of resistance to

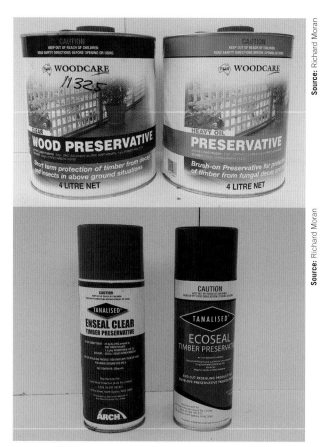

FIGURE 10.7 Rot stop wood preservative and end seal

fire. When timber is heated the temperature rises steadily to about 100°C, where it remains until the considerable amount of water vapour the timber contains is evaporated. The temperature rise then continues without much further effect until, at about 200°C, flammable gases begin to form. By 250°–300°C, ignition occurs if sufficient oxygen is present. At 350°–450°C, depending on the density of the timber, spontaneous ignition occurs.

While small pieces of timber burn through quickly, a large piece of timber will develop a layer of char, which tends to insulate its interior. Large timbers (for example, 250 mm × 250 mm) with equal flat faces may stand up to fire better than steel in the same conditions. Timber may be further protected by treating with mineral salts, which will stop the timber burning after the source of ignition is removed, or the application of intumescent paint, which forms a gas on the surface when heated to remove the presence of oxygen.

Cutting

Timber may be cross-cut (across the grain) or ripped (along the grain) by using a hand or power saw (see Figure 10.8 and Figure 10.9). Blunting of these tools may be caused by the high silica content in the timber of most hardwoods, which acts like an abrasive. Hardened or tungsten teeth on the blades will help overcome this problem.

During the cutting process, especially when cutting across the grain with a power saw, fine splinters and sawdust may be thrown up into the operator's face; therefore, it is essential that wrap-around safety goggles or glasses be worn to prevent possible permanent eye injury.

Handling

Care needs to be exercised when handling timber, especially in its rough-sawn (RS) state, as it will still have furred edges. This may cause the painful entry of splinters, resulting in possible skin infection or serious damage. Also, some species of timber, mainly unseasoned eucalypt hardwoods, will stain the skin a bluish colour during handling due to the moisture and extractives contained in it. Normal washing of the hands followed by rubbing with a cut lemon should remove most of the stain.

Both of these problems can be overcome by wearing leather gloves (see Figure 10.10), at least during manual handling procedures. Most long lengths of timber can be carried safely and comfortably by placing them on the shoulder and wrapping the hands over the top. To prevent physical injury, always use your legs to do the lifting and keep your back straight (see Figure 10.11 and Figure 10.12).

FIGURE 10.8 Cross-cutting timber with a hand saw

FIGURE 10.9 Safe use of a power saw to rip timber

FIGURE 10.10 Wearing leather gloves to prevent splinters

Another way to handle large quantities of timber is by mechanical means, through the use of cranes (usually a truck-mounted crane referred to as a hiab; see Figure 10.13) or a forklift truck.

Stacking

Timber in its green or unseasoned form is very unstable, and this can cause problems with its shape during the

FIGURE 10.11 Lifting correctly to prevent back injury

FIGURE 10.12 Carrying timber safely and comfortably

drying-out or seasoning process (see Figure 10.14). It is therefore necessary to stack timber lengths with evenly sized and spaced pieces of timber, called 'gluts',

FIGURE 10.13 Hiab lifting a sling of timber

between each layer to allow for even air circulation. Timber should also be stacked with the longest lengths at the bottom and the shorter lengths on the top.

Storing

Timber should be stacked neatly off the ground and protected from the elements with a waterproof covering if stored outside (see Figure 10.15). If the timber is ordered as a 'sling', then it should be left strapped, kept dry and stored on gluts off the ground until it is ready to be used. This applies to any timber product, such as wall frames and roof trusses. Remember, moisture over long periods of time will seriously degrade timber.

Environmental issues

The environmental issues that need to be considered for timber are disposal and personal hygiene considerations for people handling treated pine.

 When disposing of treated timber, it must not be burnt, particularly CCA-impregnated timbers. When treated timber is burned, the chemicals are released into the atmosphere as gases and fumes.

In enclosed environments, the gases from treated timber can have serious health effects if inhaled. There are also concerns relating to the leaching of these chemicals into the groundwater if they are placed into unlined landfill.

Untreated and unpainted timber should be separated from general waste on the construction site. Waste timber should be collected and recycled rather than being used as landfill.

Because of the chemicals used in treated pine, people handling these timbers must wash their hands prior to eating.

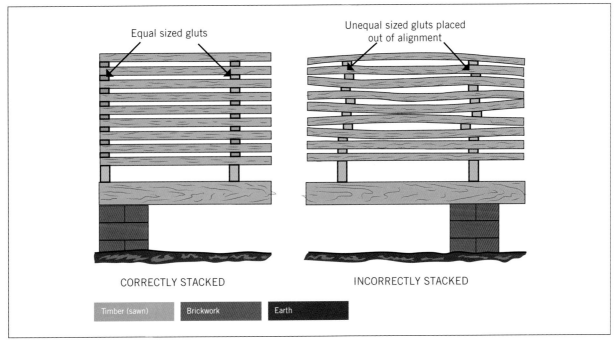

FIGURE 10.14 Bowing and bending caused by poorly aligned gluts

FIGURE 10.15 A sling of framing timber stored out of the weather

LEARNING TASK 10.2 STACKING TIMBER

Using timber of the same sectional size, but different lengths, correctly stack the timber as if it is to be picked up using a forklift truck or a hiab crane.

COMPLETE WORKSHEET 1

Engineered timber products

Engineered timber products have many forms and a wide variety of uses. The most commonly used products are:

- particleboard
- fibreboard
- hardboard
- plywood
- laminated veneer lumber (LVL)
- cross-laminated timber (CLT)
- Glulam (glue-laminated) beams
- 'I' beams
- wood plastic composites.

These products are either slices of timber or timber particles combined with a binder (resin or adhesive) and formed into sheets or sections that usually have a higher strength-to-weight ratio than the timber it came from. In the case of LVL, CLT, Glulam beams and 'I' beams, these products can be ordered in any lengths, only being limited by the method of transporting the material to the site (for example, semi-trailer or shorter table-top truck).

Manufacture

Particleboard, fibreboard and hardboard sheets are formed using a range of different-sized wood fibres, mainly Radiata pine forest thinnings and sawmill residues (for particleboard and fibreboard) and mainly eucalypt chips for hardboard, which are bonded together using a variety of resins, such as urea-formaldehyde, melamine-formaldehyde and phenol-formaldehyde.

The glued wood particles are placed in spreaders. These spread the particles onto mats on a transfer device such as a belt or metal caul plate. The mats are then transferred to the hot press, where they are compressed under high pressure to the thickness required. The high temperature of the press cures the resin, forming a solid board product, and the rough panels are trimmed and cut to size as required. Boards are usually sanded prior to sale or prior to prefinishing, with various surface and edge treatments available.

Some common trade names for these products include:
- particleboard – pyneboard, pineboard, chipboard
- fibreboard – MDF, Fibron
- hardboard – Masonite, Burnieboard, Weathertex.

Plywood, a glue-laminated wood product, is an engineered panel constructed of thin sheets or veneers that are rotary-peeled from the log. The veneers range in thickness from 0.8 mm to 3.2 mm. The number of veneers used varies according to the thickness required and the intended use of the sheet. As a guide, the minimum number of veneers is three and the maximum 11, hence the names 3 ply, 5 ply and so on. The number of veneers is always an odd figure to allow both faces of a sheet to have grain running in the same direction (see Figure 10.16). Each veneer has its grain running at 90 degrees to the previous one, and these veneers are hot-pressed using phenolic resins of varying types to suit an internal or external use of the ply sheet. The resulting composite material is stronger than solid timber, due to the long grain of each veneer being laminated at 90 degrees to the other.

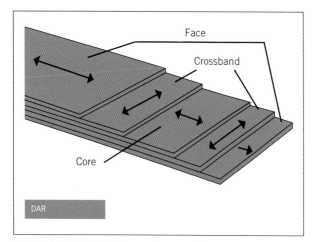

FIGURE 10.16 Typical plywood composition

Plywood products are also available in 'I' section profiles. They are used in both domestic and commercial construction for beams and various other common framing members (see Figure 10.17).

Uses

Manufactured timber products can be used in all areas of building construction and fit-out, such as structural particleboard flooring (wet and dry areas), particleboard shelving with edge strips, fibreboard cupboard doors and tops, fibreboard cabinets, fibreboard enamelled furniture, hardboard strip cladding, hardboard wall-frame wind bracing, plywood flooring, plywood decorative wall lining, plywood wall-frame wind bracing and plywood 'Plasply' formwork, to mention just a few.

FIGURE 10.17 hyJOIST®

Durability

As the basic ingredient of these materials is timber, the durability rating is similar. The durability may be enhanced by additives to the resins, which can be sprayed onto the particles. Fire retardant, insecticide and fungicidal chemicals may be added in small quantities. Paraffin wax also can be added in small quantities, either as an emulsion or sprayed in the molten state, to provide water resistance and to control swelling caused by temporary wetting. Some plywoods, especially those used for structural formwork, can be purchased with the surface pre-sealed with materials such as phenolic film, as used on Plasply. Alternatively, for a very high-quality surface, the raw plywood can be faced with GRP (fibreglass). This surface protection increases the durability level and therefore saves cost over time.

Fire ratings

Many factors need to be taken into consideration when determining the fire rating of a material. For example, jarrah has a below-average or poor ability to resist ignition when exposed to flame; it has a fair resistance to spreading the flame over itself; it gives off a medium or fair level of heat when burning, which lowers the hazard to materials around it; and it produces little smoke when burning, which lowers the risk of smoke damage to other materials around it. When all these factors are averaged out, jarrah would be classified as having a fairly good fire rating, even though it burns readily.

Therefore, for ease of identification and rating of materials, a rating of good, fair or poor is given to individual materials based on their ability to resist ignition, spread of flame, heat produced and smoke developed. Some timber materials are rated in Table 10.2.

Cutting

All manufactured timber products can be cut using conventional hand and power saws or large, accurately cutting panel saws. It is recommended that a tungsten

TABLE 10.2 Ignition performance of tested materials

Timber	Resistance to ignition	Resistance to spread of flame	Heat generated	Amount of smoke created
Jarrah	Poor	Fair	Fair	Low
Particleboard and fibreboard	Poor	Poor	Poor	Low
Plywood	Poor	Poor	Poor	Low

carbide-tipped saw blade be used to allow for the dulling effects of the resin within the product.

When cutting thin sheet materials with a hand saw, the saw should be used at a fairly flat angle and the offcut side of the sheet slightly lifted with one hand to prevent the saw jamming in the kerf. This method also helps to prevent excessive splitting or chipping of the underside of the sheet (see Figure 10.18).

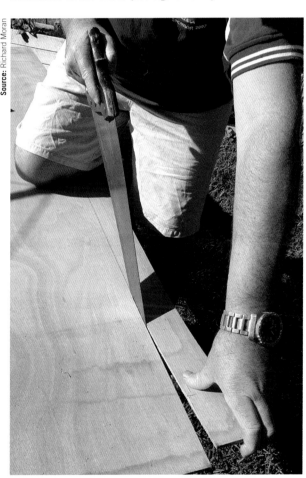

FIGURE 10.18 Correct method for cutting of sheet material

 Due to the fibrous nature of these products, small particles and splinters are thrown out when using power saws; therefore, appropriate eye protection must be worn to prevent possible injury.

Handling

Care needs to be exercised when handling and moving sheet material as the edges or the face may be damaged.

In many instances, sacrificial sheets are placed on the outside of the sheet packs to protect the face of the first and last sheets. Some of the sheet material may be light in weight but will bend and break easily, while other sheets are thick and heavy, requiring two people to carry them. If the sheets are in a pack, it may be easier and safer to mechanically move the pack using a forklift or pallet truck (see Figure 10.19). Single sheets must be handled manually using correct lifting techniques and using hooked handheld carrying devices to make it easier and safer for the handler (see Figure 10.20). This prevents fingers from being squashed when the sheet is set down edge first.

FIGURE 10.19 Mechanical handling of sheets stacked on a pallet

Stacking

Manufactured timber product sheets should always be laid flat and stacked off the ground on packing timbers, called gluts, or on pallets that allow the sheets to remain dry with easy access underneath for mechanical lifting purposes. Where there are high levels of moisture, sheets should be covered to avoid damage caused by moisture absorption. The stack can be mechanically lifted, using a forklift or pallet truck, to load it onto a delivery vehicle or to move it to another covered position.

HANDLE CARPENTRY AND CONSTRUCTION MATERIALS

LEARNING TASK 10.3 MATERIAL RESEARCH ASSIGNMENT III

Prepare a folder with a cover sheet containing the following details:
- your name
- your class/group name or code
- name of the assignment ('Material research assignment III')
- your teacher's name
- the due date.

Part I
Refer to the section above beginning at 'Engineered timber products', then provide the following details:
1 a brief description of each material
2 a brief description of the manufacturing process
3 at least two uses of the material
4 how the material should be safely handled during use or preparation for use.

The materials to be identified and described are:
- particleboard
- fibreboard
- hardboard
- plywood
- laminated veneer lumber (LVL)
- cross-laminated timber (CLT)
- Glulam beams
- 'I' beams
- wood plastic composites.

Produce half a page for each of these materials.

Part II
Include a manufacturer's brochure and/or details obtained from the internet, which relate specifically to each of the materials listed above. These details should be attached to the assignment at the end of the description of each material.

FIGURE 10.20 Manual handling of sheet material

Storing
The best way to store any manufactured timber products is flat on pallets or gluts – covered and out of moist areas.

On the job site always choose an area where the material will not be damaged by passing machines, vehicles or people.

Environmental issues
Manufactured timber products are environmentally efficient, in that they are made from materials that were previously waste, but are now recycled into useful products. This is an environmentally good practice.

There is, however, a darker side to the story that relates to how these materials are currently disposed of. These materials are often being used for landfill or are being burnt.

Therefore the chemical binders and formaldehydes used in producing the products are either released into the atmosphere or into the ground. Recycling this material into second- and third-generation materials is being investigated.

There is another consideration, too – when these products are cut using a circular saw or electric planer, dust particles and chemical fumes are released. Over time these may affect the health of the carpenter.

 COMPLETE WORKSHEET 2

Concrete components
Concrete is basically artificial rock, and when it is cast into formed moulds in a plastic or wet state it will set in the shape of the mould with characteristics similar to those of natural solid stone.

Concrete has been around since the Roman Empire (first-century BCE to fifth-century CE), when the Romans used a primitive form of it to construct vast structures and works of engineering; first using stones and then bricks to form a shape that was filled with a lightweight concrete made from porous volcanic rock. It has evolved over the centuries, but was not widely used until an improved cement was developed in the early

1800s. Since then, concrete has been used in many shapes and forms and combined with steel reinforcement. Today it is one of the most commonly used building products.

On its own, concrete is very strong in compression (when it is being squashed or crushed), but is very weak in tension (when it is being stretched or pulled apart). It consists of four main parts, which when combined create an artificial stone with a high durability level. The components are:
- cement
- fine aggregate – sand in Australia
- coarse aggregate – usually stone or rock fragments
- water.

Cement

This material acts like a glue, bonding all the particles together. The first type of cement, invented by the Romans, was prepared from volcanic ash and lime called pozzolana, while the modern-day cement, developed by Joseph Aspdin in England in 1824 and called Portland cement, is made from limestone and clay crushed and mixed together with water and burned at a very high temperature, where it fuses together to form marble-sized 'clinkers'. It is then ground finely to form the common grey powder known today as Type GP – general-purpose Portland cement (previously known as Type A cement). It is commonly available in 20 kg bags with 50 bags to the tonne. There are two types of general-purpose cement:
- Type GP – general-purpose Portland cement
- Type GB – general-purpose blended cement.

All cement must comply with the requirements set out in Australian Standard 3972 Portland and blended cements.

Cement storage

Cement may be delivered to the site in bags or in bulk form. It will deteriorate quickly if allowed to absorb water or moisture from the ground or atmosphere, so bags should be stored off the ground, covered with a tarp or plastic sheet and preferably placed in a shed, while bulk cement should be stored in watertight bins or hoppers.

Fine aggregate (sand)

Sand is the name given to fine particles of stone less than 5 mm in size. Natural sands are usually hard and durable because they are the remnants of decomposed rock that have withstood the physical and chemical ravages of time. Sand for concrete use must be a clean, washed, regulated mixture of coarse and finer grains made up of round and sharp-edged particles to create a good workable mix. Some types of sand available for concrete work are:
- pit sand – mined from the sand dunes of beaches, which must be free of salt and clay
- river sand – smooth and coarse particles dredged from rivers and banks
- beach sand – generally not used due to a high content of salt and shells
- crusher fines – produced as a by-product of rock crushing. Particles are rough and splintery in shape, so only small quantities are used.

Coarse aggregate (blue metal)

This consists of crushed rock such as basalt, granite, diorite, quartzite and the harder types of limestone (see Figure 10.21). Special types of coarse aggregate, such as blast furnace slag, expanded shale and clay, may also be used. A good coarse aggregate would be:
- dense and hard, but not brittle
- durable and chemically inert
- clean, with no silt, clay or salt
- rough and of various sizes over 5 mm
- non-porous to help prevent water penetration of the finished concrete.

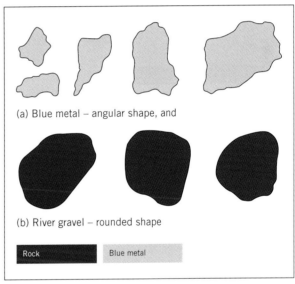

FIGURE 10.21 Material used for coarse aggregate in concrete

The average size of coarse aggregate for general domestic work is up to 20 mm; for most structural building construction it is up to 75 mm; and for massive structures, like dams, up to 150 mm.

Water

As a guide, only water that is suitable to drink (potable water) should be used for concrete mixing. The water used should be free of problem-causing substances such as acids, alkalis, oils, sugar and detergents. Water containing decayed vegetable matter should be avoided, as this may interfere with the setting of the cement within the mix. Seawater may be used but is not recommended, as it causes corrosion of steel reinforcement and forms surface efflorescence (a white salt powder on the surface).

Aggregate storage

Aggregate should be stored separately in metal or concrete storage bins and covered with a tarp to keep it dry. This prevents inaccurate weight batching caused by the presence of excessive water, as well as avoiding the material being blown around the site by the wind, or washed away with the rain (see Figure 10.22).

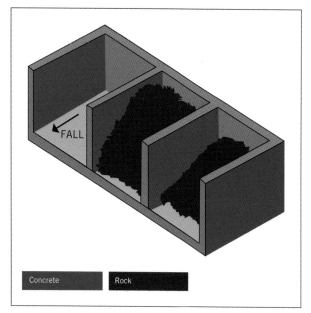

FIGURE 10.22 Storage of aggregates

Reinforcement

Reinforcement is used within the concrete for the purpose of giving the 'reinforced concrete' strength when tensile and shear forces are applied to it. Reinforcement may be divided into two categories:
- bars
- steel fabric sheets.

Bars

Generally reinforcement bars are constructed from hot-rolled 500 MPa or 500 N deformed steel bars; ranging in size from 12 mm to special sizes of 50 mm in diameter (see Figure 10.23), and round bars ranging in size from 4 mm to 12 mm in diameter. There are also now available lightweight fibreglass reinforcement rods ranging in size from 6 mm to 25 mm.

Steel fabric

This is a mesh of steel wires placed at right angles and welded together by an electrical process to form reinforcing sheets. There are two types of steel fabric sheets. The first is trench mesh, which is a sheet 6 m long by 200 mm (3 bar) up to 400 mm (5 bar) used for strip footings (see Figure 10.24). The second type of sheet is 6 m long by 2.4 m wide, and is used in reinforced concrete slabs, paths and driveways.

FIGURE 10.23 Deformed reinforcement bars

Durability

As mentioned above, reinforced concrete as a completed product is extremely durable, and concrete buildings will last many years. The weak link is the reinforcement used in the concrete. If the steel reinforcement is exposed to moisture and air it will corrode or rust and will result, over time, in the failure of the reinforced concrete component. Therefore, it is important to have the reinforcement completely buried in the concrete to an engineer-specified depth referred to as cover. In some locations that are highly corrosive (for example, by the sea) the engineer may specify galvanised or even stainless steel reinforcement. With the advent of the new fibreglass reinforcement, buildings in these highly corrosive environments may soon be using fibreglass reinforcement, because it is not affected by salt and is cheaper than galvanised or stainless steel reinforcement.

Strength

Reinforced concrete is considered to be the strongest building material available today, and this is evidenced by the fact that all mega-high-rise buildings throughout the world are built using reinforced concrete.

FIGURE 10.24 Trench mesh reinforcement

Fire rating

Concrete serves as a good insulator to many materials in a fire. It won't ignite, spread flame, generate heat or create smoke, and is therefore classed as having a good fire rating. Due to its structure, however, it does expand when exposed to extreme heat and may explode and shatter if this occurs. Steel also has a good fire rating, but at its critical temperature of 550°C it will soften sufficiently to expand and buckle, and allow a reinforced or prestressed concrete member to fail.

Handling

On large-scale constructions the quality of the final concrete product must be assured to a set standard.

For this reason, all handling of concrete components is carried out by mechanical means so engineers can be assured that the concrete will meet the required specification. However, on small-scale tasks, such as a slab for a small shed, handling can be performed manually: concrete materials are mixed by hand or in a tilting drum mixer. In this case the cement (which comes in 20 kg bags), the sand and the coarse aggregate must be stored on site prior to mixing the components into concrete. When handling bags of cement, it must be remembered that these were altered from 40 kg to 20 kg to reduce the incidence of back injuries. Another consideration when handling cement is to remember that wet cement will affect the skin and the eyes and should be washed off as soon as possible. Sand and gravel is brought to a small site in a tip truck or the back of a utility or table-top truck, and may need to be shovelled into place. Refer to Chapter 11 for a rundown of tools commonly used in concreting.

Bulk steel reinforcement, due to its mass, is usually mechanically handled by crane or hydraulic lifting arms. Individual bars or sheets can be manually handled, but some care needs to be taken. Due to its mass and length, most steel will need two people to move it (see Figure 10.25). Another difficulty is that steel reinforcement may have a textured surface (caused by deformation patterns created during manufacture) and could be rusty due to being stored outside or being placed on the ground.

FIGURE 10.25 Sheet reinforcement

It also has many sharp edges, especially when it is cut, and may inflict minor or even severe wounds during handling. The surface can even harbour a bacillus called tetanus, which is a microscopic vegetable organism causing muscle spasm and lockjaw.

Therefore, it is recommended that leather gloves be worn during both handling and cutting. The fibreglass reinforcement has stiff fibres protruding from the rods, so it is highly recommended that leather gloves are worn when handling these reinforcement rods.

Storage

Steel reinforcement should be stored off the ground on timber gluts so that it doesn't become contaminated with vegetable matter, animal matter, oils or mud. It can have a small amount of rust on its surface as this actually aids bonding, but should never have flaky rust, as this will continue to flake and cause spalding of the concrete surface, which will lead to 'concrete cancer' (see Figure 10.26). Spalding or spalling refers to concrete that has become pitted, flaked or broken up. This could be caused by poor installation, stress and environmental factors that damage concrete. You can address the spalling problem immediately when the concrete is poured.

Fibreglass reinforcement should be treated similarly to steel reinforcement, with the additional requirement

FIGURE 10.26 Concrete cancer

that it should not be exposed to direct sunlight for extended periods of time.

Environmental issues

Concrete is in many ways similar to bricks, masonry and stone in that the ingredients need to be mined. Similarly to brick manufacture, cement production uses vast amounts of electrical energy, but concrete, like brick, has a high life expectancy and will last many years. Concrete is able to be crushed and recycled as rubble to be used in the next generation of concrete.

COMPLETE WORKSHEET 3

Paints and sealants

Paint has existed in a simple form as a mixture of earth pigments and animal fat, applied with the fingers, or plant stems crushed at one end to form a fan shape, for many thousands of years. In Australia, these materials and methods have been used for over 50 000 years. Modern paint has its beginnings around the 1920s. The mixture consists basically of pigment, binder and solvent. The pigment is mined or manufactured in the form of a fine powder that gives paint its colour and opacity (meaning it doesn't allow light to pass through it). The binder is the liquid part of the paint, which may be an oil, such as linseed, or a resin. The binder holds the pigment together and enables the paint to stick to a surface. Solvents are used to dissolve the resins or thin the oils, to make paint workable. Solvents are an extract of petroleum oils, such as mineral turpentine.

Paints for building purposes are available in two forms:

1. oil paints – solvent-thinned. Oil paints contain pigments, binders, solvents, driers and extender pigments. They are designed to soak into the surface, lock into it and provide a key for the next coat. Thinning and washing up is carried out using mineral turpentine, sometimes called turps.
2. latex paints – water-thinned. Latex paints contain pigments, binders, solvents, driers and extenders. The main differences between oil and latex paints are the limited surface penetration of latex paints and the use of water instead of turps for thinning. Clean-up is also carried out using water.

The range of latex paints available is the same as for oil-based paints, but the latex range also includes acrylic and vinyl paints, such as 100% acrylic paints, vinyl paints, vinyl/acrylic paints and vinyl/latex paints. The variation in this range is due to differences in the binder.

Use and application of paints

Generally, oil-based and water-based paints can be used either internally or externally. Oil-based paints are preferred externally for their ability to soak into materials and seal them from the effects of weathering, but in recent times water-based paints have been used externally for their ease of application, low cost and easy clean-up properties. The range of uses for both types is as follows:

- primer – penetrates and seals the surface and provides adhesion for the next coating
- sealer – designed to stop suction or absorption at the surface and provides a base for the next coat
- undercoat – adds film thickness and contains more pigments for grain/texture filling and provides good sanding ability for a smooth finish
- flat finish – not designed for external use, as it attracts dust and dirt; mainly used for general ceiling finishes
- satin finish – mainly used internally for areas where regular washing is required, such as in bathrooms, laundries and kitchens; can be used on walls and ceilings
- gloss finish – can be used internally or externally where a hard, shiny, durable finish is required for appearance and ease of cleaning.

All types of paint can be applied by using a brush, roller, paint pad or sponge, sheepskin or synthetic paint mitten and by spray-gun (see Chapter 11 for a rundown of tools commonly used for painting). The surface, position, cost, use, texture and durability will determine the appropriate application method (see Figure 10.27).

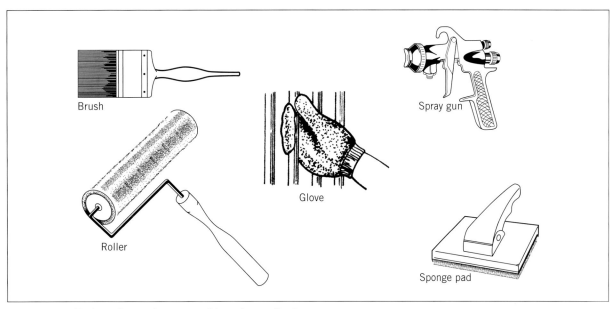

FIGURE 10.27 Tools and/or equipment used for paint application

Good brushes and rollers are expensive, so the utmost care should be taken to ensure they are kept in good condition (see Figure 10.28). They should be cleaned after every use in the appropriate solvent recommended by the manufacturer, and be ready for use when next required. After cleaning, they should be stored either flat or hung in such a way that the bristles are not damaged, and in a position where they cannot be contaminated by dust, corrosive liquids or powders.

Fillers

Fillers are smooth paste mixtures used for filling slight surface imperfections. They range from the plaster/cellulose-based (Polyfilla type) to thinned brush or spray fillers. Fillers are used to form a blemish-free surface ready to receive a paint or stain finish.

Fillers can be applied with a spatula or putty knife, a tube and gun as with common caulking or joint/gap compounds, by rubbing into the grain of timber with a cloth, or brushed or sprayed on.

Sealants

Sealants are substances that block the passage of liquids or gases through the surface of a material or joints or openings in or between materials. There are several different types of sealant, including polyurethane, acrylic and silicone, and each has different qualities including adhesive properties, waterproofing, thermal insulation, acoustic insulation, elasticity and fire retardance. An example of where a sealant may be used for its elasticity and water-proofing properties is the mastic joint between two panels of brickwork in an articulated masonry wall.

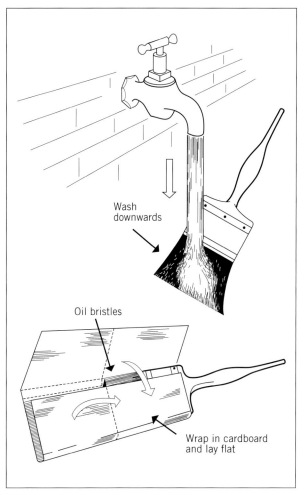

FIGURE 10.28 Cleaning and proper storage of brushes

Durability

The durability of paint is dependent on the strength of the coating. There are three important factors that cause the surface coating to break down: these are a combination of UV light, heat and moisture.

Durability of paint is affected by several items including the surface preparation, quality of the binders and the quality of the pigments.

Sealants by their nature must be durable as they are required to be exposed and able to cope with movement, heat, cold and moisture changes. If a sealant fails, the joint or surface that it is covering will be left unprotected.

Fire rating

As paints contain at least some oil and solvents they tend to be highly flammable and are, therefore, given a poor fire rating. There are, however, some specialist paints available that are used as fire retardants: that is, the gases given off are in the form of bubbles or a foam when the surface is heated and these act to smother any flame. These are called intumescent paints. Wallpaper has characteristics similar to paper and, therefore, attracts a poor rating.

Some sealants are designed to be fire rated (see Figure 10.29) while others are not.

FIGURE 10.29 An example of a fire-rated sealant

Handling

Paints are supplied in 1 L, 2 L, 4 L, 10 L and 20 L cans or drums. The larger the drum, the heavier it will be. Using a hand trolley to move several drums at one time is better than multiple trips, or even trying to lift and move heavy drums. Fillers and sealants usually come in tubes and it is important to clean the skin of any residue after use.

Storage

Careless storage of paints and sealants can prove to be expensive and, in some cases, dangerous. Paints containing solvents or volatile thinners should be stored in a cool, dry area, preferably below 15°C, with tight-fitting lids. Brushes and other materials deteriorate when subjected to damp conditions. To prolong the life and shape of brushes, they should be thoroughly cleaned in the recommended solvent, washed with warm soapy water, dried and wrapped in a cardboard liner and laid flat on a shelf. Rollers and spray guns should also be thoroughly washed and dried before storage.

COMPLETE WORKSHEET 4

Steel components

Steel is a combination of iron, carbon and manganese and is easily worked into shape from a molten state.

Mild steel may be melted and poured into billets, or short thick sections, that can be cooled, reheated and rolled into shape. As the steel is rolled and worked it gains a finer grain structure owing to the removal of impurities like slag. Steel can be produced with a remarkable range of properties through the addition of only 2% or 3% of other elements. Steel is strong in both tension and compression, which makes it a very versatile material.

Types and uses

Steel may be used in many structural and ornamental areas within a building (see Table 10.3 for an outline of the types of steel and their uses). Steel may be welded, bolted, riveted, bent, crimped or clipped together to form a variety of building elements such as reinforcement, door frames, wall frames, roof frames, roof sheeting, brickwork lintels, handrails, fasteners, beams, columns, gutters and downpipes or furniture (see Figure 10.30).

Cutting

Steel can be cut using a hacksaw (manual or mechanical), tin snips, bolt cutters (see Figure 10.31), hand shears, oxyacetylene torch or a metal angle grinder.

 Care should be taken when cutting with abrasive discs or an oxy-torch, as red-hot pieces of metal may be thrown into the eyes. Safety face shields or goggles should be worn during these operations. Also, razor-sharp edges are formed when cutting sheet and cold-formed materials, so leather gloves should be worn.

Fire rating

Steel is classified as having a good fire rating, but, as mentioned previously, at its critical temperature of 550°C it will soften and may buckle and collapse. Stainless steel has a low thermal expansion rate and would be a preferred material to use in high-heat-exposure situations.

Handling

Most heavy steel items are lifted using a crane, but manual handling can also occur. Where items are too heavy for one person to lift or hold in position, chain

TABLE 10.3 Types and uses of steel

Type	Uses
Carbon steel	Usually classified according to the carbon content. Up to a point, the more carbon used, the stronger, harder and stiffer the steel is. However, as the carbon content is increased, the steel becomes less able to be stretched out into wire as it becomes more brittle.
Alloy steel	Alloy steel is carbon steel that has elements such as manganese, nickel, chromium, vanadium and copper added to improve the steel's properties. Alloy steel may have strength, hardness and improved resistance to corrosion.
High-strength, low-alloy steel	Sometimes called weathering steel, it has up to 40% higher strength and resistance to corrosion than carbon steel. Thinner, lighter sections may therefore be used for exposed structural positions.
Stainless steel	There are many different types of stainless steel, but the basic ingredients are carbon steel with around 12% chromium. This provides enough of an inert film to prevent the steel from corroding or staining.
Structural steel	Structural steel is hot-rolled into sections, shapes and plates not less than 3 mm thick. These elements make up the structural frame of a building rather than ornamental elements such as stairs, grates, etc.
Reinforcement steel	Reinforcement steel may form part of some of the previously mentioned types: • A quenched and tempered deformed bar is a low-carbon steel that obtains its strength by a heating and water-cooling treatment. • A micro-alloyed deformed bar is a low-carbon steel that obtains its strength from the addition of small amounts of vanadium during the smelting process. • Other bars, rods and coils obtain their strength and various properties by modification of the hot-rolling process.

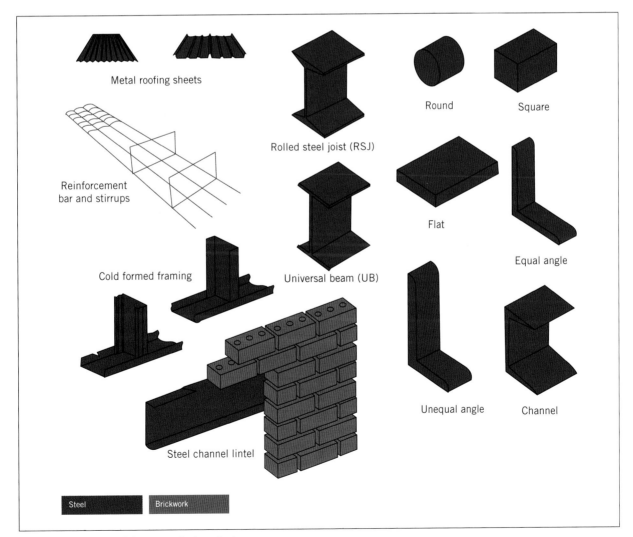

FIGURE 10.30 Some of the many steel products

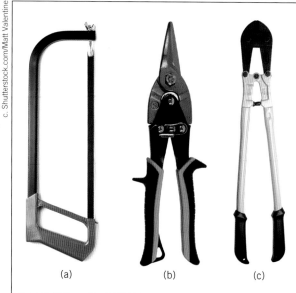

FIGURE 10.31 Various tools used to cut steel: (a) hacksaw, (b) straight-blade tin snips, (c) bolt cutters

Source: a. Shutterstock.com/luckyraccoon; b. Shutterstock.com/Yanas; c. Shutterstock.com/Matt Valentine

blocks and hoists may be used to gain a mechanical advantage.

Storage
Uncoated or unprotected steel items should be stored under cover until ready for use. Some steel products, such as stainless steel door-push plates, have a protective plastic coating, which prevents the surface from being spoiled or marked prior to installation.

Environmental issues
Steel is one product that can be reused and recycled. Scrap steel can be placed into a smelter and reformed into other metal objects.

LEARNING TASK 10.4

MATERIAL RESEARCH ASSIGNMENT IV

Prepare a folder with a cover sheet containing the following details:
- your name
- your class/group name or code
- name of the assignment ('Material research assignment IV')
- your teacher's name
- the due date.

Using Table 10.3 as your basis, obtain photos (either from the internet or photos you have taken yourself) of the six different varieties of steel. Then write why each type of steel was used in the particular location (approximately three pages in total).

 COMPLETE WORKSHEET 5

Insulating materials

The purpose of a building is to provide sheltered protection from the elements and possible intruders.

Apart from the structural and security benefits of the building enclosure, the thermal and acoustic properties of the materials used must be considered when designing the structure. The mass and density of walls, floors and ceilings will aid with the insulation process, but will also increase the weight and cost of the building.

Therefore, lightweight materials with good insulating properties are preferred. The conduction of heat and sound occurs when energy waves are passed from atom to atom through a material. The lower the density of the material, the lower the conductivity of heat and sound will be. Air has the lowest density of all the materials present in a structure, and if it can be trapped motionless within a material it will improve the insulation properties of that material. Mineral wool (blown or expanded rock), organic or wood fibre (cellulose or cork), foamed plastic (polystyrene foam) and air-entrained concrete, as well as many other types of insulation, all work on the principle of trapped air. They are usually treated to improve fire and insect resistance. Double-glazing of doors and windows works on the same principle of trapping air to lower sound transmission from outside the building.

Types and uses
See Table 10.4 for an outline of the types of insulating material and their uses. There are many other insulating materials available on the market for use in buildings for specific purposes; for example, reflective film can be applied to windows and glass doors to prevent heat conduction through the glass (see Figure 10.32). Some insulation materials (for example, asbestos and urea-formaldehyde foam) are no longer used due to their hazardous nature and links with cancer.

Fire rating
Insulation materials vary in their inherent capacity to resist the effects of fire. Materials such as glass fibre, mineral fibres, foamed concrete and cellulose insulation have a good fire rating. Materials such as wood fibre insulating board, polystyrene foam and other plastic foamed products all have a poor fire rating. Materials with a poor rating can be offset by using combinations of materials; for example, by facing polystyrene with metal.

TABLE 10.4 Types and uses of insulating materials

Type	Uses
Blankets or batts	These may be of fibreglass, rock wool and organic or inorganic materials, and are used to insulate ceilings by being placed between joists or under roof sheeting in the form of a roll or blanket.
Blown or poured-in insulation	Several types of loose insulation can be used, such as perlite and vermiculite (lightweight volcanic minerals), glass fibre and cellulose fibre. These are blown or poured between ceiling joists to any desired thickness. The lightweight open-bodied nature provides good thermal and acoustic properties.
Foil	Shiny aluminium may be used on its own from a roll or as a backing to blanket insulation and plasterboard. Although aluminium is normally a good conductor of heat, its shiny surface will reflect back into the air up to 95% of the heat that reaches it. This makes it a good insulator.
Mineral fibre	Glass, molten rock or slag from a furnace is blown with steam or high-pressure air to create mineral fibres that trap air. It can be spun to form batts, blankets or loose material laid between ceiling joists.
Polystyrene	This is a foamed plastic material used for many purposes, such as for hot or cold food product containers, cold-storage compartments, insulation sheets up to 50 mm thick in brick veneer cavities, and sandwich panel (thick polystyrene centre faced with steel or timber sheets).
Wood fibreboards	Soft, open-bodied boards or pre-prime soft pulp boards such as Caneite, often used for pin boards, provide good thermal and sound absorption properties due to the open weave of the cane fibres. They are used mainly for acoustic tiles in suspended ceilings.

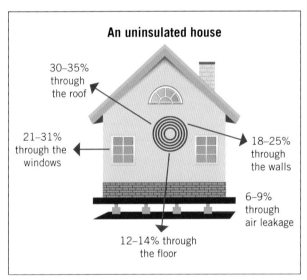

FIGURE 10.32 Typical heat loss of an uninsulated brick veneer cottage in temperate regions of Australia

Handling

Many insulating materials do not present handling problems, although great care should be taken when handling any fibre-based material.

> Because minute fibres become airborne during handling and installation, it is necessary to wear a respirator with a high-efficiency particle filter. Fibres that cannot be seen with the naked eye cause damage by setting off cell reactions in the lungs, which may end up as cancer. Also, fibres in the eyes and on naked skin can cause severe irritation, so safety goggles and leather gloves should be worn, especially when handling glass fibre materials.

Storing

In the event that insulation needs to be stored prior to being installed, it is important to keep the insulation material away from moisture and any active work area. By its very nature most insulation material will absorb moisture quite readily, resulting in poor performance of the insulation.

Environmental issues

Insulation is required in all new homes today to reduce the loss and transfer of heat and cold, and to reduce traffic noise. The biggest issue for builders is what to do with the waste. At this time, there is no reuse or recycling technology and so insulation waste generally goes to landfill.

COMPLETE WORKSHEET 6

Plasterboard sheeting

Plasterboard was introduced into Australia in the early 1960s and is the modern version of the original plaster sheet known as fibrous plaster, which is plaster of Paris reinforced with a hemp fibre, coconut fibre or horse hair.

Plasterboard is a strong, stable panel product, consisting of a core of gypsum (calcium sulfate dihydrate) sandwiched between two paper surfaces. It is made by feeding wet plaster between the two sheets of paper as they travel along a conveyor, allowing the gypsum plaster to set, then cutting it into set lengths.

Panels emerge with the face, back and long edges covered with paper. The long edges are also tapered,

> **LEARNING TASK 10.5**
>
> **MATERIAL RESEARCH ASSIGNMENT V**
>
> Prepare a folder with a cover sheet containing the following details:
> - your name
> - your class/group name or code
> - name of the assignment ('Material research assignment V')
> - your teacher's name
> - the due date.
>
> Obtain photos either from the internet or photos you have taken yourself of the following six different insulation materials:
> - fibreglass batts
> - polyester batts
> - polystyrene
> - rock wool batts
> - spray-in cellulose insulation
> - wool batts.
>
> Write the advantages and disadvantages of using each of these products (approximately three pages in total).

FIGURE 10.33 Setting plasterboard sheets with base coat

finished with several softer top or finishing coats prior to sanding smooth. Sheets are available in widths of 1200 mm and 1350 mm, lengths from 2400 mm up to 6000 mm and thicknesses of 6.5 mm to 25 mm.

Types and uses

The types of plasterboard available are outlined in Table 10.5.

Cutting and fixing

Plasterboard sheets may be cut with a hand saw, keyhole saw or jigsaw, but the usual method for cutting is to score through the surface with a utility knife (see Figure 10.34), bend the offcut back at 90 degrees and enabling joints between sheets to be butted together, reinforced with a perforated paper or cotton tape, filled with a hard base coat (see Figure 10.33) and then

TABLE 10.5 Types and uses of plasterboard

Type	Use
Cornices	Available in 55 mm, 75 mm and 90 mm coved profiles of 10 mm thickness to finish the connection between walls and ceilings.
Extra strength	10 mm thick sheets specially formulated to span greater fixing centres on ceilings, without sagging. Common trade names are Unispan and Spanshield.
Fire-resistant	Available in 13 mm, 16 mm and 25 mm thick specially processed, glass fibre-reinforced, mineral core sheets. It may be used for walls, ceilings, partitions, lift wells, stairwells, shafts and ducting where fire may spread. A common trade name is Fyrchek.
Flexiboard	6.5 mm thick sheets specially formulated to construct curved walls, ceiling arches and feature panels with ease.
Foil-backed	Available in 10 mm and 13 mm thick sheets for walls and ceilings where a vapour barrier or thermal insulation is required.
Glass fibre sheeting	Reinforced gypsum core using glass fibre. This provides a stiff body capable of withstanding impact and offers some fire resistance. It may be used to line the corridors of hospitals where trolleys and mobile beds may constantly collide with walls. A common trade name is Plasterglass.
Special panels	Available in 13 mm thick cut-to-size panels to suit grid-system suspended ceilings. They are also available in sculptured panels for decorative wall applications.
Square edge	Available in sheets of 10 mm and 13 mm thickness for use in office partitions where joints may be covered with aluminium, vinyl or timber mouldings. This enables quick and easy installation where removal or alteration may be necessary.
Standard tapered edge	Available in sheets of 10 mm and 13 mm thickness for use on internal walls and ceilings. A common trade name is Gyprock™.
Water-resistant	Specially processed plasterboard available in 10 mm thick sheets. The core and lining board facing are treated to withstand the effects of moisture and high humidity. It is used for lining walls and/or ceilings of bathrooms, kitchens, laundries, garages and ceilings of walkways and verandahs. Common trade names are Aquachek and Watershield.

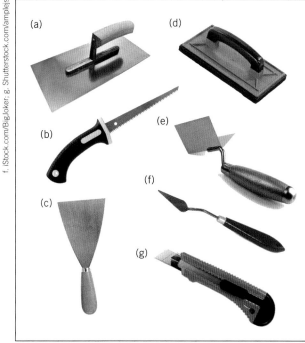

FIGURE 10.34 Tools used for plasterboard work: (a) flat steel trowel, (b) keyhole saw, (c) broad knife, (d) hand sander, (e) corner tool, (f) small tool, (g) utility knife

Handling

When manoeuvring plasterboard, always lift, handle and carry the sheets on edge. All lifting should be done with a straight back, bent knees and, at the least, depending on the size of the sheet, in pairs (see Figure 10.35).

FIGURE 10.35 Always handle plasterboard at the edges to avoid sheet breakage.

then run the utility knife through the backing paper still attached.

An alternative method is to again score the surface with a utility knife, bend the offcut back at 180 degrees, then pull forward quickly to snap it off cleanly. This method, however, takes practice and experience to master effectively.

Sheets are butt-jointed with walnuts of stud adhesive spaced 230 mm apart. Wall sheets are nail fixed, using ring-shank lattice head nails around the perimeter, and temporarily held in the centre by nails driven through scrap blocks of plasterboard. Ceiling sheets are nailed around the perimeter and double-nailed, not less than 50 mm apart, at 300 mm centres. Screws may also be used, normally when fixing into metal studs or furring channels. All joints are taped and set with external corners reinforced with a perforated metal angle prior to setting. End joints of sheets should be staggered so that a straight joint does not occur in the full height of walls or full width of ceiling.

Fire rating

Generally plasterboard has a fair fire rating. It has a low spread of flame and a low smoke development index due to the non-combustible gypsum core, but it has a high ignition index due to the paper facings. Special fire resistant plasterboard, such as CSR Fyrchek, has an excellent fire rating and is widely used in fire-prone building situations. It will provide fire-resistance ratings from half an hour to four hours for walls, and up to two hours for ceilings. Double sheeting will increase the ratings substantially.

Storage

All materials should be kept dry, preferably by being stored inside the building. If it is necessary to store plasterboard outside, it should be stacked flat, off the ground, properly supported on a level platform and protected from the weather with plastic sheeting or a similar waterproof covering. Care should be taken to avoid sagging and damage to edges, ends and surfaces.

LEARNING TASK 10.6 GYPROCK™

Visit the Gyprock™ website (http://www.gyprock.com.au). Read and become familiar with the website and then answer the following questions:
1 Where is the closest retailer of Gyprock™ to you? Write the company name and address.
2 What are the important events in relation to Gyprock™ that occurred in the following years?
   a 1947
   b 1992.
3 Which 'Gyprock™ specialty plasterboard' would be best suited for use in bathrooms or other wet-area rooms?

Asbestos

Asbestos is a naturally occurring mineral and can typically be found in rock, sediment or soil. It has strong fibres that are heat resistant and have good insulating properties. You can't see asbestos fibres with the naked eye and, because they are very light, they can be blown long distances by the wind.

Asbestos becomes a health risk when its fibres are released into the air and breathed in. Breathing in asbestos fibres can cause asbestosis, lung cancer and mesothelioma. The risk of lung cancer from inhaling asbestos fibres is greater if you smoke.

Those who get health problems from inhaling asbestos have usually been exposed to high levels of asbestos for a long time. Symptoms don't usually appear until 20 to 30 years after initial exposure.

A total ban on asbestos came into effect in Australia on 31 December 2003. It is illegal to make it, use it or import it from another country under the Commonwealth Customs Regulations 1956.

Workers must not handle asbestos unless they have been trained and hold a licence that is current and appropriate for the type of work being done.

Each state and territory in Australia has an Environmental Protection Agency (EPA) that provides guidance on the removal and disposal of asbestos in their jurisdiction (see Table 10.6).

Uses

Asbestos was once used in Australia in more than 3000 different products including fibro, flue pipes, drains, roofs, gutters, brakes, clutches and gaskets.

Because of its properties, which are described as being either 'non-friable or 'friable', asbestos was seen as being very useful for building products.

Friable asbestos is a material containing asbestos that, when dry, is in powder form or may be crushed or pulverised into powder form using your hand. This material poses a higher risk of exposing people to airborne asbestos fibres. Friable asbestos was commonly used in industrial applications (see Figure 10.36 and Figure 10.37) rather than the home, although loose-fill asbestos has been found in homes in NSW and the ACT, where it was sold as ceiling and wall insulation.

Non-friable or bonded asbestos products are solid and you can't crumble them in your hand – the asbestos

FIGURE 10.36 Friable asbestos coating around pipework

FIGURE 10.37 Friable asbestos coating around structural steel that has been coated on site

has been mixed with a bonding compound such as cement (see Figure 10.38 and Figure 10.39). If non-friable asbestos is damaged or degraded it may become friable and will then pose a higher risk of fibre release.

Durability

Asbestos is a very durable material in each of its different forms, which is one of the reasons it was used in so many different building products. The longevity of asbestos is an ongoing issue as it does not break down and deteriorate in either the built or natural environment.

TABLE 10.6 Environmental Protection Agency state guidelines

State	Website
ACT	https://www.worksafe.act.gov.au/health-and-safety-portal/safety-topics/dangerous-goods-and-hazardous-substances/asbestos
New South Wales	https://www.epa.nsw.gov.au/your-environment/waste/industrial-waste/asbestos-waste
Northern Territory	https://ntepa.nt.gov.au/your-environment/asbestos
Queensland	https://www.asbestos.qld.gov.au
South Australia	https://www.epa.sa.gov.au/environmental_info/waste_recycling/disposing-waste
Tasmania	https://epa.tas.gov.au/regulation/waste-management/controlled-waste
Victoria	https://www.epa.vic.gov.au/for-community/environmental-information/waste/about-asbestos/asbestos-epa-role
Western Australia	https://www.commerce.wa.gov.au/worksafe/asbestos

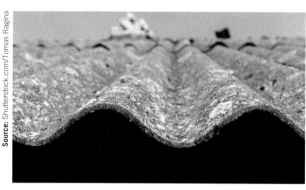

FIGURE 10.38 Non-friable asbestos roofing

FIGURE 10.39 Non-friable asbestos sheeting

Fire rating

The fire rating of asbestos products is very high, and these products have traditionally been used in situations where heat is generated in a building. Asbestos-based sheeting was also heavily used for cladding on buildings due to its high fire rating.

Cutting

Asbestos products should not be cut under any circumstances. The risk of releasing fibres into the atmosphere is too great. If any asbestos-based products need to be cut or broken down to smaller sizes, it must be done under strict guidelines where the material is watered down to minimise any fibres becoming airborne (see Figure 10.40).

Handling

The handling of any asbestos-based product needs to be conducted by a professionally trained and licensed person. There are specialised PPE requirements that need to be used when handling any asbestos-based products (see Figure 10.41).

 It is important not to disturb any material that is suspected of containing asbestos until it has been clearly identified by a person with training and experience in asbestos products.

FIGURE 10.40 Asbestos being wetted down

FIGURE 10.41 Handling asbestos using appropriate PPE

Stacking

Depending on the situation, asbestos products that are removed from a structure are usually sealed up in plastic and relocated to an appropriate storage facility. The material is usually wetted down then wrapped in plastic and loaded into a drum or bin. It is important that this type of work is conducted by fully licensed people who use the appropriate PPE and methods of handling (see Figure 10.42 and Figure 10.43).

FIGURE 10.42 Asbestos material being wrapped in plastic by licensed personnel with appropriate PPE

FIGURE 10.43 Asbestos loaded in bin with plastic lining ready to be wrapped

Storage

The Australian Government's Asbestos Safety and Eradication Agency has a website where you can search for facilities that accept asbestos waste: https://www.asbestossafety.gov.au/removal-and-disposal/search-disposal-facilities. These facilities ensure that the asbestos is stored in an appropriate manner so that the material can no longer pose a threat to the wider public.

LEARNING TASK 10.7 ASBESTOS

Part I
Visit the Asbestos Safety and Eradication Agency website (https://www.asbestossafety.gov.au/removal-and-disposal/search-disposal-facilities). Search the website and identify the closest disposal facility to you for both residential asbestos and commercial asbestos.

Part II
Search the internet and find two asbestos removal companies in your area.

 COMPLETE WORKSHEET 8

Glass

Glass is a naturally occurring material that can be created by heat from volcanic activity. The first known glass factories date back to approximately 1400 BCE. The evolution of glass production as we know it today has been due to the ability to create furnaces with sufficient heat to melt all of the ingredients into one consistent material.

The raw materials used in glass come down to three ingredients: sand (silica), soda (sodium oxide) and lime (calcium oxide). There are other additives that can be added to the manufacturing process that will produce different requirements such as colouring.

The use of glass in buildings is known as glazing and the tradespeople who work with glass in the construction industry are known as glaziers.

Several different types of glass are manufactured and used for different purposes. The following are the most common types of glass:

- *Annealed* or *float glass* is the most commonly used glass. 'Float' refers to part of the manufacturing process, where molten glass is floated on a bed of molten metal, producing a flat, uniform surface. When the glass is broken it has jagged sharp edges that are very dangerous.
- *Toughened glass* is strengthened by being rapidly cooled by chemical immersion or air jets during the production process. When it is broken it fractures into small irregular pieces.
- *Laminated glass* is made of two layers of glass that are sandwiched together with a plastic layer in between. This layer holds the glass in place when broken, making it a highly safe product. Such glass is used in areas where there is a chance of human contact.
- *Patterned glass* is produced by the glass being run through rollers to produce a desired impression in the glass surface. The different finishes serve both aesthetic and privacy purposes (see Figure 10.44).

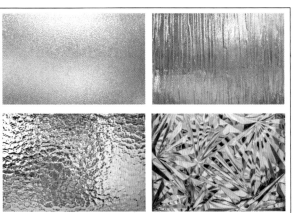

FIGURE 10.44 Images of different patterned finishes on glass

Glass blocks offer another use of glass in construction. They are typically used to create a wall that provides a decorative element, either externally or internally (see Figure 10.45). The installation of glass blocks is usually done by a bricklayer rather than a glazier.

FIGURE 10.45 A range of different finishes to glass bricks

Uses

Glass is typically used in residential construction by being integrated in doors and windows and, in some circumstances, as decorative elements such as room dividers or splashbacks for kitchens (see Figures 10.46–10.49).

FIGURE 10.48 Frameless glass shower screen

FIGURE 10.46 Glass window

FIGURE 10.49 Glass splashback for a kitchen

The location and the use of windows and doors will determine the type of glass used, as specified by Australia's National Construction Code (NCC). Its main aim is to maintain a high level of safety in situations where human impact with glass may occur. The commercial and high-rise sector also use glass as a curtain wall for the overall exterior to a structure.

The use of glass in both domestic and commercial construction is dangerous and glass products should only be used by appropriately trained professionals.

FIGURE 10.47 Glass door

Durability

Glass is a very durable construction element and has the ability to withstand severe weather conditions. The main effect on durability of glass is the risk of mechanical impact. The durability of glass that is used in today's construction industry also takes into account ultraviolet shielding, soundproofing and, importantly, the mechanisms that hold the glass in position in a building.

Fire rating

Australia's National Construction Code states that glass must be held in place by non-combustible frames to stop the spread of fire in a building. This is especially the case when used in high-rise buildings for curtain walls.

Even though glass is considered to be a non-combustible material, if exposed to sufficient heat it will break and melt.

Cutting

Cutting of glass is undertaken by tradespeople who have specialised tools to minimise the potential of damage to the glass. The finish to the cut edge of a pane of glass can have an effect on the structural integrity of the glass. The common process for cutting float glass is to score one surface, which creates a weakness in the surface where the glass will crack when sufficient pressure is applied. For other types of glass, such as toughened glass, there are special requirements for cutting and these are undertaken by the manufacturer (see Figure 10.50).

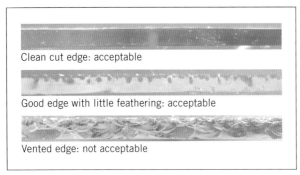

FIGURE 10.50 Good and bad outcomes of glass-cutting

Handling

Handling of glass varies depending on the size and location which the glass is being used for. Small pieces of glass can be easily handled by a single person. However, when laminated glass is being used for commercial applications, mechanical means need to be applied. Among these mechanical means may be cranes and vacuum lifting apparatuses (see Figures 10.51–10.53).

FIGURE 10.51 Plastic pump-up vacuum cup, 200 mm

FIGURE 10.52 Double pad vacuum lifter, lifting capacity 60 kg

FIGURE 10.53 Glass being lifted using a crane and specialised equipment

It is important that the appropriate PPE, such as eye protection and wrist guards, is used when handling glass (see Figure 10.54).

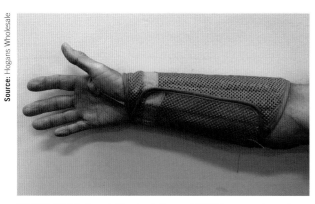

FIGURE 10.54 Specialised wrist guard PPE for use when glass is being handled

Stacking

Glass should be stacked in a manner so as the glass is upright and on a slight lean to prevent it from falling over. For products that include glass, such as windows and doors, you need to ensure that there is no direct contact where pressure may be applied directly to the glass at a single point.

Storage

Glass and glass products need to be stored in a dry and safe area away from the chance of any impact. Glass needs to be stored in a location that is free from debris and away from high traffic areas. It is also recommended that there is separation between sheets of glass to allow for air to flow freely between the sheets to prevent moisture building up and possibly staining the glass.

COMPLETE WORKSHEET 9

Adhesives

The growth of the modern-day plastics industry has resulted in the development of many new adhesives from synthetic resins. The newer manufactured adhesives can be used on materials such as glass and metals. Traditional glues are still used in today's industries, though mostly on natural materials such as timber.

The use of adhesives and glues has become an integral part of the evolution of building techniques, from the application of Laminex on kitchen bench tops to the use of construction adhesives to lay flooring and floor coverings.

Some of the advantages of modern-day adhesives are shorter times for adhesives to reach their potential strength and, in many cases, higher bonding strengths.

Uses

Adhesives and glues are used across several different areas in the construction industry (see Figures 10.55–10.58). For example, they can be used to join:

- Gyprock™ to timber or metal wall frames
- floor tiles to concrete floors
- manufactured boards and natural timbers for staircases
- timber skirting to brick walls.

FIGURE 10.55 Liquid nails for timber to brickwork

FIGURE 10.56 Stud adhesive used for Gyprock™

HANDLE CARPENTRY AND CONSTRUCTION MATERIALS

FIGURE 10.57 Adhesive for gluing tiles to concrete floors

Many products that are manufactured for building construction use adhesives, from structural components such as laminated veneer lumber (LVL) beams as floor joists, to cabinetry using particleboards, medium density fibreboard (MDF), plywood and hardboards.

Durability
Depending on the type of adhesive that is being used, durability can be high. Modern research and development has produced adhesives that function well in applications where they are exposed to the extremes of outdoor weather. Similarly, there are adhesives used to fix floor finishes that perform well despite the constant stress and strain of high-traffic areas.

Fire rating
When they are in a fluid state, before they are combined to create a final product, many adhesives are highly flammable. Once the adhesive has cured, the fire rating improves and most materials develop a high fire rating.

Cutting
Construction products that use adhesives are cut in the same manner that is used to cut the main material. So, in the case of timber products containing adhesives, a saw, router or plane can be used. Rubber-based adhesives will need to be cut with more of a slicing action by a single sharp blade.

Handling
The handling of adhesives depends on the type of adhesive being used. Where there is a chemical reaction required for an adhesive to cure, it is important that the appropriate PPE is used. Adhesives need to be treated as a potentially dangerous substance. Several adhesives come in applicators that limit the chances of exposure to the operator and accidental misapplication on the target product.

Stacking
Adhesives come in a range of different containers, depending on the characteristics of the adhesive. If the adhesive comes in small packages, these are usually bundled together in a larger container. In some instances, large quantities are stored in plastic containers that interlock when stacked together.

Storage
The storage of adhesives again varies depending on the type of adhesive. One phrase is common when it comes to adhesives – 'pot life'. This essentially means that adhesives can only remain in their pots (containers) for a certain amount of time. The pot life of any adhesive is specified on the container in which it is stored.

COMPLETE WORKSHEET 10

Transporting material

Materials need to be moved to and from a construction site as well as around the site. There are many mechanical aids to assist this, including cranes, forklifts and pallet trolleys. Materials can also be moved by hand (manual handling).

Cranes
The crane is commonly used to move large components around a job site, such as packs of wall frames and trusses (including moving individual trusses into place), large structural steel components, and packs of floor and roof sheets (see Figure 10.58).

FIGURE 10.58 Crane lifting trusses

Forklifts
Forklifts can be split into two categories: all-terrain and hard-surface trucks. It is important that an all-terrain

forklift is used for the delivery of bricks onto all building sites, as these machines are designed for this purpose and, if used correctly, are less likely to roll over (see Figure 10.59). Hard-surface trucks have smaller wheels and are designed to work on level or near-level concrete or pavement; for example, factory units loading material onto a truck.

FIGURE 10.59 All-terrain forklift trucks

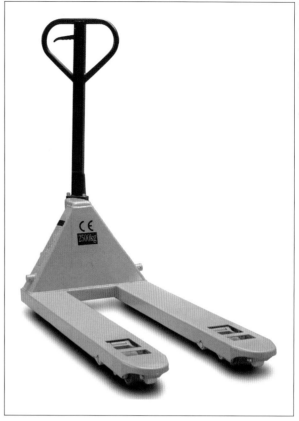

FIGURE 10.60 Rubber-tyred pallet trolley with hydraulic lift

Pallet trolleys

Pallet trolleys and platform trolleys (see Figure 10.60 and Figure 10.61) are used to safely move heavy loads on level surfaces. The pallet trolley has a hydraulic lifting mechanism that allows it to slide under timber pallets or raised loads. It has a lifting and carrying capacity of 1800 kg. Platform trolleys are fitted with four castors for easy steering and are used to transport items like bags of cement, drums or boxes to a capacity of 500 kg.

Safety and maintenance of trolleys
- All moving parts and wheels should be oiled regularly.
- Hydraulic lift equipment should be checked for leaks on a regular basis.
- Do not use these trolleys on uneven floors or unpaved ground.

Manual handling

The oldest method of moving an object is simply to pick it up by hand and move it. Among the most common injuries in the construction industry are back-related injuries, including muscle strain, pulled or torn muscles, bulging discs, ruptured discs and dislocated or broken bones. Correct lifting technique may not stop all back injuries, but it will reduce their likelihood and frequency.

If there is a single person lifting an object, follow the steps outlined in Figure 10.62.

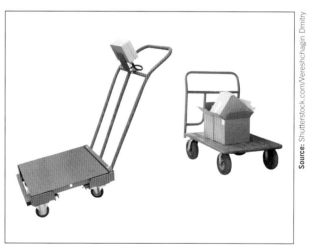

FIGURE 10.61 Rubber-tyred metal-framed platform trolley

If there are two people lifting an object, each person should follow the steps outlined in a single-person lift but with the addition that they either carry at the ends of the load or on the same side of the load.

Remember, if you don't need to manually lift something – don't. Drag it, push it or use a machine to do the task for you.

HANDLE CARPENTRY AND CONSTRUCTION MATERIALS

FIGURE 10.62 Correct manual lifting technique

LEARNING TASK 10.8 STRAPPING TIMBER

In Learning task 10.1 you were asked to correctly stack timber for pick-up by forklift or hiab crane.

In this activity you are required to steel band or polypropylene strap the timber so it will not move during transport – see the image below for further information.

FIGURE 10.63 Polypropylene banding being used to strap timber

 COMPLETE WORKSHEET 11

Clean up

All job sites, regardless of size, should be left in a clean manner that is clear of any excess material and ready for the next trade or for final use by the end user. From a business and individual perspective it is part of professional practice.

Any material that can be reused or recycled should be managed separately from general waste, and this applies to both new construction and demolition of existing structures. A good example of the recycling of a construction material is that of plasterboard. Instead of going to landfill, it can be recycled to be manufactured into a range of different products. Regyp is a Queensland company that collects and reuses gypsum to create products used in industries such as:

- *agriculture* – fertilisers for use with wheat, cotton, canola, potatoes, sugar cane, lettuce, pasture, mushrooms, grapes, dairy and sorghum, as well as golf courses
- *industrial* – plasterboard manufacture and cement manufacture
- *civil works* – stabilisation, flocculation and rehabilitation works
- *aquaculture* – fish farm flocculation.

There are financial benefits from recycling as well as environmental considerations. Dumping waste into landfill is becoming more expensive. Adopting practices to reduce tip fees in a construction project may assist you to be more cost competitive than other construction firms bidding for the same project.

When cleaning up, consideration needs to be given to the type of material being managed; that is whether it is hazardous (such as some chemicals) or non-toxic (such as timber or tiles).

Many construction projects will have a waste management plan, resulting from the local government approvals process or through the tendering process of the project. The intent of waste management plans is to reduce wastage and promote resource recovery where possible.

Dust suppression is an issue with several different materials. Where the cutting of material is required, there is a higher chance that the cutting process will create harmful airborne particles. On large construction sites, consideration needs to be given to managing earth works.

SUMMARY

In Chapter 10 you have learnt how to manually handle construction materials and components safely; store construction materials and components; apply environmental management principles associated with construction materials; and how to prepare material for mechanical handling. In addition, the chapter has:
- considered physical and economic factors related to construction trade materials
- identified the uses, durability, handling and manufacturing considerations of timber and timber products
- covered the uses of concrete components and cement, and identified reinforcement and durability considerations
- discussed different types of paint and sealants, and their use and application
- identified storage, handling and stacking requirements, and identified the fire rating and environmental considerations of construction trade materials.

REFERENCES AND FURTHER READING

Asbestos Safety and Eradication Agency, *Asbestos Waste in Australia*, 2015, https://www.asbestossafety.gov.au/sites/asea/files/documents/2017-10/ASEA_Report_Asbestos_waste_in_Australia_final_ACC_1.pdf.

Asbestos Safety and Eradication Agency, *Guidance for Importers and Exporters*, https://www.asbestossafety.gov.au/importing-advice/guidance-importers-and-exporters.

Australian Building Codes Board, *National Construction Code*, https://ncc.abcb.gov.au.

Australian Glass and Window Association, *Glass Types*, https://www.agwa.com.au/AGWA/Content/Consumers/Considerations/GlazMat/Glass%20Types.aspx?WebsiteKey=62f4db16-4aac-49fe-a32c-e7df232050c1.

Davis Glass, https://www.davisglass.com.au/home.

G. James, www.gjames.com.au.

Good Environmental Choice Australia, *Adhesives, Fillers and Sealants*, 2017, http://geca.eco/standards/click-to-view-afsv4-0-2014.

GTS Glass, https://www.gtsglass.com.au.

National Occupational Health and Safety Commission, *Code of Practice for the Safe Removal of Asbestos*, 2005, https://www.safeworkaustralia.gov.au/system/files/documents/1702/saferemoval_ofasbestos2ndeditionnohsc2002_2005.pdf.

New South Wales Technical and Further Education Commission, *Topic 2.8 Plastics CPCCBS6001*, 2015.

New South Wales Technical and Further Education Commission, *Topic 2.7 Glass*, 2015.

NSW Government, *NSW Asbestos Waste Strategy 2019–21*, https://www.epa.nsw.gov.au/-/media/epa/corporate-site/resources/waste/19p1900-nsw-asbestos-waste-strategy-2019-21.pdf.

Regyp, https://www.regyp.com.au.

Safe Work Australia, *Asbestos*, https://www.safeworkaustralia.gov.au/asbestos.

WA Department of Training and Workforce Development, *Apply Sheet Laminates by Hand: Learner's Guide*, 2013, https://www.dtwd.wa.gov.au/sites/default/files/teachingproducts/BC2017_CCBY.PDF.

WA Department of Training and Workforce Development, *Hand Make Timber Joints: Learner's Guide*, 2013, https://www.dtwd.wa.gov.au/sites/default/files/teachingproducts/BC2019_CCBY.PDF.

YouTube, *Lindsay Wall's Asbestos Warning to Tradies*, https://youtube/LBePOQyu3SE.

Relevant Australian Standards
AS 1288 – 2006 Glass in buildings – Selection and installation
AS 1478.1 Chemical admixtures for concrete, mortar and grout – Admixtures for concrete
AS 1684.2 Residential timber-framed construction – Non-cyclonic areas
AS 3972 Portland and blended cements
AS/NZS 4667 – 2000 Quality requirements for cut-to-size and processed glass
AS/NZS 4668 – 2000 Glossary of terms used in the glass and glazing industry

GET IT RIGHT

The photo below shows the storing of materials on a job site. Identify how the materials could be better stored and provide reasoning for your answer.

Source: Richard Moran

WORKSHEET 1

Student name: _____

Enrolment year: _____

Class code: _____

Competency name/Number: _____

To be completed by teachers

Student competent ☐

Student not yet competent ☐

Task

Read through the sections from the start of the chapter up to 'Engineered timber products', then complete the following questions.

1. Name the two factors that affect the selection of the material used in a building and provide an example of each.

   _____
   _____

2. State the four main chemical components of timber.

   _____
   _____
   _____
   _____

3. A tree increases in size by adding growth rings and a new layer at the very tips of branches. This means that the tree grows horizontally, but gives the impression of growing vertically. (Circle the correct answer.)

   TRUE   FALSE

   _____

4. Name three types of hardwood.

   _____
   _____
   _____

5. Name three types of softwood.

   _____
   _____
   _____

6. Timber and timber products may be used for a variety of purposes. List all the uses you can think of.

   _____
   _____
   _____
   _____
   _____
   _____
   _____

7. What is the name of the chemical used to improve the durability rating of timber, and which turns it a typical green colour?

   _____

8. What does the durability of timber relate to?

   _____
   _____

9. Timber is usually classed into groups ranging from good to poor, with Class 1 being the best and Class 4 being the worst. Identify the following timbers by durability class:

   Blackbutt   _____
   Brush box   _____
   Douglas fir   _____
   Grey gum   _____
   Radiata pine   _____
   Spotted gum   _____
   Tallowwood   _____
   Tasmanian oak   _____
   Western red cedar   _____
   White cypress pine _____

10. Complete the statement:

    Timber is strongest along the straight length of its _____.

    _____

11 All timbers will eventually burn in a fire, but some timbers have a natural resistance to fire. Name three of these timbers.

_____

_____

12 What may cause the blunting of saws and other cutting tools when cutting through timber?

_____

_____

13 Describe how timber should be stored on a job site.

_____

_____

WORKSHEET 2

Student name: _____

Enrolment year: _____

Class code: _____

Competency name/Number: _____

To be completed by teachers

Student competent ☐

Student not yet competent ☐

Task

Read through the sections from 'Engineered timber products' up to 'Concrete components', then complete the following questions.

1 List six products that are engineered timber products.

_____
_____
_____
_____
_____
_____

2 List three uses for engineered timber products.

_____
_____
_____

3 What is the easiest and safest way to move a pack of plywood?

_____
_____

4 How should engineered timber products be stored?

_____
_____

5 What are the environmental benefits of using engineered timber products?

_____
_____

WORKSHEET 3

Student name: _____

Enrolment year: _____

Class code: _____

Competency name/Number: _____

To be completed by teachers
Student competent ☐
Student not yet competent ☐

Task

Read through the sections from 'Concrete components' up to 'Paints and sealants', then complete the following questions.

1. Complete the following statement:

    Concrete is very strong in _____ but is very weak in _____.

2. What are the four materials mixed together to make concrete?

    _____

    _____

3. Water used for making concrete should be free of which problem-causing substances?

    _____

    _____

4. Name the two categories of concrete reinforcement.

    _____

    _____

5. What will happen to steel reinforcement that is exposed to moisture and air?

    _____

6. What is the technical term used to describe reinforcement that is buried in concrete to an engineer-specified depth?

    _____

7. Describe how bags of cement should be stored on the job site.

    _____

    _____

WORKSHEET 4

To be completed by teachers

Student competent ☐

Student not yet competent ☐

Student name: _____

Enrolment year: _____

Class code: _____

Competency name/Number: _____

Task

Read through the sections from 'Paints and sealants' up to 'Steel components', then complete the following questions.

1 What are the two forms of paints for building purposes?

_____

_____

2 Which two liquids are used for thinning paint and washing up?

_____

_____

3 What are fillers?

_____

_____

4 Name the three different types of sealants mentioned in the text.

_____

_____

5 It is possible to have a sealant that is fire rated. (Circle the correct answer.)

TRUE   FALSE

_____

6 What size containers is paint supplied in? (Circle the sizes that are not correct.)

1 kg   2 L   4 kg   6 L   10 L   15 kg   20 L

HANDLE CARPENTRY AND CONSTRUCTION MATERIALS

WORKSHEET 5

To be completed by teachers
Student competent ☐
Student not yet competent ☐

Student name: _____

Enrolment year: _____

Class code: _____

Competency name/Number: _____

Task

Read through the sections from 'Steel components' up to 'Insulating materials', then complete the following questions.

1 Name three of the six different types of steel.

_____
_____
_____

2 Which two forms of PPE should be worn when cutting steel with an angle grinder?

_____
_____

3 List the six tools that can be used to cut steel.

_____
_____
_____
_____
_____
_____

4 Complete the sentence:

Most heavy steel items should be lifted using a _____.

_____

WORKSHEET 6

To be completed by teachers
Student competent ☐
Student not yet competent ☐

Student name: _____

Enrolment year: _____

Class code: _____

Competency name/Number: _____

Task

Read through the sections from 'Insulating materials' up to 'Plasterboard sheeting', then complete the following questions.

1   Complete the sentence: Apart from structural and security benefits of the building enclosure, the _____ and _____ properties of the materials used must be considered.

2   Name four different materials that may be used as insulation.

    _____

    _____

    _____

    _____

3   What three items of PPE are required when handling insulation material?

    _____

    _____

    _____

4   What generally happens to waste insulation material?

    _____

HANDLE CARPENTRY AND CONSTRUCTION MATERIALS

WORKSHEET 7

To be completed by teachers
Student competent ☐
Student not yet competent ☐

Student name: _____

Enrolment year: _____

Class code: _____

Competency name/Number: _____

Task

Read through the sections from 'Plasterboard sheeting' up to 'Asbestos', then complete the following questions.

1 What two thicknesses (in mm) is standard tapered edge plasterboard available in?

   _____

2 List four hand tools commonly used when working with plasterboard.

   _____

   _____

   _____

   _____

3 Complete the following:

   Plasterboard will provide fire-resistance ratings from _____ to _____ hours for walls, and up to _____ hours for ceilings. Double-sheeting will _____ the ratings substantially.

4 How many people are needed to lift a long length of plasterboard?

   _____

5 How should plasterboard be stored?

   _____

   _____

WORKSHEET 8

To be completed by teachers
Student competent ☐
Student not yet competent ☐

Student name: _____

Enrolment year: _____

Class code: _____

Competency name/Number: _____

Task

Read through the sections from 'Asbestos' up to 'Glass', then complete the following questions.

1. When did a total ban on the use of asbestos come into effect across Australia?

   _____

2. Name and describe the two different ways asbestos can be produced for use in the construction industry.

   _____

   _____

3. Health effects are increased from inhaling asbestos if you are a smoker. (Circle the correct answer.)

   TRUE   FALSE

   _____

4. You do not need to be a licensed person to handle asbestos products. (Circle the correct answer.)

   TRUE   FALSE

   _____

5. Which agency provides guidance when it comes to the management of asbestos?

   _____

WORKSHEET 9

Student name: _____

Enrolment year: _____

Class code: _____

Competency name/Number: _____

To be completed by teachers

Student competent ☐

Student not yet competent ☐

Task

Read through the sections from 'Glass' up to 'Adhesives', then complete the following questions.

1. What are three different elements in a building that are made from glass?

   _____
   _____
   _____
   _____

2. Why is it desirable to have a clean cut edge on glass?

   _____
   _____

3. What PPE is used by glaziers to prevent their arms from being cut?

   _____
   _____

4. Identify two different mechanical means of handling glass.

   _____
   _____

WORKSHEET 10

To be completed by teachers
Student competent ☐
Student not yet competent ☐

Student name: _____

Enrolment year: _____

Class code: _____

Competency name/Number: _____

Task
Read through the sections from 'Adhesives' up to 'Transporting material', then complete the following questions.

1 Provide three examples of where adhesives or glues are used in the construction industry.

   _____

   _____

   _____

2 What does the term 'pot life' mean?

   _____

   _____

3 Adhesives should be considered as what type of substance?

   _____

WORKSHEET 11

To be completed by teachers
Student competent ☐
Student not yet competent ☐

Student name: _____

Enrolment year: _____

Class code: _____

Competency name/Number: _____

Task

Read through the sections from 'Transporting material' up to 'Clean up', then complete the following questions.

1 List three mechanical aids that may be used to move materials around a building site.

   _____
   _____
   _____
   _____
   _____

2 List three safety and maintenance items for trolleys.

   _____
   _____
   _____

3 Complete the sentence:

   Forklifts can be divided into two categories: _____ and _____ trucks.

4 List the basic steps for lifting something correctly (for one person).

   _____
   _____
   _____
   _____
   _____
   _____

5 Complete the sentence:

   Remember if you don't need to manually lift something, _____, _____ it or use a _____ to do the task for you.

HANDLE CARPENTRY AND CONSTRUCTION MATERIALS

EMPLOYABILITY SKILLS

Using the following table, describe the activities you have undertaken that demonstrate how you developed employability skills as you worked through this unit. Keep copies of material you have prepared as further evidence of your skills.

Employability skills	The activities undertaken to develop the employability skill
Communication	
Teamwork	
Planning and organising	
Initiative and enterprise	
Problem-solving	
Self-management	
Technology	
Learning	

USE CARPENTRY TOOLS AND EQUIPMENT 11

This chapter covers the outcomes required by the unit of competency 'Use carpentry tools and equipment'. These outcomes are:
- Plan and prepare.
- Select, check and use tools and equipment.
- Clean up.

Overview

In the construction industry you will be called on to use and operate a wide range of tools and equipment. In order to work efficiently you must be able to identify, use and maintain these tools correctly. Selection of the right tool will assist in producing a higher quality of work with maximum efficiency.

Hand tools have long been used, and are still used, to form functional items from raw materials, with a fairly high degree of accuracy. Hand tools are used for simple tasks and where power tools would be of no advantage. Correctly used hand tools show the ability and skill of the person using them. Hand tools have an important role in setting and marking out the work to be undertaken.

Since the Industrial Revolution (from the late 1700s), machinery or powered tools have gradually taken the lead in making tasks quicker, easier, cheaper and more accurate than ever before. Modern technology has given us tools and machinery that have made our lives and work easier, with the ability to create structures only dreamed of in times gone by.

The modern construction worker makes use of many electrical, pneumatic, fuel cell and battery-operated tools on projects that range from simple residential work to impressive landmark structures seen in most capital cities.

Although power tools are generally considered by many as superior to hand tools, it should not be forgotten that safe use, regular maintenance and an appreciation of function comes from an understanding of using hand tools properly.

Employers value people who fit into their workplace, solve day-to-day problems, manage their time and are keen to continue learning. These types of skills are known as employability skills. The employability skills you develop while working through this unit will be assessed at the same time as the skills and knowledge.

It is important to show evidence that you have developed these skills. Your trainer or assessor will discuss with you how to record your employability skills. The following table provides some examples of things you might do to develop employability skills in this unit.

EMPLOYABILITY SKILLS

Employability skill	What this skill means	How you can develop this skill
Communication	Speaking clearly, listening, understanding, asking questions, reading, writing and using body language.	• Liaise with others when organising work activities. • Interpret plans and working drawings.
Teamwork	Working well with other people and helping them.	• Work as part of a team to prioritise tasks. • Offer advice and assistance to team members and also learn from them.
Planning and organising	Planning what you have to do. Planning how you will do it. Doing things on time.	• List tasks in a logical order of action. • Select appropriate tools and materials.
Initiative and enterprise	Thinking of new ways to do something. Making suggestions to improve work.	• Initiate improvements in using resources. • Respond constructively to workplace challenges.
Problem-solving	Working out how to fix a problem.	• Participate in job safety analyses. • Check and rectify faults in tools and equipment if appropriate.
Self-management	Looking at work you do and seeing how well you are going. Making goals for yourself at work.	• Maintain performance to workplace standards. • Set daily performance targets.
Technology	Having a range of computer skills. Using equipment correctly and safely.	• Apply computer skills to report on project work. • Follow manufacturers' instructions.
Learning	Learning new things and improving how you work.	• Use opportunities for self-improvement. • Trial new ideas and concepts.

Planning and preparing to use tools and equipment

Prior to using any tool on the job – whether it be a hand tool, power tool or pneumatic tool – it is important to prepare and organise the task to be done. There are five things to consider before starting to use the machine or tool:
1. Is the work area safe, clean and tidy, with adequate lighting and ventilation?
2. Has the most appropriate tool been selected to do the task?
3. Is the tool ready for use, with the appropriate cutter or blade?
4. Is the tool safe to use?
5. Is the person using the tool familiar with the tool and able to operate it safely?

LEARNING TASK 11.1 GRINDING AND HONING

Using sharp tools for the job is critical to ensure an accurate, neat and professional finish. All tools need some maintenance; however, tools with cutting edges require special attention to ensure efficient use.

Ask your trainer or employer to demonstrate how to correctly grind and hone various tools that have cutting edges.

Now practise the correct techniques yourself to:
- grind and hone wood chisels
- grind and hone plane blades.

Identifying and selecting hand, power and pneumatic tools

Upon identifying a tool, you will then need to learn what it is designed for, its correct use and any important safety aspects. The following examples outline some common tools that you are likely to use in the construction industry.

Hand tools

Hand tools are used for different tasks and with different materials. For example, a carpenter uses a claw hammer to hit nails into timber, whereas a bricklayer uses a lump hammer and bolster to break bricks in half. To simplify the hand tools available, we will discuss them in groups associated with the material or tasks they perform.

Woodworking tools

The tools listed below are generally used by a carpenter on a construction site.

Squares

There are several different kinds of squares and each kind has its own use but all are equally useful when called upon.

- A combination square is the most versatile type of square. It has a sliding blade and is used for marking 45 and 90 degrees; it can also be used to mark parallel lines from an edge (see Figure 11.1).

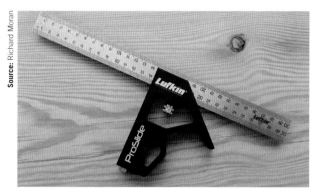

FIGURE 11.1 Combination square

- A roofing square has a 600 mm blade and a 400 mm tongue, and is primarily used with the aid of buttons or fences for calculating angles for roofing and stairs (see Figure 11.2).
- A quick/pocket square is designed to provide a quick means for development of various cuts on roofing members (see Figure 11.3).
- A try square is used for measuring 45- and 90-degree angles (see Figure 11.4). It consists of a metal blade fixed at a right angle.

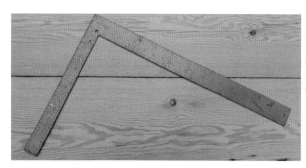

FIGURE 11.2 Roofing square

FIGURE 11.3 Quick/pocket square

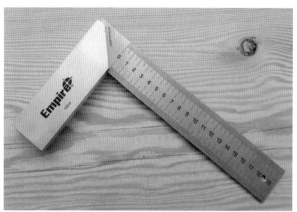

FIGURE 11.4 Try square

Hammers

There is a variety of shapes and sizes of hammers available to suit specific tasks or operations.

- Solid timber mallets are used in joinery work for hitting or striking chisel handles (see Figure 11.5); timber-handled chisels might be damaged if a metal hammer is used on them.
- Claw hammers are used to drive and/or extract nails (see Figure 11.6). They have a metal head which is hardened and tempered, and a timber, steel or fibreglass handle that is usually fitted with a leather or rubber grip. Head sizes are available from 225 g to 910 g. It is necessary to wear eye protection when using a claw hammer due to the risk of small pieces of metal flying off nails during the driving process.

FIGURE 11.5 Timber mallet; the mallet head may be made of brush box and the handle of spotted gum

FIGURE 11.6 Claw hammers

Plasterboard hammers

Plasterboard hammers differ from carpenter's hammers in that the head is domed to avoid damaging the plasterboard. They also may have a hatchet-style end for rapid trimming of plasterboard (see Figure 11.7).

FIGURE 11.7 Plasterboard hammers

- Warrington or cross pein hammers have a small driving face with a wedge-shaped top used to start small nails, like panel pins, held between the fingers (see Figure 11.8).

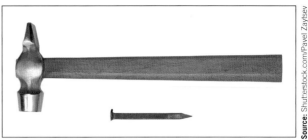

FIGURE 11.8 Warrington hammer

- Sledgehammers have a double-faced steel head with a long handle (see Figure 11.9). Available in sizes from 1.8 kg to 12.7 kg, they are used for demolition work or for driving pegs, and may have a timber, steel or fibreglass handle.

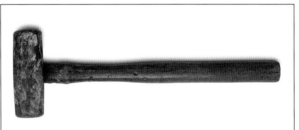

FIGURE 11.9 Sledgehammer

Nail punches

Nail punches are used to drive nails below the timber surface to allow putty to fill the hole and give a smooth finish with no visible nail showing. They are available in a variety of sizes and lengths. Larger versions are called floor punches (see Figure 11.10).

Wood chisels

Wood chisels have a sharpened steel end or blade with a timber or hardened plastic handle designed to be struck with a mallet or hammer (see Figure 11.11). Chisels are used to remove sections of timber to produce joints, sockets or holes.

Hand planes

Hand planes are tools that are either pushed or pulled along the surface of a piece of timber to scrape away or remove timber to a set line or intended surface. There are many planes including jointer or try planes, jack planes, smoother planes and block planes, to name just a few (see Figure 11.12).

Bench grinders

Bench grinders have a variety of purposes including to grind plane blades and chisels and to sharpen drill bits, when grinding at the correct angle of 25–30 degrees. When grinding, the steel becomes hot and needs to be cooled with water. Once the correct angle has been achieved, the cutting edge needs to be honed at 30–35 degrees on a stone (see Figure 11.13).

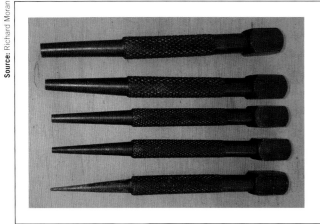

FIGURE 11.10 Nail punches and floor punches

FIGURE 11.11 A selection of chisels used to cut into timber

Oil stones

Stones are used to hone the cutting edge to produce a fine sharp edge. These include oil stones and diamond stones.

Oil stones (see Figure 11.14) are divided into two main groups:
- Natural stones are imported, with 'Norton Washita' being the most common. They are white in colour and very durable.
- Artificial stones have an abrasive component, which does the actual cutting, and a bond component, which supports the abrasive grains while they cut. They are made from aluminium oxide and silicon carbide.

Diamond stones

Diamond stones are manufactured from industrial grade diamonds that are not useful for jewellery. They are bonded to a nickel backing to provide a long-lasting flat surface (see Figure 11.15).

Hand saws

Hand saws are tools with a spring-steel blade with teeth set along the bottom edge. There is a timber or plastic handle set on one end that allows the user to push and then pull the blade across the timber, producing a clean cut. There is a range of different hand saws available including the rip saw, crosscut saw, panel saw, tenon saw and pad saw (see Figure 11.16).

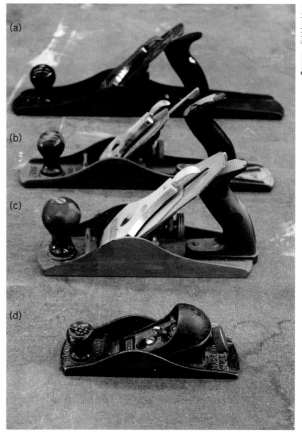

FIGURE 11.12 (a) Jointer or try plane, (b) jack plane, (c) smoother plane, (d) block plane

FIGURE 11.13 Bench grinder with guide

FIGURE 11.14 Oil stones

FIGURE 11.15 Diamond stones

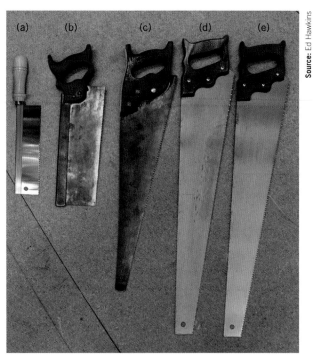

FIGURE 11.16 (a) Pad saw, (b) tenon saw, (c) panel saw, (d) crosscut saw, (e) rip saw

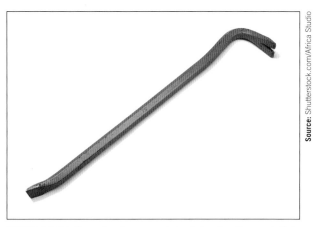

FIGURE 11.17 Round or hexagonal steel shank with specialist ends

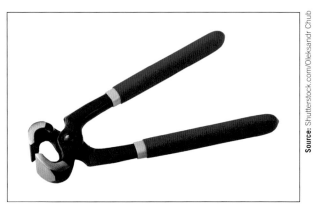

FIGURE 11.18 Pincers

Pinch or wrecking bars

Pinch or wrecking bars have a flat chisel-like end for levering and a straight shank with a hooked end claw for pulling nails (see Figure 11.17). They are used as a lever for prising framework apart, lifting heavy objects into place, de-nailing timber and general demolition work. The larger bars are around 800 mm long and the smaller bars, known as jemmy bars, are around 300 mm long.

Pincers

Pincers have a pair of circular jaws for holding and removing nails or fasteners (see Figure 11.18), and can have a small claw for levering and extracting tacks on the end of one handle. They are very handy for grabbing and holding during fixing operations.

Tapes and rules

For information regarding tapes and rules, such as the four-fold rule (see Figure 11.19), refer to Chapter 3.

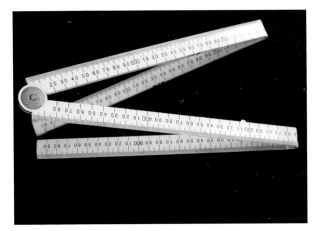

FIGURE 11.19 Four-fold rule

Marking gauge

This tool is used to impress or mark a line into a piece of timber. It is used by setting the pin the required distance from the block. The pin is then drawn along the timber, marking a line parallel to the selected face or edge of the timber (see Figure 11.20).

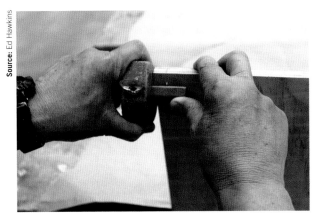

FIGURE 11.20 Marking gauge

Utility or safety trimming knife (Stanley knife)

A utility knife is often used for sharpening pencils as well as for accurate marking out; a pencil may leave a 0.5–1 mm line, whereas a knife blade leaves a much finer cut into the timber. A utility knife blade must be able to be folded or retracted when not in use and the knife should be stored in a secure place when not in use (see Figure 11.21).

Note: do not carry this tool off-site on your person as it is considered a weapon in many states and territories.

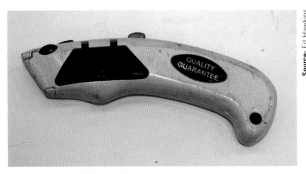

FIGURE 11.21 Utility knife

Planks

Planks may be used with ladders and scaffolding and are made of solid timber (see Figure 11.22) or boxed aluminium (see Figure 11.23). Timber planks are a minimum size of 225 mm × 38 mm (Oregon) and usually up to 3.6 m in length (they may be 32 mm thick in hardwood).

FIGURE 11.22 Typical timber plank

FIGURE 11.23 Typical aluminium plank

Aluminium planks are 225 mm × 50 mm and are available in standard lengths of 3.0 m, 4.0 m, 5.0 m and 6.0 m; they can also be custom-made to length.

Saw stools

Saw stools may be made on site from timber or purchased pre-made with pressed metal legs and a timber head (see Figure 11.24). They are also available in folding form for easy transport and storage.

FIGURE 11.24 Typical timber saw stool

Saw stools may be used to create a bench by placing one or two planks on top to cut timber and other materials.

Clamps

Clamps are available in a range of types and sizes, from the traditional heavy duty 'G' clamps and 'F' clamps to quick-action clamps, long sash clamps and the very small and lightweight spring clamps (see Figure 11.25).

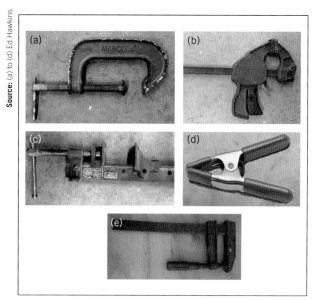

FIGURE 11.25 (a) G clamp, (b) quick-action clamp, (c) sash clamp, (d) spring clamp, (e) F clamp

Clamps are used to hold the material being worked, allowing the operator of the hand or power tool to use both hands.

COMPLETE WORKSHEET 1

Metalworking tools

The tools listed below are generally used on construction sites that involve working with metals.

Spanners

Spanners are used to tighten or loosen nuts, bolts and hex head screws. They are available in several styles, materials and sizes, including imperial and metric.

Styles include single open-end, double open-end, podger (for scaffolding, rigging and formwork centre adjustments), double-ring end, ring and open-end, and a variety of shaft shapes ranging from straight to almost semicircular. They may be made from drop-forged steel or chrome vanadium steel and are available in a host of socket sizes and types. Adjustable or shifting spanners are also very common. Spanners are useful to carpenters for fitting and adjusting surface and/or cavity sliding doors (see Figures 11.26–11.33).

FIGURE 11.26 Drop-forged single open-end spanner

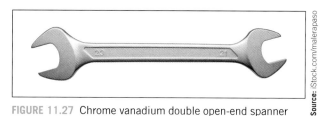

FIGURE 11.27 Chrome vanadium double open-end spanner

FIGURE 11.28 Podger for scaffolding and formwork centre adjustments – may be used for levering

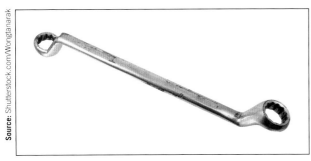

FIGURE 11.29 Double-end ring spanner

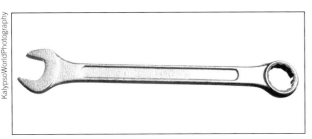

FIGURE 11.30 Ring and open-end combination

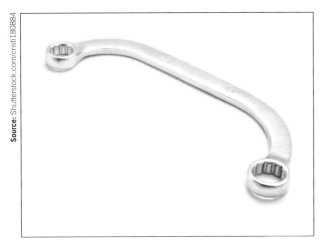

FIGURE 11.31 Half-moon ring spanner

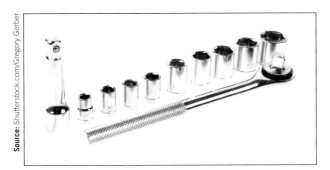

FIGURE 11.32 Square-drive ratchet handle and socket

Pliers

Pliers have a variety of shapes and sizes for particular jobs. They can be used for holding, cutting, banding, twisting and stripping wire. Circlip pliers are specialist pliers used to open or close spring clips (see Figure 11.34).

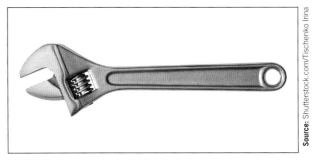

FIGURE 11.33 Adjustable shifting spanner

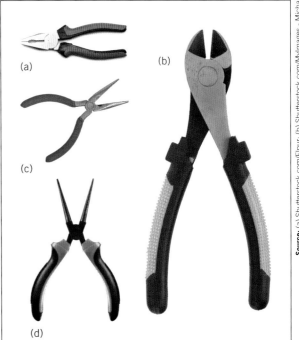

FIGURE 11.34 (a) Insulated combination pliers, (b) insulated diagonal cutters, (c) needle-nose pliers and (d) external straight circlip pliers

Tin snips

Tin snips are used for cutting straight or curved lines in thin sheet metal such as barges, cappings and corrugated iron, and for cutting banding straps found on slings of timber, pallets of bricks, etc. (see Figure 11.35). Small tin snips, also called jewellers' snips, are useful for trimming thin sheet material and cutting small-radius curves, arcs and circles (see Figure 11.36).

Aviation snips

Aviation snips (also called tin snips) are the best hand tools for cutting sheets of metal. When choosing a snip for a particular task, your choice will depend on whether the waste will be on the right or left hand side: red cuts left, yellow cuts straight or left and right, and green cuts right (see Figure 11.37).

Pop rivet guns

Hand pop rivet guns are used with the aid of a rivet to secure two pieces of sheet metal together such as gutters and flashings (see Figure 11.38).

USE CARPENTRY TOOLS AND EQUIPMENT **325**

FIGURE 11.35 Tin snips with a straight edge blade for general cutting

FIGURE 11.36 Jewellers' snips for curved work

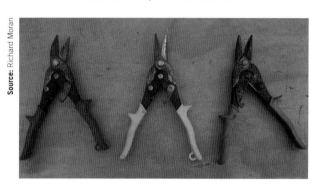

FIGURE 11.37 Red, yellow and green aviation snips

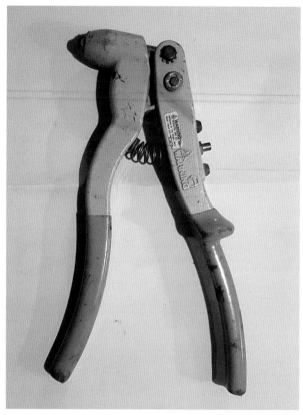

FIGURE 11.38 Hand pop rivet gun

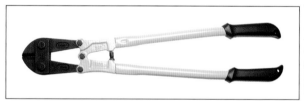

FIGURE 11.39 Bolt cutters

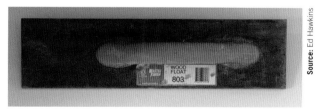

FIGURE 11.40 Wood float – available in various lengths and widths

Bolt cutters

Bolt cutters are used for heavy cutting of rods, bars and thick-gauge wire, the main use being to cut reinforcement bars and steel fabric in preparation for concrete work. They have long handles for leverage with centre cut jaws for general cutting (see Figure 11.39).

Concreting tools

Listed below are some of the tools that are generally used when working with concrete.

Trowels and floats

Wooden floats are used for smoothing and finishing wet concrete and for bedding ceramic tiling to give a smooth, textured or non-slip finish (see Figure 11.40). They are also used to apply the cement render to masonry walls. Steel trowel finishes are usually smooth and dense to provide a hard-wearing, easy-to-clean surface (see Figure 11.41). Wood floats are also useful for floating up the surface of sand bedding prior to laying pavers to ensure that all hollows are filled and the surface is even.

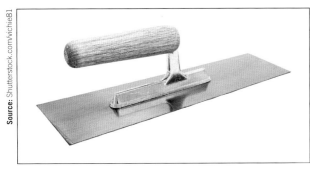

FIGURE 11.41 Steel float – available in a variety of shapes and sizes

Concrete screeds

Concrete screeds are used to screed concrete to ensure a flat, level concrete surface. They are available in various lengths (see Figure 11.42).

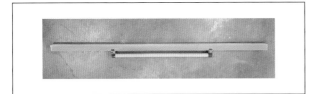

FIGURE 11.42 Concrete screed

Bull floats

Bull floats are used to smooth concrete surface after the concrete has been screeded and before the concrete has been finished (see Figure 11.43).

Edging tools

Edging tools are used by concreters to give a neat finish around the edges of concrete (see Figure 11.44).

Hoses

Hoses are used for a variety of purposes on a building site. They may be used to deliver oxygen, acetylene, abrasive for sand blasting or compressed air, or simply to supply or remove water. Hoses are flexible enough to coil or roll up but have the strength to withstand rough treatment. They are normally made of a reinforced plastic or rubber material (see Figure 11.45), or from a tightly woven fabric, as used for fire hoses.

Painting tools

Putty knives

Putty knife blades vary in length from 100 mm to 150 mm and have a variety of shapes (see Figure 11.46). They are used to fill nail holes, cracks and surface imperfections. They may also be used to patch or reglaze windows.

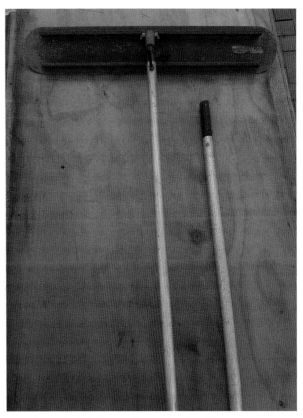

FIGURE 11.43 Bull float and extension handle

FIGURE 11.44 Concrete edging tools

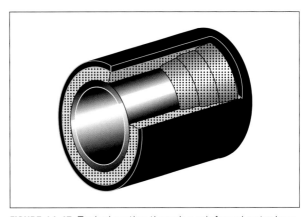

FIGURE 11.45 Typical section through a reinforced water hose

USE CARPENTRY TOOLS AND EQUIPMENT 327

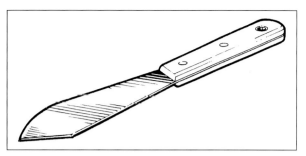

FIGURE 11.46 Putty knife

Filling knives

Filling knife blades vary in width from 50 mm to 150 mm and may be used to apply oil- or water-based fillers to open-grained timbers or shallow holes in the surface of a variety of materials. A broad knife may be used to fill or patch wider cracks and areas with a flat, smooth surface (for example, plasterboard) (see Figure 11.47).

FIGURE 11.47 A broad knife and a filling knife

Hacking knives

Hacking knife blades are usually 100 mm to 125 mm long with tapering sides and a thick edge on one side to allow them to be hit with a hammer (see Figure 11.48). They are used to remove old, hard putty from a window sash to enable the removal and replacement of glass.

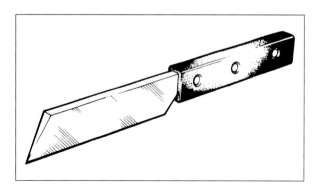

FIGURE 11.48 Hacking knife

Shave hooks

Shave hooks are available in a variety of head sizes to suit the surface they are used on (see Figure 11.49). They

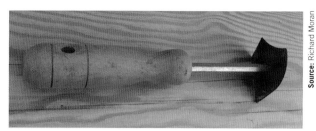

FIGURE 11.49 Shave hook

are used in conjunction with a blow torch or liquid paint removers to scrape old paint from ornamental beadings or mouldings and to take out cracks in cornices prior to filling.

Paint brushes

Paint brushes are made up of a handle (usually hardwood); the stock, which holds the filling or bristles; a setting of epoxy resin or vulcanised rubber to bind the filling together at the end; and the filling itself (see Figure 11.50). The filling is made from pure bristle or animal hair; a synthetic fibre like nylon; a natural fibre such as grass or straw; or a mixture of bristles and fibres.

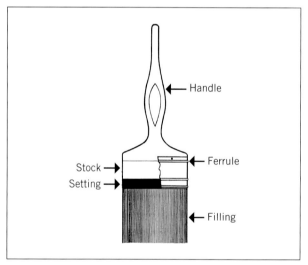

FIGURE 11.50 Standard-type brush

Paint brushes may be used to apply oil- and water-based paints to a variety of surfaces. The size and type of filling, together with the length of the handle, will determine the specific task the brush is designed for; for example, staining, cutting in, flat surfaces, applying a textured finish and even applying diluted acid to brickwork to remove mortar smears.

Rollers

Rollers consist of a central core of heavy-duty cardboard tube impregnated with phenolic resin to resist solvents and water, and are covered with a selection of fabrics. This cover, called the nap or pile, may be made from natural fibres such as wool, mohair or cotton, or synthetic fibres such as acrylic, polyester or nylon.

Rollers are used to paint a wide variety of internal and external surfaces such as walls, ceilings, furniture, corrugated shapes, pipes, textured surfaces and wire fences. The roller tube is fitted over a wire frame with a handle, and paint is applied to the roller by rolling it in a ribbed rolling tray to evenly distribute the paint on the nap (see Figure 11.51). A variety of fittings, extension arms and cleaning devices are available.

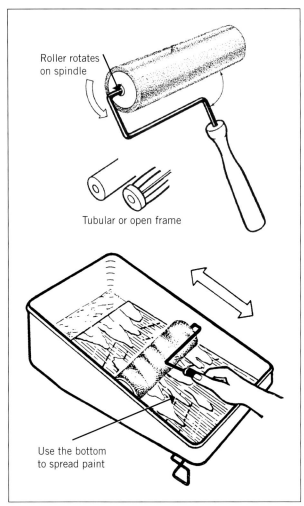

FIGURE 11.51 Roller, roller frame and metal or plastic tray

Abrasive papers

A surface-coated abrasive may be defined simply as an abrading medium (grain) bonded to a flexible backing. A common size of abrasive paper sheet is 275 mm × 225 mm. This is cut into six pieces for use with a cork rubbing block. When using abrasive paper on timber surfaces by hand, care should be taken to work only along the direction of the grain, using a uniform pressure throughout.

The grade of paper used will vary according to the original condition of the timber and the fineness of finish required. Papers of 60- to 80-grade would generally be suitable to remove marks left by machines or hand tools, and finishing work can be done with grades of 100 to 180. The higher the number, the finer is the paper's grit (see Figure 11.52).

FIGURE 11.52 Abrasive papers

As well, there are different grit materials that may be adhered to the paper or cloth backing. Natural abrasive grains include:
- flint (cream to grey)
- emery (black)
- garnet (red).

Manufactured abrasive grains include:
- aluminium oxide (reddish-brown)
- silicon carbide (blue-black).

Excavation tools

Tools such as crowbars, picks, shovels and mattocks (see Figure 11.53) may be used by carpenters from time to time in the preparation of the site to allow site set-out activities to proceed. They may also be used to clean up, dig holes for fence posts, move rubbish, dig and trim trenches, grub out tree roots or break up old concrete paving.

Light digging, shovelling, cleaning out and spreading tools

After breaking compacted soils or materials into manageable sizes, shovelling or cleaning-out tools are employed to move loose materials clear of the work area or to load the spoil directly into a barrow or transporter for disposal.

The most suitable hand tool for this purpose is the long-handled shovel, of either round-nose or square-mouth design, depending on the type of material being moved. The square-mouth shovel is best suited to granular materials such as sand or gravel, while the round-nose is a universal design and performs well in materials of uniform or irregular shape and size. Long-handled shovels allow for a more upright posture. This helps to prevent back strain and enables extra body weight to be applied with the assistance of the more efficient leg muscles. A short-handled, square-mouth shovel allows more control when cleaning out.

Trimming, detailing and finishing tools

As noted previously, several of the basic breaking and cutting tools and some shovelling and cleaning-out tools

FIGURE 11.53 Hand tools for breaking, cutting and grubbing (a) Crowbar, (b) Fork, (c) Mattock, (d) Pick, (e) Spade, (f) Long-handled round-mouth shovel, (g) Spud bar

are suitable for trimming and finishing operations. For example:

- A cutter end or grubbing mattock is an ideal tool for trimming a flat surface such as a pavement base or trench bottom. A square-mouth, short-handled shovel is a good trimming and detailing tool for medium-density moist soils.
- A crowbar or spud bar is suitable for trimming the sides of trenches or footing pads.
- A spade will also give good results when trimming or squaring an excavation, although it is not efficient for shovelling or cleaning out loose materials from corners and edges of work (see Figure 11.54).

FIGURE 11.54 (a) short-handled square-mouth shovel, (b) long-handled square-mouth shovel, (c) short-handled round-mouth shovel, (d) long-handled round-mouth shovel

Although trimming and detailing are considered finishing operations in the excavation or filling process, the extent of this work will vary according to the degree of accuracy required for each project. For example, a stout steel garden rake of about 500 mm width is sometimes used to spread and level out granular filling and sand base course or screed bed under a slab. The only group of tools not included in the main use groups are special-purpose tools, such as trenching shovels, post-hole shovels, auger post-hole diggers and intermediate or narrowed-down long-handled, round-nose shovels (see Figure 11.55).

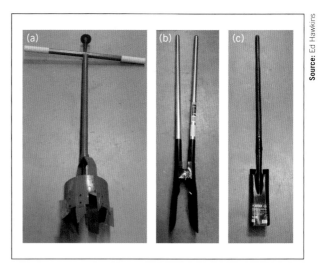

FIGURE 11.55 (a) auger post-hole digger, (b) post-hole shovel, (c) trenching shovel

Spirit levels

Spirit levels are usually made from timber or aluminium, and have several glass or plastic tubes containing a liquid with a trapped air bubble inside. The spirit levels are available in lengths of around 250 mm and up to 2.4 m, with a standard length of around 800 mm, which is suitable for checking door heads are level. A straight edge may also be used. Straight edges may be metal (aluminium) or well-seasoned timber (Western red cedar is suitable), have two straight, parallel edges, and are commonly 2.4 to 3.0 m in length. Care should be taken to prevent damage to the edges, and they should not be left exposed to the weather or stored in a manner that allows them to bow or twist.

Used together, levelling may be carried out by spirit levels and straight-edges in the horizontal plane (see Figure 11.56). Refer to Chapter 9 for information on the use of levels in bricklaying and blocklaying, and Chapter 12 for their use in tiling.

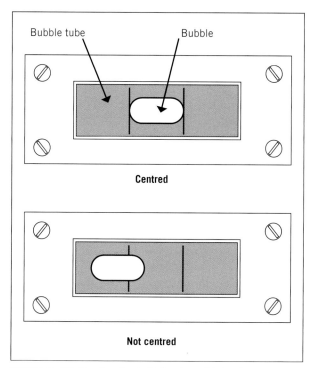

FIGURE 11.56 Reading the bubble of a spirit level

COMPLETE WORKSHEET 2

Power and pneumatic tools

The first thing to consider is what is meant by 'power' and by 'pneumatic'. Power means the type of energy that drives or moves the tool or machine. There are now five 'power' sources in relation to hand-operated power tools:

- 240-volt AC electrical
- 12–36-volt DC electrical
- gas – uses explosive gas (butane or propane) with a battery for ignition purposes
- nitrogen/battery – uses a sealed tank of compressed nitrogen gas with a battery
- explosive powered tool (EPT) – uses a 0.22 rifle cartridge.

There is also another form of 'power', which is 'pneumatic' power. Pneumatic simply means compressed air. For example, car and truck tyres are filled with compressed air. So any tool that uses compressed air, not an explosive gas, is in fact a pneumatic tool. This means that a tool that is directly or indirectly connected to an air compressor is a pneumatic tool.

240-volt AC electrical power

It is important from a safety perspective to be aware that 240-volt AC can kill an unsuspecting person. So the first line of defence against electrical shock from the mains service, electrical leads or powered tools is the installation of an earth leakage device or residual current device (RCD) or core balance unit. This unit is either fixed to the temporary power board, or is a portable unit that plugs into the general-purpose outlet (GPO) on the mains board or generator and thus becomes a mini-distribution board with four or six outlets (see Figure 11.57). This device senses the smallest differential in supply and demand, and then isolates the power in milliseconds.

FIGURE 11.57 An RCD-protected portable power board

All power from a supply source, whether a temporary builder's service or a permanent installation, should pass through one of these units installed immediately adjacent to the power source. As an additional measure, users of the temporary power supply should wear heavy rubber-soled shoes, which will give maximum protection against electrocution.

Extension leads

Each state and territory has its own code of practice or guidelines for electrical installation. These are all based on Safe Work Australia's *Model Code of Practice:*

Managing electrical risks in the workplace, which requires portable electrical equipment and leads to be regularly inspected by a licensed electrician. The purpose of inspection is to identify any defects in the leads or equipment that are potential hazards, and to reject or advise on repair or replacement where possible. The electrician must tag the equipment as evidence that the inspection has been carried out.

These tags (see Figure 11.58) are clamped to the item after inspection and approval and contain the following information:
- name and licence number of the inspecting electrician
- date of inspection
- name of plant inspected (for example, power drill)
- due date for re-inspection.

FIGURE 11.58 Electrical tags

All extension leads should be heavy-duty, sheathed for construction and have a rating of 10 A. An extension lead in constant use and drawing near to its full current rating will generate heat and suffer a voltage drop. To minimise damage to the lead and possible hazard to users in this situation, any extension lead of 15 A rating must not be longer than 40 m if it has 2.5 mm² conductors.

Every lead should be unwound from its spool or uncoiled fully when in use as, when a lead is coiled up, any heat generated cannot be exchanged efficiently and risks damage or personal hazard to workers.

All leads used on building and construction worksites must have plugs and sockets of clear plastic to aid inspection for the correct termination of conductors and to identify loose or burnt terminals and wires. Moulded, non-rewireable plugs are also satisfactory.

Leads running from the temporary power supply or from any permanent source should be elevated clear of the ground over the distance from power source to worksite. This is achieved by the use of stands with weighted or broad-based frames and vertical masts with hooks or slots that support the lead without damage.

The minimum height for the lead is 2.1 m from the ground. If the lead can be run adjacent to a wall or other rigid structure, it can be attached at the required height to serve the same purpose.

Any damage to an extension lead, flexible lead to a power tool, or other piece of equipment means it must be taken out of service as soon as the damage is detected. Patching leads, repairing plugs or sockets, applying bandages or trying to camouflage a fault or damage to a lead is not acceptable and breaches national, state and territory WHS legislation.

Circular saw

One of the most widely used electrical tools on the job is the circular saw (see Figure 11.59). The circular saw is identified by the diameter of the blade fitted. There are five sizes available:
- 160 mm (6¼ inch) – battery saw available in this size
- 185 mm (7¼ inch)
- 210 mm (8¼ inch)
- 235 mm (9¼ inch) – most used and the preferred saw for carpenters
- 270 mm (10½ inch).

FIGURE 11.59 Circular saw

The motor of the machine is generally universal and the power source is generally 240-volt AC supply, although there are now good quality 36-volt battery-powered saws available.

These saws may be used for:
- ripping
- grooving
- trenching
- cross-cutting
- rebating
- compound cutting.

Routine maintenance
- Always check the motor housing and the blade for damage prior to using the saw.
- Most blades today are single-use replaceable items, but blades can be sharpened or touched up for reuse many times. Tungsten-tipped blades may be sent to a saw doctor for special sharpening if required.
- Dust, vacuum or blow the body of the saw and around the motor housing so that any build-up of dust doesn't inhibit airflow into the motor.

- Check for worn carbon brushes and replace them as required. Note: many new saws are now using brushless motors.
- When not in use, store saws in a dry, dust-free box to ensure long life.

Safety
- Always wear eye and hearing protection while operating saws.
- Never operate these tools with loose clothing, long loose hair or necklaces, as these may be caught by the blade.
- The material to be cut must be firmly secured before cutting.
- Ensure that the extension lead and lead of the saw are well behind the saw during cutting operations.

Drop saw and compound mitre saw

These saws are also widely used electrical tools and allow for a high degree of accuracy in cutting. The motor housing and the blade is set on a hinge or pivot directly above a turntable to enable the saw to be turned and allow for cuts from 0 degrees (square across the timber) to generally 45 degrees.

If the saw only has this action it is referred to as a drop saw (see Figure 11.60), but if the saw incorporates a slide allowing for the motor housing of the saw and the blade to be pulled forward, this is referred as a sliding saw. Finally, if the motor housing and the blade are able to be tilted to 45 degrees this is referred to as a sliding compound mitre saw (see Figure 11.61).

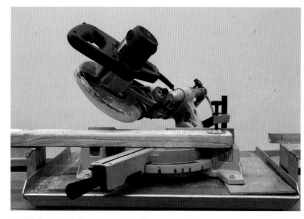

FIGURE 11.61 Compound mitre saw set to cut a compound mitre cut

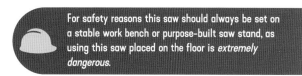

For safety reasons this saw should always be set on a stable work bench or purpose-built saw stand, as using this saw placed on the floor is *extremely dangerous*.

Other safety considerations when using this tool are:
- Never try to cut timber along its length (ripping the timber) as the blades of this saw are designed for 'cross cutting' only; if the operator attempts to rip timber, the blade may pull the timber through the saw and also pull the operator's hand beneath the spinning saw blade.
- Never try to cut a piece of timber that is not secured against the back fence or stop of the turntable. As the timber cuts, the blade will pull the timber back away from the operator and slam it against the fence, possibly jamming the saw blade in the timber.
- Never cut wide pieces of timber by cutting into the timber and pulling the blade towards you. Always start the cut on the edge of the timber closest to you and push the motor housing and the blade away from you.

Routine maintenance and safety

The information for circular saws in the previous section also applies to drop saws and compound mitre saws.

Electric planers

Planers are used to remove excess timber while leaving a smooth finish. The depth of cut can be adjusted to a maximum depth of 3 mm. Planers can range from 82 mm common to 155 mm wide timber. Edges can be chamfered using the groove on the bottom of the planer. There are two types of blades: double edge disposable or those that can be resharpened multiple times (see Figure 11.62).

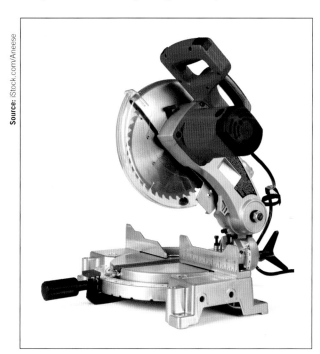

FIGURE 11.60 Drop saw

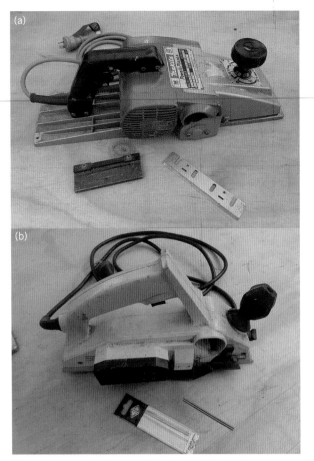

FIGURE 11.62 Electrical planers and blades: (a) 155 mm, (b) 82 mm

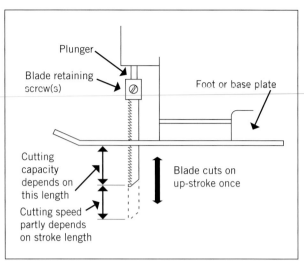

FIGURE 11.63 Cutting action of a jig saw

FIGURE 11.64 Jig saw cutting a piece of ply

Routine maintenance
- Always check the motor housing, planer lead and blade for damage prior to using the planer.
- Most blades today are single-use replaceable items, but some blades can be sharpened or touched up for reuse many times.
- Dust, vacuum or blow the body of the planer and around the motor housing so that any build-up of dust does not inhibit airflow into the motor.
- Store in a dry, dust-free environment to prolong the life of the planer.

Safety
- Always wear eye and hearing protection when using the planer.
- Always unplug planer from power source before adjusting and replacing blades.
- Never put fingers near the blades.
- Never hold the piece of timber being planed – ensure the material is firmly secured instead.

Jig saw

The portable jig saw is used to cut concave and convex shapes in thin materials (see Figure 11.63 and Figure 11.64). It is useful on site due to its portability where access to a band saw is not possible. These saws are available in a wide range of sizes and styles, from light-duty single-speed to heavy-duty variable-speed. They may be used to cut curved shapes or along straight lines in materials such as ply, hardboard, soft and hard metals and cardboard. Care should be taken when cutting brittle material, as the blade cuts on the up-stroke, which may cause surface chipping of the material.

Routine maintenance
- Replace blades if excessive pressure has to be applied when cutting or if blades are burnt or bluing.
- Ensure that air vents in the machine casing are not blocked, as this will cause overheating.

Safety
- Never put fingers or hands under the material while cutting.
- Make sure sufficient clearance is available under the piece being cut.
- Drill a hole in the material first so that the blade may be inserted. Do not use the blade to start a cut in hard materials, as this may shatter the blade. Plunge cutting is permissible only in soft materials.

Sabre saw

Another saw with a similar action to the jig saw is the sabre saw (see Figure 11.65). This is designed for two-hand operation to cut steel pipe, steel plate and floor panels, or for awkward setting positions.

FIGURE 11.65 Cordless and powered sabre saws

Portable router/trimmer

The portable electric router is a high-speed spindle moulder and shaper, which may be used in the workshop or on site. The versatility of the router lies in the variety of bits and cutters that are designed for its use, and unlike the joinery shop fixed bench-type spindle moulder, it is portable (see Figure 11.66).

FIGURE 11.66 (a) Plunge router, (b) standard router, (c) trimmer router

It has a high-speed revolving cutter (2000–27 000 rpm) that gives a very neat, clean cut. There is a wide range of routers available, with differing power and speed ratings to suit a variety of work. They are usually purchased with accessories such as an adjustable fence and a template guide for cutting grooves and rebates, trimming and edging, and also for joints.

Routine maintenance

- Always check the motor housing, router/trimmer lead and blade for damage prior to using the router.
- Check for damaged cutters.
- Dust, vacuum or blow the body of the router and around the motor housing so that any build-up of dust does not inhibit airflow into the motor.
- Store in a dry, dust free environment to prolong the life of the router.

Safety

- Always wear eye and hearing protection.
- Unplug the router/trimmer from power source before replacing cutters.
- Hold router firmly when switching on, as torque may twist the machine out of your hands.
- Never put fingers near the cutters.
- Never start the router with the cutter in contact with the job as it will kick and cause damage to the job and possibly injure the operator.
- Never hold the piece of timber – ensure material is firmly secured and make sure two hands are on router during use.

Electric and pneumatic drills

Drills cover a range of tools that provide rotation for boring, cutting and fastening operations. Most drills are electrically operated (240-volt or rechargeable battery type); however, they can also be operated by air or by flexible drive from a petrol motor.

Drills are classified according to their size, use, power source and speed, as well as the type of boring or cutting 'bit' they drive and the material they bore into. Sizes range from palm-size pistol-grip through to fixed machine tools for workshop use. Some drills deliver medium- to high-frequency impact or hammer forces, as well as rotating for boring into concrete, masonry or rock. Manufacturers' information charts will assist in selecting the right drill for a specific task.

The cutting tools, bits or drills that actually do the boring are as specialised as the drills that drive them in terms of speed of rotation, diameter of the hole and best sharpening angle for the cutting edge.

Common drill types include the following:
- pistol-grip drill – can be used one-handed for small-diameter holes in soft materials like timber and aluminium
- pistol-grip with side handle – designed to be used with both hands to prevent the drill from twisting if it jams; used for larger holes or heavy drilling (see Figure 11.67)
- side handles with breast plate – used mainly for drilling steel or large-diameter holes when more pressure needs to be applied.
- angle-head drill – used for drilling in confined spaces or awkward positions.
- impact drill/driver – used for drilling holes in masonry materials with a tungsten carbide-tipped bit (see Figure 11.68). The impact as well as the rotation allows for easy drilling into these materials. Impact drivers are fitted with a hexagonal chuck that allows the fitting of screwdriver and hex-head bits quickly and easily to the driver. The impact action with the turning action drives screws in timber (see Figure 11.69).

FIGURE 11.67 Pistol-grip drill with side handle drilling steel

FIGURE 11.69 Battery-powered impact driver

FIGURE 11.68 Tungsten carbide-tipped drill bit (masonry bit)

- pneumatic drills – used for very heavy-duty work, as found on a large building site or mine. They may be used to drill holes in concrete for reinforcing steel starter bars, into solid rock to receive long rock bolts, or even underwater where the use of electricity is not possible. They are powered by compressed air.
- power screwdriver – this operates on similar principles to drills, and some drills may actually be used as screwdrivers due to their variable speed and reverse functions. It comes with a variety of bits and fittings to accommodate most head shapes.

Drill bits must be securely held by the chuck on the drill. Chucks may be key operated (see Figure 11.70) or simply tightened by hand if a keyless chuck is fitted (see Figure 11.71).

Routine maintenance
- Any electrical problem, including replacement of leads, must be dealt with by a licensed electrician.

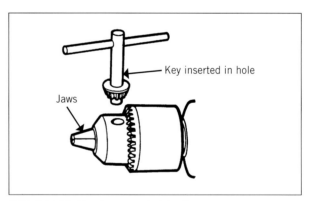

FIGURE 11.70 Chuck operated with a key

- Internal greasing of gears, internal cleaning and so on should be carried out by a qualified service person.
- Brushes should be replaced as necessary and checked for contact.
- Pneumatic hoses should be neatly coiled when not in use, with ends capped to prevent foreign matter from entering. Store hoses off the floor away from acids, oils, solvents and sharp objects.

Safety
- Always wear eye protection as material can be thrown from the drill bit into the operator's eyes.

FIGURE 11.71 Keyless chuck type

- Never attempt to open or close a keyed-chuck with the power still connected.

Portable electric sanders

The main types of sander in use are the drum, belt, disc and orbital sanders (see Figures 11.72–11.74). Drum sanders are traditionally used for sanding timber floors

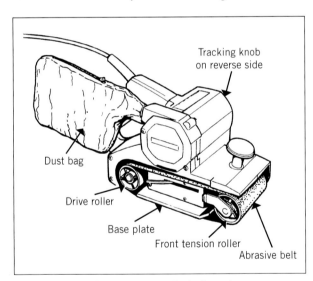

FIGURE 11.72 Main components of a belt sander

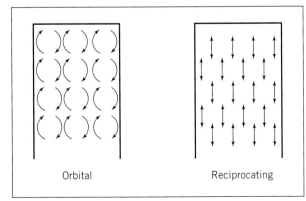

FIGURE 11.73 Sander actions

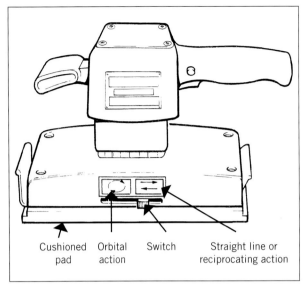

FIGURE 11.74 Main components of an orbital sander

to a smooth finish. Belt sanders are used for removing large amounts of timber very quickly and, if used correctly, are capable of producing a much flatter surface than a disc sander. Belt sanders are also used for sanding timber floors along the wall line where drum sanders are unable to go. Disc sanders are generally used for removing material quickly – usually paint from timber or steel. Care should be taken as these sanders can easily gouge timber. Finally, orbital sanders, which come in a range of shapes and sizes, are designed to finish the surface of the timber or base material prior to painting or finishing.

Papers for machine use are available as sheets, rolls, discs and belts. Abrasive grains or grits used for coating the paper may be of natural stone, or are manufactured in an electric furnace under temperature conditions up to 2300°C.

Sanding should always be in the direction of the grain, as cross-grain sanding will tear the timber fibres and leave marks. The dust bag should be fitted at all times to reduce dust levels in confined spaces.

Orbital and reciprocating sanders are not designed to remove large quantities of timber like the belt sander, but are mainly used to finish off prior to painting or staining. They are both fitted with a cushioned base pad and may also be fitted with a dust bag. Both machines are safe to use as the movement of the base is limited to little more than a vibration.

Routine maintenance
- For belt sanders – always fit the belt with the indicator arrow facing the correct direction, otherwise the join in the belt might come apart.
- For belt sanders – adjust the travel of the belt sideways with the adjustable tracking knob.
- For belt sanders – keep a firm grip with both hands when in use, as most belt sanders exert a high level of forward thrust.

- For all sanders – keep housing vents free of dust and check and replace carbon brushes as required.
- For all sanders – keep vents on the motor housing free of dust to prevent clogging and overheating.
- Check and replace carbon brushes as required.

Safety
- Always wear a respirator when using sanders in a confined location.
- Dust extraction bags should be used with all sanders (where provided) or connect the sander to a vacuum cleaner to reduce the amount of fine dust, which creates an unhealthy work environment.

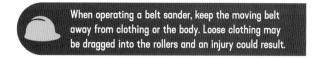

When operating a belt sander, keep the moving belt away from clothing or the body. Loose clothing may be dragged into the rollers and an injury could result.

Powered nailers and explosive-powered tools

Construction workers on site are required to fasten different materials together, such as putting together a timber frame, securing skirting boards to a wall or even securing a timber batten to a steel or concrete column. All these actions may be done using a powered device that drives in a nail.

The first and most common device is what is commonly known as a 'nail gun'. This may be powered in the following ways:
- pneumatic – compressed air
- gas-powered – often referred to as a gas gun
- nitrogen/battery-powered – some manufacturers refer to these devices as nitrogen-powered and others simply call it a battery gun.

All of these devices have a common factor – they are designed with all the same features. Some will even accept the same nails, which are specially manufactured to load into a clip (see Figure 11.75) on the nailer rather than being loaded individually.

FIGURE 11.75 Nails being loaded into nailer

The purpose of a nailer is to drive nails quickly and efficiently into and through timber to fasten materials together. All fasteners work on the same basic principle:

a sealed chamber with a piston (similar to a car motor) is filled with compressed air or an exploding or compressed nitrogen gas, driving a piston that pushes the nail out of the nailer and into the material (see Figure 11.76 and Figure 11.77).

FIGURE 11.76 Compressed air nailer

FIGURE 11.77 Gas nailer

Explosive-powered tools (EPTs) are often referred to by their manufacturer's name; for example, Ramset™ gun or Hilti gun. These nailers are specialty tools and are designed to nail timber to steel or masonry; they are not to be used to nail timber to timber.

The most common tool is now the low-velocity or indirect-acting explosive-powered tool (EPT). This uses a 0.22 cartridge to drive a piston down, pushing a hardened nail down the barrel of the tool, through the timber and into the steel or masonry (see Figure 11.78).

FIGURE 11.78 A stripped-down indirect-acting EPT

The high-velocity or direct-acting explosive-powered tool (EPT) uses a 0.22 cartridge to directly push a hardened nail down the barrel of the tool, through the timber and into the steel or masonry (see Figure 11.79). This tool is being phased out and many building sites will not permit their use because of the danger involved. These tools in untrained hands *have killed and will kill*.

FIGURE 11.79 A direct-acting EPT

EPTs in general are considered dangerous and it is a requirement that danger warning signs be placed to notify others on the site that an EPT is being used (see Figure 11.80).

FIGURE 11.80 Explosive-powered tool danger warning sign

Routine maintenance

- For all nailers – regularly inspect for damage to the casing.
- Some compressed-air-type nailers require a few drops of oil to be put into the air intake to lubricate the piston seals. (Note: read the manufacturer's booklet, as putting oil into synthetic-sealed nailers will damage these seals.)
- Never exceed the maximum allowable pressure for a compressed air nailer as set by the manufacturer – generally around 120 psi.
- Always place a nailer in a secure, water-resistant box when you have finished using it.

Safety

- Always wear eye and ear protection when using this tool.
- Never point these tools at anyone at any time – loaded or unloaded.
- Never walk around with your finger on or over the trigger.
- Always disconnect the power source of the tool if leaving it for a period of time.
- Always lock away an EPT when not in use.

LEARNING TASK 11.2

SAFE AND EFFECTIVE USE OF HAND TOOLS

Undertake a practical exercise to demonstrate safe and effective use of hand tools. Here is a list of potential projects that you can complete using a variety of carpentry hand tools:
- carry-all toolbox
- carry-all fastener box
- timber mallet
- cross-legged stool.

Identifying, selecting and using plant and equipment

On any construction site today in Australia, workers are operating plant and equipment. For the purpose of this book and for simplicity, the plant and equipment covered will only include:

- 240-volt power supply – mains and generators
- compressors
- pneumatic and electric jackhammers and breakers
- concrete mixers
- concrete vibrators
- wheelbarrows/brick barrows
- industrial vacuum cleaners
- industrial work platforms
- ladders and trestles.

Safety

When refuelling petrol-driven power tools:
- only refuel after the power tool has been switched off and has stopped running
- always use a suitable funnel when refuelling
- avoid splashing fuel on hot parts and electrical components.

240-volt power supply – mains and generators

A builder has two choices when it comes to obtaining temporary electrical power on site. The builder may use the mains (240-volt) supply to a single board mounted on a pole (see Figure 11.81) containing one or more general-purpose outlets, although on larger sites the supply is likely to be both 240 volts (single-phase) and 415 volts (three-phase) for heavy equipment. The power may be reticulated to several points around the site and involve main and sub-boards. All work associated with the application, installation of pole and board and reticulation of power is the domain of the licensed electrical contractor. The builder simply engages the contractor, signs the agreement for supply on the application and plugs into the service when it is installed, inspected and connected by the supply authority.

For short-term power supply on site, a generator may be used. It is normally a petrol-powered, air-cooled portable unit used to supply 240-volt power to operate electric-powered tools (see Figure 11.82).

FIGURE 11.82 Portable site generator

Generators range from small, extremely quiet and lightweight 1-kVA units to large, on-site 8-kVA generators. Most carpenters using a generator will have a 2.5-kVA, 3.5-kVA or 5.0-kVA generator, depending upon the type of work and the tools that need to be powered by it. For example, a carpenter might want to buy a generator to operate all these tools simultaneously: a circular saw, impact drill and angle grinder.

Circular portable saw	4 amps
Impact drill	2 amps
Angle grinder	4 amps
	= 10 amps

$$\text{Formula:} \quad kVa = \frac{\text{amps} \times \text{volts}}{1000}$$

$$10 \text{ amps} \times 240 \text{ volts} = \frac{10 \times 240}{1000}$$

$$= 2.4 \text{ kVa}$$

In this case, the minimum capacity of the generator would be 2.4 kVA. A reserve of 1–2 kVA is advisable for start-up of tools; therefore, the nearest standard size would be 3.5 kVA.

Note: 1 kVA = 1 kilovolt-ampere or 1000 volt/amperes per hour – 1 amp

= 1 ampere × 240 volts

= 240 watts

Therefore, a tool rated at 960 watts = 960 W ÷ 240 V = 4 amperes.

FIGURE 11.81 Typical builder's temporary power pole and board

Portable site generators are an essential unit on new, large housing estates where mains power is not yet available and in country or isolated areas where access to electricity is not possible. Most units are compact enough to fit into the back of a utility or trailer.

In some instances large diesel-powered generating units for operating heavy equipment with high current demand are available, mounted on the back of a truck or purpose-made trailer (see Figure 11.83).

FIGURE 11.83 Trailer-mounted diesel generator

Routine maintenance

With any mechanical equipment, maintenance is vital for the continued running of the machine as well as the safety of those who are operating it. Listed below are some of the potential checks that need to be done to keep the machine running.
- Check the oil level before use each day when using a four-stroke engine.
- Check the petrol level each day and make sure the correct oil/petrol mix is used in two-stroke engines.
- If the generator fails to start, check the spark plug. Clean and reset the plug or replace it as required.
- Clean the air filter regularly to prevent a build-up of oil choking the airflow.
- Replace the pull start cord when it becomes frayed or breaks.
- Check power-point outlets for wear or damage.
- Make sure equipment is cleaned after use and before storage.

Compressors

A site compressor is usually portable (see Figure 11.84) and is used to operate nail guns, paint spraying equipment, needle guns for paint removal, small sand-blasting guns and small air tools generally, and also to supply air for respiratory masks via an air purifier.

Just like generators, compressors come in a range of sizes and configurations. The differences are dependent on the following:

LEARNING TASK 11.3
HAND AND PORTABLE POWER TOOLS: QUIZ

Search online for a copy of the CFMEU's *OH&S Bulletin* 'Hand and Portable Power Tools' and use this document to complete the following questions.
1. True or false? Tools found to be defective should be reported and replaced immediately.
2. What are the two most common injuries that can happen when using power tools?
3. True or false? Before servicing a tool you should always have the power source connected to the tool.
4. How should a tool be pulled up or lowered while working on an elevated work platform?
5. What is the best way to carry knives or sharp objects?

FIGURE 11.84 Portable site compressor

- L/s (litres per second) or cfm (cubic feet per minute) is the amount of air being brought from the compressor to the tool. Tanks with higher cfm ratings deliver more air. This is the measure of capacity; for example:
  - 1.2 L/s (2.5 cfm) or (70 L/minute)
  - 4.6 L/s (10 cfm) or (283.2 L/minute)
- kPa (kilopascals) or psi (pounds per square inch) is the pressure exerted by the compressed air on the walls of its container. Generally, the higher the number, the higher the pressure, resulting in greater force pushing the nail.
- Trailer-mounted site compressors (see Figure 11.85) are available for use with a breaker or rock drill for demolition work and for breaking up firm soil.

Larger units are available to operate a range of air tools and sand blasters. Jobs include large excavation, breaking concrete and demolition, and they can also act as a stand-by in case a factory compressor breaks down or there are power restrictions.

FIGURE 11.85 Trailer-mounted site compressor

Petrol- or diesel-powered site compressors supply 47 L/s or 118 L/s.

Routine maintenance
- An operating compressor unit should be set as level as possible in a clear area, where cool air that is free of dust is available.
- Prior to starting, oil, water and fuel levels must be checked to avoid the possibility of engine wear.
- When starting up the plant, the air cocks should all be opened. After the engine has been started and had sufficient time to warm up, the air cocks may be closed.
- When stopping the plant, the air in the receiver should first be released by opening the cocks; then the engine itself should be stopped by allowing it to lose speed gradually.
- Pneumatic tools should be handled carefully and serviced frequently to help ensure trouble-free use of the plant.
- Prior to coupling up the tools to the air line and putting them into service, any moisture or dirt in the line should be blown out to prevent it entering the working parts of the tools.
- All bolts should be tightened up regularly and the shanks of chisels and points should be of the correct length and cross-section.
- Air lines should be checked to ensure all joints are secure and so there's a minimum of air loss at these points.

Careful maintenance and attention to these points will greatly lengthen the life of equipment and ensure the best working efficiency.

Pneumatic and electric jackhammers and breakers
Electric demolition hammers (see Figure 11.86), commonly called 'Kango hammers', are efficient tools for light demolition of brickwork, concrete and hard-

FIGURE 11.86 Electric demolition hammer with moil point

packed clay. They are also available with a rotary hammer drill facility for boring holes into rock or reinforced concrete.

These electric hammers are available in 8.3 kg, 10 kg, 15 kg and 33 kg weights. They may use moil points, chisels or clay spader attachments and/or 13 mm to 38 mm diameter drill bits.

Pneumatic jackhammers or breakers (see Figure 11.87) are available to carry out similar work to the smaller demolition hammers. They are used for chipping, drilling and scabbling. Small points, chisels and drill bits ranging from 12 mm to 25 mm may be used with this 7-kg hammer.

Routine maintenance
- Safety goggles and ear muffs should be used, as well as a silencer, to cut excessive noise.
- Moil points, chisels and other attachments need grinding and hardening on a regular basis to allow efficient operation.
- Loosely coil air hoses when not in use and cap ends to prevent foreign matter from entering.

Concrete mixers
There are four types of concrete mixers available. They range in size and shape; each has its advantages and all are used in the construction industry. Refer to Chapter 9

FIGURE 11.87 Air breaker used for heavy work

for information on the use of mixers for bricklaying and blocklaying.

Tilting mixer

This small-sized mixer is generally used by bricklayers on building sites to mix mortar (see Figure 11.88). They can be powered by one of three sources – electric motor, petrol or diesel engines – though the smallest mixers are not usually available with diesel engines. These mixers are mounted on a steel chassis with the power unit at one end. The power is usually transferred from the engine to the pinion shaft by a roller chain and sprocket or a V-belt drive, and the drum revolves at about 20 rpm. The whole steel mixing drum and drive is mounted on a steel trunnion, which is tipped by a hand-operated lever. Being small, this mixer is able to be moved easily from one building site to another.

FIGURE 11.88 Tilting drum mixer

Horizontal drum mixer

The horizontal drum mixer has a larger capacity than the tilting drum type, and is of much heavier construction (see Figure 11.89). These machines have two openings: one for loading and the other for discharging or emptying. The machine is fitted with a hopper into which the dry materials are placed. The hopper is elevated to load the dry materials into the drum, water is added and the material is then mixed as it slowly works through the machine and out the other end. The advantage of this type of mixer is that it is continually mixing concrete; there is no stopping and starting as it is one process.

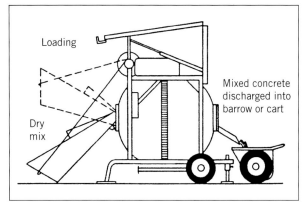

FIGURE 11.89 Mobile horizontal drum mixer

Inclined drum mixer

The inclined drum is the conical-shaped drum that is usually fitted onto a truck (see Figure 11.90). The drum has only one opening and the mixing blades are arranged so that the drum will discharge (or empty) when the rotation is reversed.

FIGURE 11.90 Inclined drum mixer with a capacity of up to 8.0 m³

Pan mixer

Pan mixers are not often seen on construction job sites as they are not really transportable. They are more often used in batching plants, as they are designed to produce the highest-quality concrete in the shortest possible time. Whereas most other types of mixer will take up to 120 seconds in normal circumstances to produce a mix, the pan mixer should not require more than 30 seconds. They are fed with material through a chute in the top of the pan and discharge through an opening in the base (see Figure 11.91). The larger models have up to three openings, which have a sliding gate cover to prevent loss during mixing.

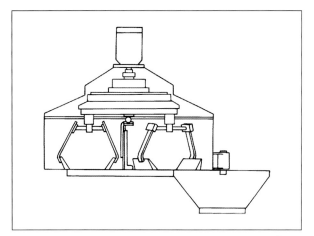

FIGURE 11.91 Mobile pan mixer with a capacity of 0.2 m³ to 1.5 m³

Routine maintenance
- Check electrical connections for proper and secure attachment. No bare wires or coloured cables should be visible at any connection point.
- Make sure all electrical connections are waterproof.
- If petrol motors are used, check oil levels as required and clean, adjust and/or replace spark plugs as required.
- Cleaning should take place as soon as possible after final use of the mixer. On very hot days this may need to be carried out as a quick wash between mixes.
- Use a stiff brush and water to clean off remaining slurry from the outside and inside of the barrel.
- If the slurry or leftover mix has started to harden inside the barrel and on the mixing blades, pieces of broken brick or dry blue metal may be placed in the barrel with water – this allows the mixing action to dislodge any stubborn remains.
- Note: Do not allow leftover slurry in the barrel to completely set prior to cleaning as this dried mortar will build up and prevent effective mixing action.

Safety
- When the drum is rotating and is being loaded with material, ensure that sufficient distance is maintained so nothing is caught in the machine.

Concrete vibrators
The poker or immersion vibrator is the most common of all the concrete vibrators; it consists of a steel poker head and a flexible drive shaft that connects the poker head to the power unit. The poker head has an internal impeller which causes the poker to vibrate (see Figure 11.92).

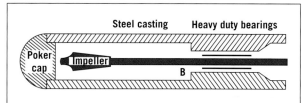

FIGURE 11.92 Section through the vibrating head of the poker vibrator

The purpose of the vibrating action is to enable trapped air and water in the concrete mix to rise to the surface. The removal of air and excess water allows the voids to be filled with a sand and cement paste, making the concrete very dense and durable.

The impeller generates vibrations as it is thrown against the casing; the casing often needs to be tapped against something to put the impeller off centre and to start it vibrating. The frequency of the vibrations varies a lot between different models and can be as low as 9000 vibrations per minute. The poker cap may be made of polyurethane material to protect the formwork. The flexible drive shaft is usually between 3 m and 6 m, but can be up to 12 m. The power unit is usually fuelled by petrol, but may also be diesel-powered, electric or pneumatic (see Figure 11.93).

FIGURE 11.93 Portable immersion vibrator (petrol-driven)

Poker vibrators have a variety of casing diameters and should be inserted at relative spacings into the concrete so as not to create air pockets or voids. Table 11.1 gives a guide to sizes and relative spacings for insertion.

TABLE 11.1 Diameter and relative spacings

Diameter of poker (mm)	Spacing when inserted (mm)
26	150
30	200
38	300
50	450
60	600
65	750
76	900
105	1000

Other vibrators include the twin screedboard vibrator (which is used for paths, roads and runways; see Figure 11.94) and the vibrating table (which is normally used for pre-cast concrete units such as step treads and concrete manhole covers). The external form of the vibrator is bolted to the outside of steel or timber forms and used where access is limited for other types of vibrator. Single-engine-mounted screedboard vibrators are available for use by two people for screeding wide concrete driveways and pavement slabs.

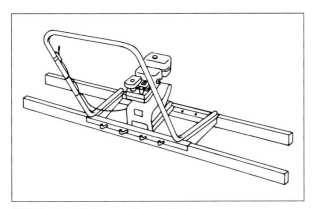

FIGURE 11.94 Twin screedboard vibrator

Routine maintenance
- Check all electrical connections for proper operation and water tightness.
- Check oil levels and clean filters regularly on petrol-driven models.
- Always clean off any concrete or cement paste with running water and a stiff brush after every use.

Wheelbarrows/brick barrows
Barrows are the traditional carriers of material on building sites. They are used to carry both wet and dry material to places that heavier and larger-capacity machines cannot reach. The single, central-wheeled type usually consists of a frame constructed of tubular steel with a bolt-on heavy steel tray (see Figure 11.95).

FIGURE 11.95 Standard builder's barrow

Wheelbarrows are sometimes galvanised with strong bracing at points of stress. The wheel has either a 355 mm × 75 mm solid rubber or 410 mm × 100 mm pneumatic tyre.

Wheelbarrows have a range of capacities, varying from light-duty 60 L barrows (0.060 m^3) up to 125 L (0.125 m^3) heavy-duty builder's barrows. A recent development is the ball barrow, which instead of the conventional wheel has a 355 mm diameter pneumatic ball and a steel body with a moulded rubber bin (see Figure 11.96). The ball barrow is lighter than a conventional barrow, weighing only 10 kg.

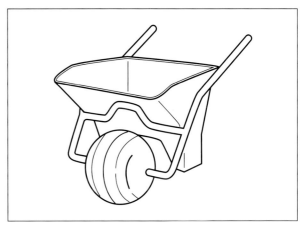

FIGURE 11.96 Ball barrow

Manufacturers claim that the ball rides lightly over rough and muddy sites where wheels would sink in. All parts are snap-fit, enabling simple on-site replacement. The body is smooth and easily cleaned, with a capacity of up to 110 litres (0.11 m^3).

The two-wheeled barrow has a tubular steel chassis with a detachable steel body that tips forward (see Figure 11.97). It is equipped with a pram handle and is available with alternative bodies.

Refer to Chapter 9 for information on the use of wheelbarrows for bricklaying and blocklaying.

Routine maintenance
- Cleaning should be carried out as soon as possible after use. A stiff brush should be used with water to

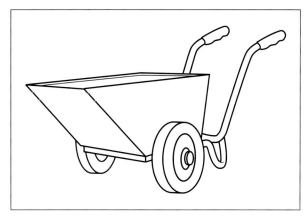

FIGURE 11.97 Two-wheeled barrow

dislodge any leftover concrete slurry as a build-up of dried slurry makes future cleaning and smooth discharging of mixes difficult.
- Regularly oil the wheel bearing.
- Regularly inflate the pneumatic tyre to the correct pressure as a tyre with low pressure is more difficult to push and to control.

Safety
- Use your legs and not your back when lifting the handles of a wheelbarrow.
- Always try to wheel over flat, rubbish-free surfaces, as any sudden jolt may injure your back.
- When constructing barrow ramps, make them the width of the barrow tray. Single planks are dangerous, as they tend to flex excessively and don't provide a surface wide enough to walk and wheel on safely. Also nail or screw low cleats across the width of the ramp to stop your boot slipping.

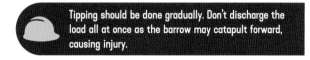

Tipping should be done gradually. Don't discharge the load all at once as the barrow may catapult forward, causing injury.

Industrial vacuum cleaners
Industrial vacuum cleaners (see Figure 11.98) are designed for wet and dry vacuuming of floors in factories, shopping centres and all commercial premises as well as building sites. They are also suited to the removal of surface water from flooded floors, cleaning ceilings and removal of water from waterlogged carpets. Capacities range from 23 litres, with a 1200-watt motor, to 210 litres, with dual electric motors. Industrial vacuum cleaners are also available in smaller back-pack models.

Specialist HEPA (high efficiency particulate air) vacuum cleaners must be used for asbestos dust removal, as well as other hazardous material. Bag removal/replacement and cleaning of these machines must be carried out within strict safety guidelines.

Routine maintenance and safety
- Always check machine leads for bare wires, cracks or cuts before use. Do not use the machine if any are found.
- Check that the switch is in the 'off' position before plugging the machine into the power point.
- Ensure that the lead is always behind the machine when operating.
- Never use extension leads.
- Never stretch the lead to its limit – use the closest power point to proceed.
- Switch the machine off at the power point before unplugging.
- Never remove the plug from the power point by jerking the cord.
- Always empty the bag after use.
- Clean filters regularly.
- Keeping the machine clean extends the useful life of the unit and also reflects well on the operator.
- Debris around the base of the machine and wheels should be removed when noticed or at the end of each use.
- Always wipe the machine over after use with a clean damp cloth and thoroughly dry it prior to storage.

Industrial work platforms
Industrial work platforms can be divided into two main groups:
- elevated work platforms (EWPs)
- temporary structures, such as scaffolding.

Elevated work platforms
Trailer lifts or cherry pickers (see Figure 11.99), scissor lifts and man lifts are all designed to allow easy access to elevated work areas. They may be used for tree lopping, factory maintenance, changing light globes/fittings in factories, painting, sign writing or reaching awkward places.
- Trailer lifts or cherry pickers will slew 360° and are powered by petrol engines. They have hydraulic outriggers for stability and the controls are mounted in the platform basket, which has a capacity of 150 kg and can lift to a height of 14 m.
- Scissor lifts are self-propelled and electrically operated or battery-powered, lift vertically only and to a height of 7.5 m.

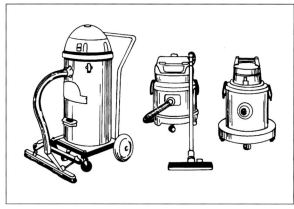

FIGURE 11.98 Various types of industrial vacuum cleaners

FIGURE 11.99 Mobile cherry picker work platform

FIGURE 11.100 Lightweight mobile scaffolding

- Man lifts are petrol-powered hydraulic lifts for one person and light equipment. They have a safe working height up to 9 m, a capacity of 136 kg, and come with adjustable outriggers.

Routine maintenance and safety
- All petrol or diesel engines must have oil levels checked regularly.
- Refer to the machine's log book for precise maintenance requirements.
- A safety harness must be worn at all times when working in the basket or on the platform (refer to your state Code of Practice for requirements).
- Platforms must not be used near overhead electrical cables.
- The operator of the trailer lift/cherry picker is required to have a certificate of competency.

Temporary structures – scaffolding
Scaffolding can be divided into different types; however, we will look solely at lightweight mobile scaffolding (see Figure 11.100). Scaffolding allows workers to operate on a safe platform above ground level. As with all work at heights, there are strict guidelines set out in the national WHS Regulations 2011, as well as Safe Work Australia's *Model Code of Practice: Managing the risk of falls at workplaces*.

According to Safe Work Australia, where work is performed using mobile scaffolds, workers should be trained to ensure the scaffold:
- remains level and plumb at all times
- is kept well clear of power lines, open floor edges and penetrations
- is not accessed until the castors are locked to prevent movement
- is never moved while anyone is on it
- is only accessed using an internal ladder.

Source: Safe Work Australia © Commonwealth of Australia. CC BY 3.0 (https://creativecommons.org/licenses/by/3.0/)

Ladders and trestles
Ladders are usually constructed from aluminium, steel, timber or fibreglass. Selecting the correct ladder for the job is an important safety consideration (see Figures 11.101–11.106). As a Safe Work Australia report of work-related fatalities in Australia from 1 July 2003 to 30 June 2011 revealed:

half of the falls that resulted in a fatality involved distances of three metres or less in the eight years 2003–11. Falls from ladders accounted for the greatest number of fatalities (37 fatalities – 16%).

Source: Safe Work Australia © Commonwealth of Australia. CC BY 3.0 (https://creativecommons.org/licenses/by/3.0/)

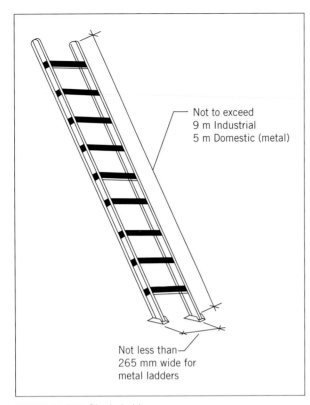

FIGURE 11.101 Single ladder

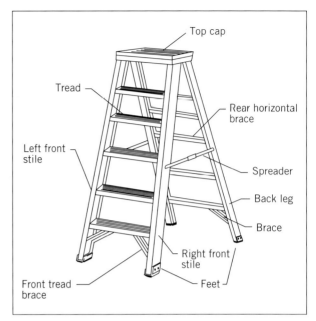

FIGURE 11.103 Step ladder

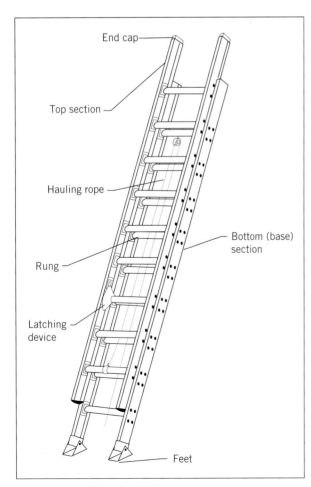

FIGURE 11.102 Extension ladder

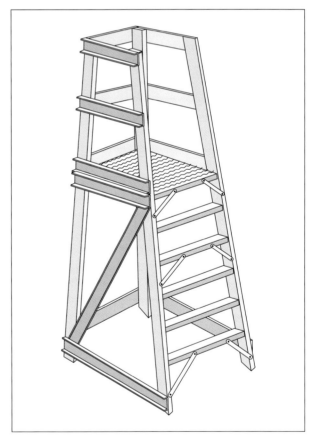

FIGURE 11.104 Platform-type step ladder

It is easier for workers to select the correct ladder when the following information is permanently marked or labelled in a prominent position:
- the name of the manufacturer
- the duty rating, in the largest lettering possible – that is, INDUSTRIAL USE or DOMESTIC USE – and the load rating in kilograms (always select industrial use)

FIGURE 11.105 Dual-purpose step/extension

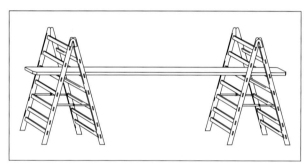

FIGURE 11.106 Trestles – used with a plank to create a working platform

- the working length of the ladder (closed and maximum working lengths for extension types)
- hazard warnings, in the largest lettering possible (particularly for metal ladders; for example, DO NOT USE WHERE ELECTRICAL HAZARDS EXIST)
- instructions for use; for example, double-sided step ladders should display the words 'TO BE USED IN A FULLY OPEN POSITION'.

Safe use

- Ladders must be in good condition, free from splits, knots and broken or loose rungs.
- Follow the 1:4 (base:height) ratio rule with ladders. For example, the foot of a 4 m ladder should be at least 1 m away from the wall against which the ladder is leaning. Make sure the top of the ladder extends at least 1 m above the landing or platform.
- The ladder should be securely fixed at the top and bottom and footed securely on a firm, level foundation.
- Never put ladders in front of doorways, or closer than 4.6 m to bare electrical conductors (sometimes it is safe to put timber ladders closer than this; e.g. 1.5 m).

An electrical current can jump from a conductor to an aluminium ladder without direct contact.

- When working with or on electrical equipment, use only wooden ladders. Do not use metal or wire-reinforced ladders when working near exposed power lines.
- One person only should be on a ladder at a time and tools should be pulled up with a rope. Workers ascending or descending should face the ladder.
- Two ladders must never be joined together to form a longer ladder.
- Ladders should not be placed against a window.
- Timber ladders should never be painted, as this could cover faults in the timber.

COMPLETE WORKSHEET 3

Using tools safely

New and inexperienced operators should be given extensive instruction on the safe use of each of the common power tools before using them. This instruction should be provided individually or at a maximum 2:1 ratio between learners and instructor.

Close supervision is also extremely important as the new and inexperienced operator gains knowledge. Remember, a 'learner' car driver does not have a single session of instruction and become totally competent; it takes time and practice.

Power tool safety checklist

Portable electric-powered tools can be a source of both mechanical and electrical hazards unless they are maintained and used correctly. The checklist in Table 11.2 contains points of general safety and ways of avoiding damage to tools and equipment or injury to the operator.

If the answer is 'Yes' to any of these questions, the worker must review work practices before using any power tools.

All power tools are potentially dangerous if not used correctly. This includes wearing or using the correct personal protective equipment.

Every time a tool is used, you will need to conduct checks for tool damage, missing guards, damaged leads, faulty triggers, poor maintenance, etc. (see Figure 11.107). These are just some of the items to be addressed prior to operating power tools. This step is especially important if the tool is unfamiliar to the operator; for example, where the tool has been borrowed or hired, or is brand new.

The following details outline the safety precautions required for power tool operation to allow preparation for safe use of the tools. The list is not comprehensive, but it does outline the main safety requirements and checks to be carried out for the safe operation of a portable power saw.

TABLE 11.2 Power tool safety checklist

General lead safety	Yes	No
Are extension leads neatly coiled when not in use to allow ease of future use?	☐	☐
Are all leads stored away from oils, solvents, acids, heat and sharp objects?	☐	☐
Are the plugs clear plastic? (Extension leads are required by law to have clear plastic plugs so that broken or exposed wires can easily be detected.)	☐	☐
Are leads carried rather than dragged? Are copper wires exposed because of dragging?	☐	☐
Are tools carried or lowered by a lanyard or the handle and not by the lead?	☐	☐
Are plugs disconnected by pulling the plug out directly and not by the lead?	☐	☐
Are leads fully rolled out and not left partially rolled up when using high-amp tools (for example, a welder)?	☐	☐
Is the outer casing of the lead intact, with no coloured cords visible along the length of the lead or at the ends?	☐	☐
Are leads free of patches or repairs with insulation tape?	☐	☐
Are leads clear of the operator's feet and the tool's cutting edge while being used?	☐	☐
Are the leads protected when they are laid across a driveway or barrow runs?	☐	☐
Are extension leads supported at a vertical height of 2.1 m above the work area and passageways up to 4 m from where the tool is to be used?	☐	☐
Are the pins firm and not loose? (Loose connections cause the plug to overheat.)	☐	☐
Tool safety	**Yes**	**No**
Is the work area tidy before, during and after using a power tool? (An untidy work area is potentially dangerous.)	☐	☐
Is the work area floor damp/wet? (DO NOT MIX ELECTRICITY AND WATER!)	☐	☐
Is the operator wearing appropriate safety glasses/goggles?	☐	☐
Are the operator and nearby workers wearing appropriate hearing protection?	☐	☐
Is the operator wearing loose clothing or jewellery? (Jewellery may become entangled in the moving parts of the tool.)	☐	☐
Is the material being worked on correctly secured? (The tool operator should have both hands free to control the tool.)	☐	☐
Is any part of the operator or assistant in a position to be hit or cut by a runaway tool?	☐	☐
Does the operator carry the tool with their finger on or over the trigger while it is still plugged in?	☐	☐

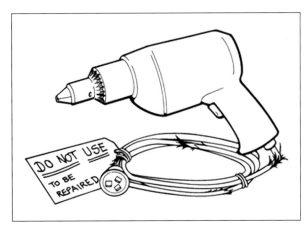

FIGURE 11.107 Labelling faulty equipment

- Check all adjustments and adjustable fittings to ensure they are tightened before use.
- Ensure that the saw is fitted with a hood guard and a returning spring guard.
- Never tie the guard back.
- Always operate the saw with both hands.
- After all cutting operations, check to see that the guard has returned before placing the saw on the ground or other surface.
- Rest the saw on a timber block when not in use to prevent damage to the guard.
- Use the correct type of blade for the work being undertaken.
- Switch off the power and remove the plug before making any adjustments to the saw.
- Keep the work area clean and clear of offcuts.
- Always wear safety glasses and hearing protection when using the saw.

To use a power tool correctly means obtaining training and instruction. It is highly recommended that you gain personal training from an experienced operator in how to operate power tools. However, you can also access supplementary training on video or DVD that will assist with the personal training. There are also video clips on YouTube that will give you a good basis for learning how to use tools correctly.

LEARNING TASK 11.4 POWER TOOL INSTRUCTION

Undertake safe power tool instruction with a teacher or instructor in relation to the following basic portable power tools:
- circular saw
- compound mitre saw
- planer
- impact drill
- compressed air nailer and compressor.

Note: For safety and compliance with WHS requirements, it is recommended that power tool instruction should be carried out at a ratio of no greater than two students per teacher/instructor for approximately four hours per session. It is also recommended that safe instruction and use be conducted as required for AQF (Australian Qualifications Framework) level 3 competencies.

LEARNING TASK 11.5 POWER TOOL USE

Use basic portable power tools, as directed by your teacher/instructor, to assist in the construction of various practical projects and/or site activities.

Note: For safety and compliance with WHS requirements, it is recommended that only one or two tools be used at any one time by a class or training group. The tools should be set up and used in a designated area clearly visible to the teacher/instructor at all times.

 COMPLETE WORKSHEET 4

Cleaning up after using tools and equipment

At the completion of the job, it is important to clean the work area of any garbage, scraps and material offcuts, as well as packing away all the tools that have been used. There are several questions to consider in the cleaning up and packing away process:
- Is the work area safe for the next team of workers on the site? (This means clean and tidy, and all offcuts, shavings and sawdust have been cleaned away.)
- Have the tools that were used been cleaned, checked for damage and then correctly packed away?
- Are there tools that need to be maintained to keep them operating properly?
- Are there cutters, bits or blades that need to be replaced before next use?

Cleaning the worksite

Brooms are the basic tools used in cleaning up and are available in a variety of sizes for specific uses and may have heads containing bristles of straw, polypropylene or animal hair (see Figure 11.108). Stiff-bristle brooms such as the yard broom and straw broom may be used on rough surfaces, and soft-bristle brooms may be used on smooth surfaces, such as lined or coated floors. Brooms may also be used to provide a textured non-slip finish to wet concrete. Broom handles are made from seasoned hardwoods that are straight-grained, strong and flexible.

FIGURE 11.108 (a) Stiff straw or millet broom, (b) stiff yard broom of straw or polypropylene, (c) broad soft-bristle floor broom of animal hair or polypropylene

Brooms form an essential part of the carpenter's equipment kit as they may be required to clean up a work area before carrying out a task, or to clean up after work has been carried out. It is each tradesperson's responsibility to clean up his or her own mess and to leave a clean, dust-free and safe work surface in line with WHS Acts and regulations.

Maintenance of tools and equipment

It is important to be able to use hand and power tools safely and correctly and part of that is the maintenance of those tools. If tools are not maintained properly they will not operate to their full potential and may become a potential hazard and result in time wastage.

Below are some points that may assist a tradesperson in maintaining some of the basic tools:

- Hammers and pinch bars should be kept dry and rust-free. Handles should be kept free of oil and grease to prevent slipping during use.
- Spanners, pliers and nips must be kept rust-free, lightly oiled at moving parts and stored in a dry dust-free container.
- Tapes, especially metal-blade retractable tapes, should be kept dry and clean before retracting the blade; otherwise rusting and jamming will occur. Also, the blade should be fed back gently without allowing the hook to hit hard against the tape's body, as it may be snapped off.
- Four-fold rules should have the knuckles oiled and be gently exercised when first used. This will prevent stiff joints from snapping.
- Concreting trowels must be thoroughly cleaned, dried and stored in a toolbox after use. Don't let cement paste harden on these tools, as this will allow a build-up to take place every time they are used.
- Straw or millet brooms should not be stored wet, as this will allow the bristles to rot.
- Hoses should be loosely coiled or wrapped on a reel to prevent kinks and tangles. Protect the hose with timber battens if it is laid across a traffic road. Take care using hoses in extremely cold weather, as plastic hoses become brittle and may snap during use.
- Saw stools should be stored out of the weather and all screwed joints should be checked for tightness on a regular basis.
- Painting tools should be scraped clean after use and shave hooks should be filed to keep the edges keen.
- Brushes and rollers should be cleaned with the recommended solvent, rinsed with warm, soapy water, dried and stored away. During use they may be wrapped in plastic to prevent drying out.
- Abrasive papers must be kept dry, especially those with animal glue holding the grit, otherwise they will clog, lose the grit or tear easily during use.
- Shovels, spades, picks, mattocks, etc. should have excess soil scraped off and then be thoroughly washed with a hose. Lightly oil exposed steel areas when these tools are not being used for extended periods of time.

SUMMARY

In Chapter 11 you have learnt how to use a selection of tools and equipment. In addition, the chapter has:
- considered the use of a variety of tools and equipment, including hand tools, power tools and pneumatic tools
- discussed what these tools are used for
- indicated how to identify, select and use a variety of relevant plant and equipment.
- discussed how to use hand tools, power tools and pneumatic tools safely and efficiently, including the need to undertake training
- identified the importance of maintaining equipment and cleaning up after tools, plant and equipment are used.

REFERENCES AND FURTHER READING

Safe Work Australia, *Model Code of Practice: Managing electrical risks in the workplace*, **https://www.safeworkaustralia.gov.au/doc/model-code-practice-managing-electrical-risks-workplace**.

Safe Work Australia, *Model Code of Practice: Managing the risk of falls at workplaces*, **https://www.safeworkaustralia.gov.au/search/site?search=+code+of+practice%2C+Managing+the+risk+of+falls+at+workplaces**.

GET IT RIGHT

The photo below shows a person changing a planer blade unsafely. Identify the unsafe practices and provide reasoning for your answer.

Source: Richard Moran

WORKSHEET 1

To be completed by teachers

Student competent ☐

Student not yet competent ☐

Student name: _____

Enrolment year: _____

Class code: _____

Competency name/Number: _____

Task

Read through the sections from the start of the chapter up to 'Metalworking tools', then complete the following questions.

1 Complete the following statement:

   Hand tools have an important role in _____ and _____ the work to be undertaken.

2 List the five things that need to be considered before even starting the tool or machine.

   _____

   _____

   _____

   _____

   _____

3 List five hammer types used for general carpentry activities.

   _____

   _____

   _____

   _____

   _____

4 List the four main types of planes.

   _____

   _____

   _____

   _____

5 Briefly describe the main use of pincers.

   _____

   _____

6 Why should a utility knife only be carried or stored in your nail belt or tool bag?

   _____

   _____

7 Identify the types of plank in the photos.

8 State the minimum section size of an aluminium plank:

   _____ mm × _____ mm

9 List the five main types of clamps that can be used.

   _____

   _____

   _____

   _____

   _____

WORKSHEET 2

To be completed by teachers
Student competent ☐
Student not yet competent ☐

Student name: _____

Enrolment year: _____

Class code: _____

Competency name/Number: _____

Task
Read through the sections from 'Metalworking tools' up to 'Power and pneumatic tools', then complete the following questions.

1. What is the name of the spanner that a scaffolder or formworker would use?

   _____

2. List five actions or tasks that may be done with pliers.

   _____

   _____

   _____

   _____

   _____

3. Tin snips are used for cutting a straight line only in thin sheet metal. (Circle the correct answer.)

   TRUE   FALSE

4. Bolt cutters are used for cutting reinforcement bars and steel fabric for concrete work. (Circle the correct answer.)

   TRUE   FALSE

5. Apart from applying putty to glaze windows, state three other uses for a putty knife.

   _____

   _____

6. What is the main use for a hacking knife?

   _____

7 What are the three main parts of a paint brush?

　

　

　

8 What is the name given to the soft covering on a paint roller?

　

9 State the three main types of natural abrasive grains used for abrasive papers.

　

　

　

10 State the two main types of manufactured abrasive grains used for abrasive papers.

　

　

11 Name two tools that may be used for trimming or detailing an excavation.

WORKSHEET 3

Student name: _____

Enrolment year: _____

Class code: _____

Competency name/Number: _____

To be completed by teachers	
Student competent	☐
Student not yet competent	☐

Task

Read through the sections from 'Power and pneumatic tools' up to 'Using tools safely', then complete the following questions.

1   List the six sources of power that may be used to drive or work a power or pneumatic tool.

    _____
    _____
    _____
    _____
    _____
    _____

2   Extension leads used on all building sites need to be inspected and tagged by a licensed electrician. State the four main items that must be included on the clamped electrical tag.

    _____
    _____
    _____
    _____

3   State the minimum height (in metres) above ground for an extension lead when used on site.

    _____

4   To ensure safe use of powered tools, why should leads not be left partially rolled up during use?

    _____
    _____

5   Why is a portable site generator essential on large new housing estates?

    _____
    _____

6 List four types of concrete mixers available for use on site.

_____

_____

_____

_____

7 What is another name for the poker vibrator?

_____

8 What do the letters HEPA stand for in relation to industrial vacuum cleaners?

_____

9 Complete the statement:

Trailer lifts or cherry pickers have a platform basket with a capacity of _____ and are able to extend to a maximum height of _____ metres.

10 According to the *Model Code of Practice: Managing the risks of falls in the workplace*, what five things should workers be trained to ensure when using mobile scaffolds?

_____

_____

_____

_____

_____

11 Fill in the missing information below.

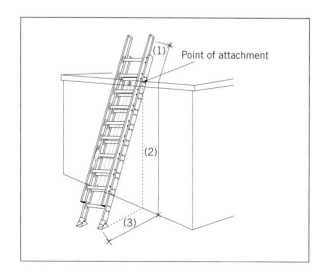

WORKSHEET 4

To be completed by teachers
Student competent ☐
Student not yet competent ☐

Student name: _____

Enrolment year: _____

Class code: _____

Competency name/Number: _____

Task

Read through the sections from 'Using tools safely' up to 'Cleaning up after using tools and equipment', then complete the following questions.

1 What are the two important things that need to happen for a new and inexperienced operator to gain knowledge and experience?

   _____

   _____

2 Extension leads are required by law to have clear plastic plugs so that broken or exposed wires can easily be detected. (Circle the correct answer.)

   TRUE   FALSE

3 What height (in metres) are extension leads to be supported at above the work area and passageways?

   _____

4 Why is it important that the operator of a tool is not wearing loose clothing or jewellery?

   _____

   _____

MY SKILLS 11

EMPLOYABILITY SKILLS

Using the following table, describe the activities you have undertaken that demonstrate how you developed employability skills as you worked through this unit. Keep copies of material you have prepared as further evidence of your skills.

Employability skills	The activities undertaken to develop the employability skill
Communication	
Teamwork	
Planning and organising	
Initiative and enterprise	
Problem-solving	
Self-management	
Technology	
Learning	

UNDERTAKE BASIC INSTALLATION OF WALL TILES

This chapter covers the outcomes required by the unit of competency 'Undertake basic installation of wall tiles'. These outcomes are:
- Prepare for basic installation of wall tiles.
- Set out basic wall tiling.
- Prepare substrate and tiles, and install tiles.
- Grout tiles.
- Clean up.

Overview

Tiles have a long history in construction, having been used as far back as the fourth millennium BCE, for both functional and decorative purposes. Roman mosaics, for example, are celebrated for their artistic qualities, and give an insight into the cultures and societies from specific regions and time periods.

Today, tiles continue to be used in multiple ways in construction, from providing a strong, hard-wearing surface that is durable for foot traffic to assisting with the control of water in wet areas of a building. Tiles also assist with the cleaning and maintenance of water areas, minimising water penetration and any associated rot or mildew.

Employers value people who fit into their workplace, solve day-to-day problems, manage their time and are keen to continue learning. These types of skills are known as employability skills. The employability skills you develop while working through this unit will be assessed at the same time as the skills and knowledge.

It is important to show evidence that you have developed these skills. Your trainer or assessor will discuss with you how to record your employability skills. The following table provides some examples of things you might do to develop employability skills in this unit.

EMPLOYABILITY SKILLS

Employability skill	What this skill means	How you can develop this skill
Communication	Speaking clearly, listening, understanding, asking questions, reading, writing and using body language.	• Listen carefully to work instructions. • Ask questions to clarify task requirements.
Teamwork	Working well with other people and helping them.	• Work individually and as a team to complete tasks in a timely manner. • Liaise with others to avoid conflicting work tasks.
Planning and organising	Planning what you have to do. Planning how you will do it. Doing things on time.	• Plan resources required for each task. • Plan sequence of activities in order to successfully complete work task.
Initiative and enterprise	Thinking of new ways to do something. Making suggestions to improve work.	• Prepare information before you discuss issues. • Suggest improvements to processes.
Problem-solving	Working out how to fix a problem.	• Manage task sequencing to cater for changes to the work environment. • Adapt to cultural differences in communications.
Self-management	Looking at work you do and seeing how well you are going. Making goals for yourself at work.	• Evaluate and improve own methods of communication. • Aim for quality results within specified task time frames.
Technology	Having a range of computer skills. Using equipment correctly and safely.	• Implement safe work procedures for machinery. • Use a range of communication tools such as two-way radios, telephone and email.
Learning	Learning new things and improving how you work.	• Research online product and work practice information. • Observe more experienced construction workers and adopt their practices to improve your work processes.

Prepare for basic installation of wall tiles

It is important that all materials for a job have been calculated in advance and delivered to the job site before work begins. It is also important that all materials are ready for installation when required so the flow of work can continue uninterrupted, particularly adhesives that have a limited time to successfully bond and secure tiles. Wastage of materials and any delays in replacing them will also add additional costs to the project.

Measures must be in place to ensure that any completed work is protected. For example, when a splashback is being installed, measures may need to be in place to protect benchtops that have already been installed.

The appropriate personal protective equipment (PPE) also needs to be available before the work commences. The work instruction will highlight any other aspects that need to be considered.

Quantities

Tiles are quantified per square metre (m^2). In some circumstances tiles are ordered and supplied as individual pieces or by the lineal metre. This is typically where there are specialised decorative tiles required, such as patterned or tiles with special finished edges and borders.

EXAMPLE 12.1 CALCULATING WALL AREA FOR TILES

The calculations below demonstrate a systematic approach to working out the wall tile area for a bathroom. As discussed in Chapter 3, where there is an irregular shape that needs to be measured, the approach is to 'cut up' the irregular shape into regular shapes (rectangles).

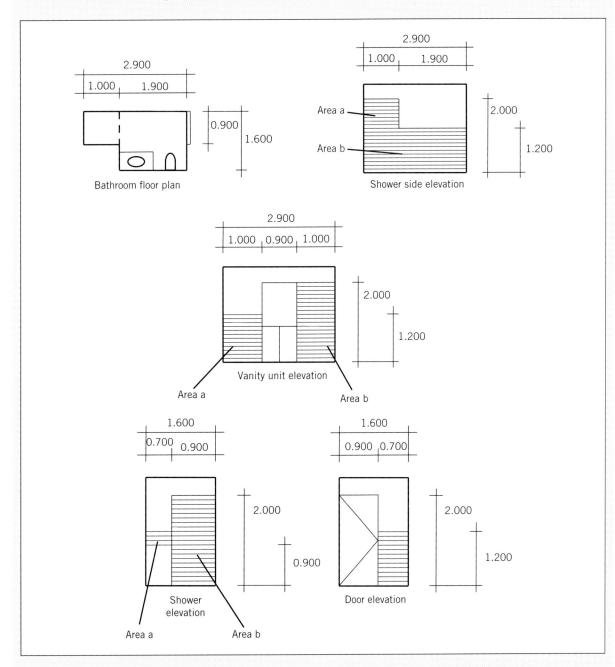

Shower side elevation
   Area a = 2.900 × 1.200 = 3.480 m²
   Area b = 1.000 × 0.800 = 0.800 m²
   Total   = 3.480 + 0.800 = 4.280 m²

Vanity unit elevation
   Area a = 1.000 × 1.200 = 1.200 m²
   Area b = 1.000 × 2.000 = 2.000 m²
   Total   = 1.200 + 2.000 = 3.200 m²

>>

>> Shower elevation
Area a = 0.700 × 0.300 = 0.210 m²
Area b = 0.900 × 2.000 = 1.800 m²
Total   = 0.210 + 1.800 = 2.010 m²

Door elevation
Total = 0.700 × 1.200 = 0.840 m²
Total area = 4.280 m² + 3.200 m² + 2.010 m²
           + 0.840 m²
         = 10.330 m²

COMPLETE WORKSHEET 1

Adhesives

Adhesive quantities are supplied in units of bags. There are different weights of bags available, though the most common size is a 20 kg bag (see Figure 12.1). Pre-mixed adhesives are available in 20-litre, 10-litre and 1-litre containers. The number of bags required for a particular job is not only dependent on the weight of the bag but also the size of the notches in the trowel that will be used to apply the adhesive.

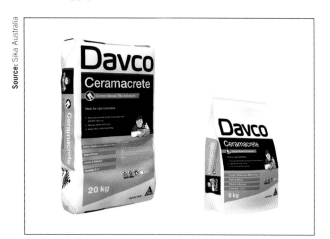

FIGURE 12.1 5 kg and 20 kg bags of tile adhesive

Each manufacturer of adhesives provides an estimate of the coverage that will be achieved through the use of their particular product. In some circumstances there are online calculators that are available to estimate the amount of adhesive that is required for a specified area. Table 12.1 lists the recommended trowel notch size for various tile sizes and the coverage of the adhesive based on the notch size.

TABLE 12.1 Recommended trowel notch size and coverage of adhesive

Tile size	Trowel size	Coverage
Up to 250 mm × 250 mm	6–8 mm notch	0.8 m²/kg
250 mm × 250 mm and over	10–12 mm notch	0.5 m²/kg

Selection of appropriate material

The two main differences between floor tiles and wall tiles are thickness and finishes. In general, floor tiles are thicker and have a form of textured finish to safeguard against slipping and provide a higher level of durability.

Wall tiles in the most part are thinner and more decorative in their finish. They can also have a high glossy finish as there is no requirement for there to be a slip-resistant features.

Tiles are made of several different materials, including ceramic, mosaic, fully vitrified, porcelain, terracotta, slate and granite (see Figures 12.2–12.8). These materials have different properties and this needs to be taken into account when determining the use of the tile, and the substrate, adhesives and grouting system to be used.

FIGURE 12.2 Ceramic tiles

FIGURE 12.3 Mosaic tiles

Adhesives

There is a wide range of adhesive products on the market for use with wall tiles. The selection of the adhesive relies on several different factors: the type and condition of the substrate, the environment factors that

FIGURE 12.4 Fully vitrified tiles

FIGURE 12.5 Porcelain tiles

FIGURE 12.6 Terracotta tiles

FIGURE 12.7 Slate tiles

FIGURE 12.8 Granite tiles

the wall tiles will be exposed to (such as whether the wall is internal or external) and if the tiling will be used in wet areas that have been waterproofed. The type of tile that is to be used will also be a contributing factor.

Before an adhesive is used, the correct sealer/primer must be selected and used to gain the maximum efficacy from the adhesive. Again, the selection of a sealer/primer will depend on the type of substrate and what preparation is required before installing the tiles.

There is such a variety of situations in which wall tiles may be installed that a clear classification structure for the use of adhesives has been created; see AS ISO

13007.1:2020 Ceramic tiles – Grouts and adhesives, Part 1: Terms, definitions and specifications for adhesives.

Classification and designation

The Standard divides tile adhesives into three types: cementitious adhesives (C), dispersion adhesives (D) and reaction resin adhesives (R). Each of these types can be further subdivided into normal adhesive (1) and improved adhesive (2) classes.

Then there is a range of different characteristics that each of these types can exhibit. These are abbreviated as single letters; for example, T denotes a slip-resistant adhesive. The overall designation of the adhesive is generated with the symbol of the type (C, D or R), followed by the class number (1 or 2) and then abbreviation of the class or classes to which it belongs (F, A, T, E, S or P). For example, the designation R1T indicates a normal reaction resin adhesive with slip resistance. The Standard lists all the combinations in table form (Based on Australian Standards).

Adhesives may cause respiratory irritation in some people and moderate inflammation of the skin either following direct contact or after a delay of some time. Long-term exposure to adhesives can lead to respiratory irritants that may result in airways disease, involving difficulty breathing and related whole-body problems. Refer to the material's SDS for specific safety information.

COMPLETE WORKSHEET 2

Tools and equipment

There is a range of tools used by tilers when installing wall tiles, including specific tools that relate to the different phases of the work: preparation, installation and clean-up. This list of tools is not extensive and only covers the minimum basic tools required to undertake wall tiling. Tilers have a wider range of tools that may be used for the different situations they will encounter.

Preparation

In this phase, the first tools used are those to clear the area of dust and debris. Next is the setting-out phase, and it is important that these tools are accurate and in good functional order and accurate as all the following works build on the setting out.

Chalk line

A chalk line is a roll of string that sits within a container that also holds powdered chalk (see Figure 12.9). As the string is pulled out of the container, the string is loaded with chalk. This is to mark a line on the substrate: the

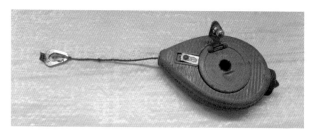

FIGURE 12.9 Chalk line

string is held into position at two ends, lifted from in the middle then allowed to flick back into position, leaving a line on the substrate. This line provides a reference point for measuring and checking that tiles are correctly positioned.

Spirit level

A spirit level is used to establish a level line or plumb line on the substrate (see Figure 12.10 and Figure 12.11). This will assist in the setting up of any temporary formwork used to support tiles. Spirit levels are also used to mark out walls and check that the wall to be tiled is plumb. Refer to Chapter 11 for information on the use of spirit levels in the construction industry.

FIGURE 12.10 Spirit level used to set out level line

Tape measure and ruler

Tape measures and rulers are used in several different situations on the job; for example:
- measuring and setting out the location of the tiles (see Figure 12.12)
- checking the size of area to be tiled
- checking the sizes of the tiles to be used
- determining the size of any tiles that need to be cut.

The ability to use a tape or ruler accurately sounds basic, but it is important that you develop appropriate skills to use tapes in a range of different situations.

FIGURE 12.11 Spirit level used to set out a plumb line

FIGURE 12.12 Tape measure used to check tiles when setting out

Pencil

The pencil is used to make appropriate marks on the substrate and any tiles that need to be cut. A hard-wearing pencil should be used as marking out on masonry surfaces may be required.

Installation

This phase of tiling is where many hand skills come into play. The ability to produce a high-quality job relies on a sound knowledge of the combination of materials and tools to complete the task. Different tiling tasks will require different skills based on the size of the tiles and the location in which they will be installed.

Notched trowel

Notched trowels are used to apply the adhesive to the substrate. There is a range of differently sized notched trowels. These sizes relate to the size of the notch, which will be determined by the requirements of the adhesive type and the size of the tile (see Figure 12.13).

FIGURE 12.13 Notched trowel used to apply adhesive

AS 3958.1-2007 Ceramic tiles – Guide to the installation of ceramic tiles specifies the size of the trowel to suit the size of the tile. For example, a 200 mm × 200 mm tiles requires an 8 mm notched trowel.

Buckets

Buckets are constantly used when wall tiling. The bucket is a multi-purpose piece of equipment that is used on site for a range of different applications, including:

- transporting tools and equipment
- mixing and temporarily storing adhesives (see Figure 12.14)
- holding water for the cleaning process
- carrying debris from the tiling activity.

FIGURE 12.14 Bucket used to temporarily store adhesive

Bucket trowel
The front edge of a bucket trowel has a flat straight edge to allow the tiler to scoop all of the contents out of the bottom of a bucket.

Straight edge
A straight edge is used as a point to work from. In many instances a straight edge is fixed in a level position for wall tiles to be installed from. These are typically made from aluminium sections (see Figure 12.15) and in some cases timber.

FIGURE 12.15 Aluminium straight edge used to support tiles

Spacers
Spacers come in different sizes and are used to maintain a consistent distance between tiles as the adhesive dries (see Figure 12.16). They are made of plastic and should be removed before grouting. There is also a range of spacers that assist in the levelling between tiles as well as spacing the tiles from each other.

Tile cutters
Tile cutters are used to make straight cuts on tiles (see Figure 12.17 and Figure 12.18) and they range in

FIGURE 12.16 Spacers used to keep tiles in the correct position

FIGURE 12.17 Tiles being cut with manual tile cutter

FIGURE 12.18 Tiles being cut with a tile-cutting machine

size from hand tools all the way up to machinery that uses water to control the dust and longevity of the cutting blade.

Other tools

Other tools used in the installation of tiles include saws that have diamond blades, drills with specialised drill bits, and tools to transport material, such as wheelbarrows. There is also a range of tools that are used in the process of grouting (discussed later in this chapter).

Clean-up

Scrapers

Scrapers are most commonly used to remove unwanted material from tiles, such as adhesives that have dried in the incorrect location, or to clear excess grout from between tiles (see Figure 12.19). Care must be taken not to damage tiles when using scrapers.

FIGURE 12.19 Paint scraper

Dustpan and broom

The clean-up process will produce both coarse and fine particles of debris and a dustpan and broom (see Figure 12.20) normally suffices to collect and remove this waste.

FIGURE 12.20 Dustpan and broom

Sponge

A sponge is used to wash down excess material from the surface of tiles, usually after the tiles have been grouted (see Figure 12.21). Different densities of sponge

FIGURE 12.21 Sponge being used to clean down tiles

materials are available for different applications throughout the tiling process.

Set out basic wall tiling

There are several considerations when setting out tiles to be attached to a wall. Typically, wall tiles are used in wet areas such as bathrooms and kitchens. In such situations, there will always be plumbing and electrical services, and the location of these services needs to be considered and allowed for when setting out tiles.

There also may be special fittings to be considered, such as soap holders and towel rails. These may affect the pattern chosen for the installation of the tiles, as the placement of fixtures can make the tiling process difficult or make the tiles look out of place.

There are several standard patterns used for wall tiles, including diagonal, diamond, pinwheel, herringbone and stretcher bond.

Diagonal is similar to the straight pattern except the tiles are laid on a 45-degree angle, turning square tiles into diamonds (see Figure 12.22). It is not uncommon for the diamond pattern to be used with a border as a feature in a kitchen splashback. The same technique can be used for an entire floor to make a small room look and feel bigger than it really is.

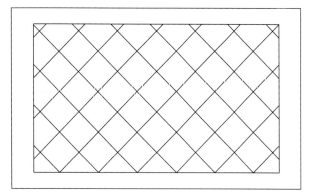

FIGURE 12.22 Diamond pattern

The pinwheel pattern, also known as the hopscotch design, uses a small square tile surrounded by much larger square tiles to create the effect of a spinning pinwheel (see Figure 12.23).

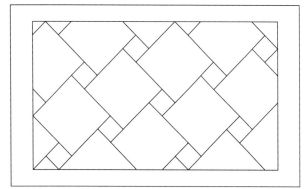

FIGURE 12.23 Pinwheel pattern

Herringbone has a 'V' shape in its pattern, so it resembles arrows pointing in a particular direction. This pattern is achieved by laying rectangular tiles in a zig-zag pattern (see Figure 12.24).

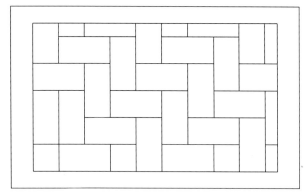

FIGURE 12.24 Herringbone pattern

Stretcher bond is usually rectangular tiles that are laid like bricks in a wall. The end of each tile is lined up with the centre of the tiles directly above and below it. This creates a staggered but cohesive look (see Figure 12.25).

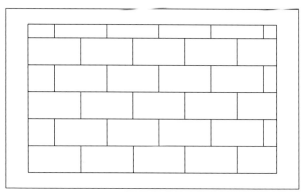

FIGURE 12.25 Stretcher bond pattern

Whichever pattern is selected, it is important that a vertical plumb line is established in the setting-out phase. This will give a reference point to ensure that the tiles in horizontal patterns are installed both level and plumb.

The plumb line will also provide a point of reference so any adjacent walls can be checked for their plumbness, which may end up having an impact on the aesthetic look of the tiles laid at the intersections of walls. Figure 12.26.

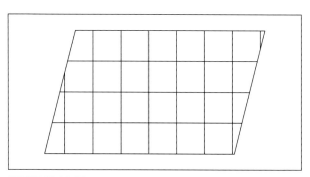

FIGURE 12.26 Tiles installed plumb to the orange reference point, although adjacent walls may not be plumb

It is also important to establish a level line around the whole of the job, which can be referred to and used as a guide. Also known as a datum, this is very important when the tiled surface continues across several walls and the consistent flow of the tiles need to be maintained. The flow of the tiles' horizontal line is preferable to the vertical line for aesthetic reasons.

Centring

A gauge rod is used to establish the position of windows, doors and bench heights so the layout can be planned to minimise any small gaps that will require tiles to be cut thinly, which can detract from the aesthetics of the finished job. A gauge rod will allow the top and bottom tiles to be evenly cut (see Figure 12.27).

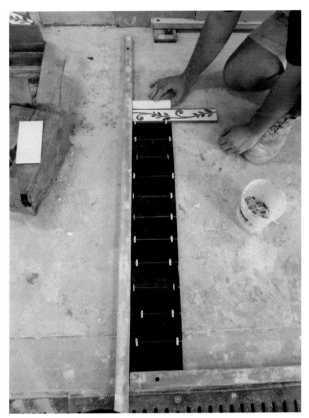

FIGURE 12.27 Tiles being laid out with a gauge rod

Consideration also needs to be given to any patterned border or decorative tiles that will meet an intersecting wall. In some circumstances the length of the patterned tiles will not match that of the main wall tiles, making synchronisation of tiles difficult (see Figure 12.28).

FIGURE 12.28 Wall junction with multiple decorative tiles with different tile lengths

LEARNING TASK 12.1 SETTING OUT A GAUGE ROD

Using a piece of aluminium or timber no less than 1.8 m in length, mark out a gauge rod to suit selected tiles. Make necessary allowances for spacers and any decorative tiles that may be used in the layout of the tiles.

Wet areas

When setting out the position of tiles for a wall, all penetrations need to be considered, especially in wet areas. Penetrations include taps, shower roses and any other areas where there is potential for water to get into the wall or behind the tiles and wall linings, leading to defects occurring over time.

Before tiling over a wet area it is important that the appropriate adhesive is used. It is equally important to ensure that the waterproof membrane maintains its integrity and is not damaged in any way during the tiling process (see Figure 12.29). If the waterproofing is damaged it may not be identified until the water area is in full use. This may lead to rectification work and potential disputes with the client.

FIGURE 12.29 Typical waterproofing applied to both floors and walls of a wet area to be tiled

 COMPLETE WORKSHEET 3

Preparation and installation

Special consideration needs to be given to the substrate before tiling. The condition of the substrate will have a direct impact on the overall finish of the tiled wall and the longevity of the final product. There are several different materials that can make up a substrate, but no matter which is selected, it must be structurally sound, plumb and as flat as possible. While the substrate may have been selected by another tradesperson, it is still the responsibility of the tiler to conduct quality checks before any work begins.

Substrate

There are several different materials and structural elements that can produce a wall. In contemporary residential buildings the walls will typically be constructed of a timber frame and lined with a sheet product. The sheet product may vary depending on the purpose of the room which is to be tiled. Common linings include plywood, fibre-cement sheeting, compressed fibre-cement sheeting and plasterboard sheeting.

Plywood

Plywood is rarely used as there are alternative materials that are more cost-effective and readily available. Plywood used as a substrate for floor and wall tiling should have a minimum thickness of 10 mm (see Figure 12.30).

FIGURE 12.31 Fibre-cement sheeting fixed to a wall to take wall tiles

FIGURE 12.30 Plywood sheeting lining a wall frame

Fibre-cement sheeting

Fibre-cement sheeting is the most common material used for wet areas in domestic construction as it is both widely available and cost-effective. Fibre-cement sheets to be used as a substrate for wall tiling should have a minimum thickness of 6 mm (see Figure 12.31). For heavy-duty commercial applications, fibre-cement sheets for wall tiling should have a minimum thickness of 9 mm.

Compressed fibre-cement sheeting

Compressed fibre-cement sheeting is more dense and can withstand additional stresses such as impacts. The sheeting is heavier and requires a higher level of fixing requirements. It is also very expensive compared to fibre-cement sheeting and requires special tools to cut it safely and effectively. Compressed fibre-cement sheets to be used as a substrate should have a minimum thickness of 9 mm for wall tiling. Figure 12.31.

FIGURE 12.32 Compressed fibre cement sheeting

Plasterboard

Different types of plasterboard are available depending on the use of the room. If the area is a wet area, there are plasterboard sheets that are specifically manufactured to maintain their integrity in such conditions (see Figure 12.33). Gypsum plasterboard sheets to be used as a substrate for wall tiling should

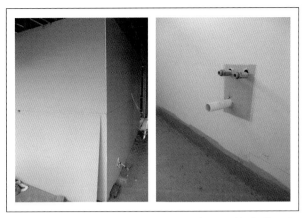

FIGURE 12.33 Plasterboard products can be used in both wet and dry areas of a building.

have recessed linerboard-bound longitudinal edges and a minimum thickness of 10 mm.

No matter which lining is selected, it is necessary to ensure that the sheeting is properly secured to the structural components of the wall. This is to ensure that the lining will not come away from the framework and that the weight of the tiles to a wall will accommodate the additional load that will be placed on the wall.

There are also specific fixing requirements for the different sheeting products that may be used as a substrate. These requirements are outlined in AS 3958.1-2007 Ceramic tiles – Guide to the installation of ceramic tiles. The manufacturer's specifications also can be used as a guide for the fixing requirements (see Figure 12.34). Such information can include the thickness of the tiles that can be used depending on the thickness of the sheeting and spacing of the wall studs (see Table 12.2). This type of information assists the manufacturer to ensure the successful use of their product.

TABLE 12.2 Manufacturer's specifications

Villaboard thickness (mm)	Maximum tile thickness (mm)	
	600 mm stud centres	450 mm stud centres
6	9	13
9	13	18
12	18	>25

Source: James Hardie Australia

It is important that the sheets are fixed as specified, as any faults that occur after the installation of the tiles may become the responsibility of the tiler rather than the installer of the sheeting. Before you install any tiles, you must confirm that the substrate has been installed as per manufacturer's specifications.

The other type of wall construction is concrete or masonry. If it is an existing wall, consideration will need to be given to any existing coatings or tiles. AS 3958.1-2007 Ceramic tiles – Guide to the installation of ceramic tiles identifies the preparation required for different substrates as well as the maximum variation on the plane of the substrate.

Assessing tile quality

Before any tiles are installed it is important that the quality of the tile is checked. There are several aspects of tiles that should be assessed by the tiler for either structural or aesthetic faults.

Structural faults include the following:

- *Incorrect size.* The size of the tiles should remain consistent.
- *Edge imperfections.* The edges of the tiles should be straight and the finishing of the glaze also should be consistent along the edges.
- *Thickness.* The tiles should maintain a consistent depth.
- *Curvature.* The tiles should be consistently flat, and not cupped or bowed.

Aesthetic faults include the following:

- *Inconsistent colour.* It is a good idea to review several tiles from different boxes to ensure that the colour is consistent and can be maintained across the whole job.
- *Textured finish changes.* Again, a review of several tiles from different boxes will ensure finishes are consistent.
- *Surface spots and blemishes.* These can include undesired bubbles and ripples in the glaze that coats the face of the tile. Again, review several tiles from different boxes.

Once a basic quality control review has been completed, the placement of the tiles in preparation for

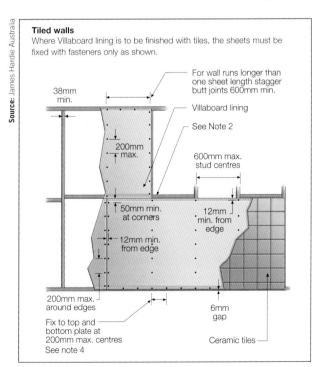

FIGURE 12.34 Example of a manufacturer's fixing requirements

laying has to be considered. The tiles should be placed close to where they will be installed. Boxes of tiles can be heavy, and handling them increases the likelihood for them to be damaged, so try to avoid double handling.

While tiles need to be easily accessible and stacked so that they are safe from damage, the tiler also needs sufficient access to the wall to apply the adhesive without the tiles presenting a trip hazard. This also means keeping the work area clean and clear of unnecessary equipment and materials (see Figure 12.35).

FIGURE 12.35 Tiles and equipment cluttering work area, leading to a poor-quality outcome

FIGURE 12.36 Finding the high point to start levelling from

It also may be necessary to prepare the tiles for installation; for example, some tiles need to be wetted down before they are laid. In these situations, additional allowance needs to be made for the time allocated to complete the work.

Install tiles

While different jobs will have different installation requirements, depending on the materials used and environment of the work site, there is a basic process that can be followed.

1. Find the lowest point on the floor line and make a mark one tile height up from the floor. This will assist with minimising the cuts required between the wall and floor junction (see Figure 12.36).

FIGURE 12.37 Setting a straight edge into position to support tiles

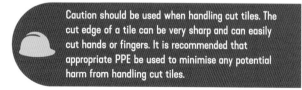

Caution should be used when handling cut tiles. The cut edge of a tile can be very sharp and can easily cut hands or fingers. It is recommended that appropriate PPE be used to minimise any potential harm from handling cut tiles.

2. Set up a temporary straight edge, making sure it is level. When you have finished tiling above the batten, remove the temporary batten. This will allow for the bottom row of tiles to be installed (see Figure 12.37 and Figure 12.38).

FIGURE 12.38 Straight edge in position ready to support tile installation

3. Thoroughly mix the appropriate adhesive for the project to a toothpaste consistency. (If the adhesive you are using is pre-mixed this step will not be necessary.)

4   Apply any primer coats to the substrate to assist in gaining the full bond of the adhesive (see Figure 12.39). Ensure that you do not apply the primer outside of the area to be tiled.

FIGURE 12.39 Applying primer to the substrate

5   Apply the adhesive using the flat side of the trowel to promote substrate contact. Then, using the recommended notch trowel at a 45-degree angle, spread the adhesive uniformly in a ridged pattern (see Figure 12.40). Continue applying in a straight pattern in a horizontal direction.

FIGURE 12.40 Applying adhesive to the substrate

6   Do not apply more than one square metre of adhesive at a time (see Figure 12.41). This will avoid the adhesive forming a skin that will reduce adhesion of the tiles.
7   Apply the tile in the predetermined position, work along the set stick and maintaining appropriate spacing, as established in the planning stage with the gauge rod (see Figure 12.42). Consistency of spacing between tiles can be maintained by using the same spacers for the whole project.
8   Check the tiling regularly with a spirit level to ensure the work is remaining straight and level

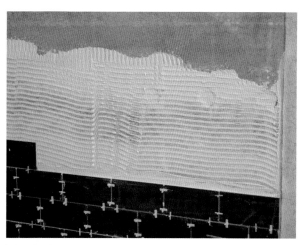

FIGURE 12.41 Limiting the amount of adhesive to prevent from drying before tiles are fixed

FIGURE 12.42 Placing tiles in the correct position as planned

(see Figure 12.43). Ensure that good contact with the adhesive is made.

Grout tiles

The grouting of tiles is a crucial step in the overall appearance and functionality of the tiled area. The adhesive bed should be thoroughly dry and fully cured, and the grout joints clean, dry and free from adhesive residues, dirt, dust and other loose debris. Cleaning the joints also includes the removal of spacers used in the installation process.

Grouts can contain strong colour pigments and this should be taken into consideration when selecting the colour. The selection of the colour will depend on the overall desired look of the tiles and the project in general.

Classification and designation of grouts is covered by AS 13007.3 – 2013 Ceramic tiles – Grouts and adhesives, Part 3: Terms, definitions and specifications for grouts

FIGURE 12.43 Checking alignment of tiles as they are progressively installed

provides a structure of the classification and designation of grout.

The Standard classifies ceramic tile grouts as one of two types: cementitious grout (CG) and reaction resin grout (RG).

Both of these types can be further subdivided into normal grout (1) and improved grout (2) classes. Then there is a range of different characteristics that each of these types can exhibit. These are abbreviated as single letters; for example, F denotes a fast-setting grout. The overall designation of the grout is generated with the symbol of the type (CG or RG), followed by the class number (1 or 2) and then abbreviation of the class or classes to which it belongs (F, W or A). For example, the designation CG1F indicates a normal fast-setting cementitious grout. The Standard lists all the combinations in table form (Based on Australian Standards).

The application of grout follows the same process, no matter the type of grout being used.

1. Grouting should only take place after the adhesive has had sufficient time to set, so the tiles will not move while grouting takes place.
2. All joints need to be clear and free from dust and debris. Any excess adhesive also should be removed, along with the spacers.
3. After mixing the grout, apply it to the wall surface with a squeegee. It is usually best to work in a diagonal motion rather than straight up and down or across (see Figure 12.44).

FIGURE 12.44 Applying grout to the wall tiles

4. Using the rubber squeegee, work the grout paste deep into the joints, again using a diagonal motion. Make sure the grout is not just sitting on top (bridging the joint) (see Figure 12.45).

FIGURE 12.45 Spreading and working the grout in between the tiles

5. Remove most of the grout with the squeegee before washing off (see Figure 12.46). It is preferable not to grout too large an area before washing it off.
6. Still working diagonally, use a sponge that has been worn in to wash off the grout. Wipe over the surface once to smooth out the joints and then do a final wipe-over to achieve a finer clean finish (see Figure 12.47).
7. The final wipe-over should leave the surface clean enough so that when the surface is dry all that is needed is a wipe-over with a clean, dry, soft rag (see Figure 12.48).

Clean up

Clean-up should occur several times during the process of tiling. The preparation of the substrate surface is the first opportunity to clean the area to be tiled. It is

FIGURE 12.46 Removing excess grout from tiles once joints have been filled with grout

FIGURE 12.47 Sponging off excess grout left on tiles

important that any loose or free material is removed. At this time it is also necessary to identify any other material that may affect the adhesion of the tiles and any staining that may affect the tile once in position.

FIGURE 12.48 Final wipedown leaves tiles clear of any grout film

The installation phase produces a higher level of debris to manage, mainly tile offcuts and excess adhesive. Care needs to be taken with the sharp edges of offcuts, both for personal protection and to maintain the integrity of waterproofing materials in wet areas. The latter can be achieved by laying down drop sheets.

The chemical components of adhesives also need to be managed – check the manufacturer's safety data sheet (see Figure 12.49). Finally, all tools used during clean-up need to be appropriate to the materials being managed, including the selection of PPE.

Methods and material for containment and cleaning up	
Minor spills	Cleaning up all spills immediately Avoid contact with skin and eyes Wear impervious gloves and safety goggles Trowel up/scrape up Place spilled material in clean, sealed container Flush spill area with water.
Major spills	Minor hazard: Clear area of personnel Alert Fire Brigade and tell them location and nature of hazard Control personal contact with the substance by using protective equipment as required Prevent spillage from entering drains or water ways Contain spill with sand, earth or vermiculite.

Source: Adapted from Parex Group (ParexGroup) Chemwatch Hazard Alert Code: 3 Davco Grey One Pot Safety Data Sheet according to WHS and ADG requirements

FIGURE 12.49 Extract from a manufacturer's SDS for an adhesive product

LEARNING TASK 12.2

INSTALL WALL TILES AND GROUT

Undertake a practical activity to demonstrate the installation of wall tiles and grout. Each student is to:
- prepare all material and substrate to install wall tiles
- set out a gauge rod as specified by the instructor
- install wall tiles in a stretcher bond pattern
- prepare the wall tiles for grouting
- apply grout to the wall area
- evaluate any aspects of the finished product that you consider not to be acceptable.

COMPLETE WORKSHEET 5

SUMMARY

In Chapter 12 you have learnt how to install wall tiles. In addition, the chapter has:
- identified the different aspects involved in preparing to install wall tiles, including selecting and quantifying tiles and adhesive for a project
- discussed the basic tools required to perform the tasks involved in installing wall tiles
- considered different types of tiles and commonly used patterns used for wall tiles
- identified the requirements for substrate elements that are to be tiled
- considered the process of installing wall tiles and applying grout to a tiled surface
- identified clean-up requirements during the tiling process.

REFERENCES AND FURTHER READING

Bunnings, *10 Tile Patterns You Need to Know*, **https://www.bunnings.com.au/diy-advice/home-improvement/tiles/10-tile-patterns-you-need-to-know**

Davco, *Data Sheets*, **https://www.davcoaustralia.com.au/resources/data-sheets**.

Davco, *SDS*, **https://www.davcoaustralia.com.au/resources/sds**.

Good Environmental Choice Australia, *Panel Boards*, **https://www.geca.eco/wp-content/uploads/2017/08/Panel-Boards-GECA-04-2011-v2i.pdf**.

James Hardie, *Internal Lining, Installation guide – Villaboard lining*, JHA-2014

PlaceMakers, *How To 20: Tiling interior walls and floors*, **www.placemakers.co.nz**.

TAFE NSW South Western Sydney Institute, *CPCCWF3003A Fix Wall Tiles*.

Relevant Australian Standards
AS 3958.1-2007 Ceramic tiles – Guide to the installation of ceramic tiles
AS ISO 13007.1:2020 Ceramic tiles – Grouts and adhesives Part 1: Terms, definitions and specifications for adhesives
AS ISO 13007.2 – 2013 Ceramic tiles – Grouts and adhesives Part 2: Test methods for adhesives
AS ISO 13007.3 – 2013 Ceramic tiles – Grouts and adhesives Part 3: Terms, definitions and specifications for grouts

GET IT RIGHT

The photo below shows a completed tiling job. Identify which work practices have not been followed to allow this to happen.

WORKSHEET 1

Student name: _____

Enrolment year: _____

Class code: _____

Competency name/Number: _____

To be completed by teachers

Student competent ☐

Student not yet competent ☐

Task

Read through the sections from the start of the chapter to 'Adhesives', then complete the following questions.

1 Why is it important that materials are ready for use when installing wall tiles?

   _____

   _____

2 Under what circumstances are tiles ordered in lineal metres?

   _____

   _____

3 Identify two reasons why tiles are used in the construction of buildings.

   _____

   _____

4 Based on the diagrams below, calculate the amount of wall tiles required in square metres. Set out the calculations in a logical sequence so that each step of the calculation process can be identified.

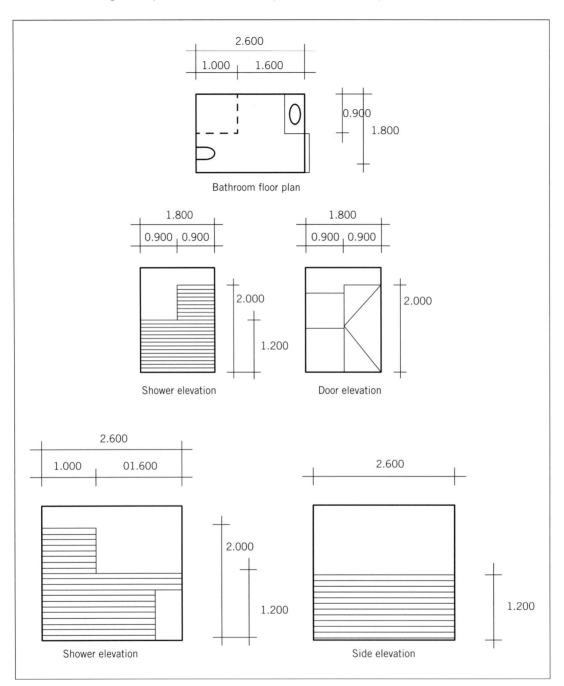

WORKSHEET 2

To be completed by teachers
Student competent ☐
Student not yet competent ☐

Student name: _____

Enrolment year: _____

Class code: _____

Competency name/Number: _____

Task

Read through the sections from 'Adhesives' up to 'Tools and equipment', then complete the following questions.

1 How do you identify the coverage of adhesive product to be used?

   _____

   _____

2 Identify two factors that will have an effect on the coverage of an adhesive product.

   _____

   _____

3 Indicate two ways in which floor tiles differ from wall tiles.

   _____

   _____

4 List three factors that affect the selection of the adhesive to be used.

   _____

   _____

   _____

5 Identify two health effects that may occur when using adhesives.

   _____

   _____

UNDERTAKE BASIC INSTALLATION OF WALL TILES

WORKSHEET 3

To be completed by teachers
Student competent ☐
Student not yet competent ☐

Student name: _____

Enrolment year: _____

Class code: _____

Competency name/Number: _____

Task

Read through the sections from 'Tools and equipment' up to 'Preparation and installation', then complete the following questions.

1 Identify two tools used in each of the three categories of tools that are used in the process of installing wall tiles. Give a brief explanation of what the tool would be used for.

Category 1:
_____
_____

Category 2:
_____
_____

Category 3:
_____
_____

2 Name two types of special fitting that may be included in tiling a wall for a bathroom.
_____
_____

3 Why is it important to establish a plumb line when planning and setting out a wall to be tiled?
_____
_____

4 Nominate two reasons why it is important to centre tiles when setting out tiles for a wall.
_____
_____

5 What are the possible outcomes if waterproofing is damaged during tiling and not discovered until the area is in full use?

_____

_____

WORKSHEET 4

To be completed by teachers
Student competent ☐
Student not yet competent ☐

Student name: _____

Enrolment year: _____

Class code: _____

Competency name/Number: _____

Task
Read through the sections from 'Preparation and installation' up to 'Grout tiles', then complete the following questions.

1 What is the meaning of the term 'substrate'?

　

2 Identify three different types of sheet substrate material.

　

3 Specify two reasons why the fixing requirements of sheet substrates are important.

　

4 What is the maximum thickness of tile that can be used on 6 mm Villaboard with studs at 600 mm centres?

　

5 Identify the two types of faults that can exist in tiles.

　

6 Identify the two faults for each of the two different types of faults.

　

UNDERTAKE BASIC INSTALLATION OF WALL TILES 393

7 Why is it recommended not to apply more than 1 m² of adhesive at a time?

_____

_____

8 When installing wall tiles why is it recommended to check for level and straightness of the tiles?

_____

_____

WORKSHEET 5

To be completed by teachers
Student competent ☐
Student not yet competent ☐

Student name: _____

Enrolment year: _____

Class code: _____

Competency name/Number: _____

Task

Read through the sections from 'Grout tiles' up to the end of the chapter, then complete the following questions.

1. Identify the two different types of grout.

   _____

   _____

2. What is the main difference between the two types of grout?

   _____

   _____

3. Should the spacers between the tiles be removed before or after applying grout to the tiled walls? Why is this so?

   _____

   _____

4. When applying and cleaning up grout, in what direction should the squeegee and sponge be moved?

   _____

5. When tiling over waterproofing, what precautions should be taken?

   _____

   _____

6. What document should be referenced when cleaning up any material that may have a chemical hazard?

   _____

   _____

7. Other than the tools directly used for clean-up, what other type of equipment is required when cleaning up?

   _____

   _____

MY SKILLS

EMPLOYABILITY SKILLS

Using the following table, describe the activities you have undertaken that demonstrate how you developed employability skills as you worked through this unit. Keep copies of material you have prepared as further evidence of your skills.

Employability skills	The activities undertaken to develop the employability skill
Communication	
Teamwork	
Planning and organising	
Initiative and enterprise	
Problem-solving	
Self-management	
Technology	
Learning	

GLOSSARY

A

alignment In correct position, usually in a straight line.

C

centre line length A reference line that is the centre or axis of an item.

charge-out fee A set amount of money charged by a tradesperson or other provider of services to attend either a private residence or a construction site at short notice.

concrete An artificial stone made up of cement, large aggregate, fine aggregate and water. Additives can be added for different purposes, such as a longer curing time or anti-freezing in cold temperatures.

controlled measures Actions taken to eliminate or minimise health and safety risks so far as is reasonably practicable.

cost analysis Breaking down costs into individual components and comparing projected amounts at the start of the project to actual amounts spent.

crane A machine that will lift and place loads of material in a horizontal and vertical direction.

D

datum Any known point, line or level from which a level line may be transferred to another position. Its elevation, or height, may be recorded and used as a permanent or temporary reference while carrying out a job.

dead load The load imposed on a building that is static, such as walls, roof and all permanently fixed elements.

decorative Serving to make something more attractive or ornamental.

durability The ability to maintain a structure or intended purpose over a given period of time under expected actions and environmental influences.

duty of care A legal and moral obligation to ensure the safety and wellbeing of others.

E

earth leakage box A device used to shut off current to prevent electric shock where there's a fault in an electrical system.

efflorescence Crystalline deposits of soluble salts on a surface from the evaporation of water.

electronic format Text or image-based content that is produced on and readable on computers or digital devices.

F

fibrous plaster Plaster reinforced with fibres such as sisal or hemp; an historical plastering system to line walls and ceilings.

financial viability Having sufficient income to meet or exceed outgoings or debts.

fire rating The duration for which a passive fire protection system can withstand a standard fire resistance test.

footing The base of a structure in contact with the ground and designed to spread and transfer loads to the supporting foundation material.

formwork An assembly or temporary construction to support and shape freshly mixed concrete until it sets and hardens.

furrow The result of dragging a trowel across the centre of a bed joint to distribute mortar across the brick.

G

gauge A consistent setting or spacing used as a guide.

greenhouse gas Any gases in the Earth's atmosphere that trap heat.

Gyprock™ Plaster sandwiched between special paper on either side. It is used to line the interior of buildings and is manufactured for different areas in a building, such as wet areas.

H

hiab A set of hydraulic arms on the back of a truck for lifting loads.

high-risk construction work Work that has a high likelihood of something going wrong, or where if something goes wrong it will be catastrophic.

I

infrastructure construction The building of assets that provide a framework for social and economic development, such as roads, airports and tunnels.

innovation The creation of something that is new.

J

job safety analysis (JSA) A written procedure developed to review work steps and their associated hazards so the correct measures can be put in place to minimise any potential harm.

K

kerf The cut made by a saw or an axe into a piece of timber.

L

labour constant The amount of time that it would take to perform a specific task; used to quantify the time required when estimating a construction project.

level line Any horizontal line that is parallel to the surface of still water.

levelling The determination and representation of the elevation of points on the surface of the Earth from a known datum, using a surveyor's level of any kind to measure the differences in elevation by direct or trigonometric methods.

live load The load imposed on a building from moving items such as people, cars and wind.

logical sequence A series of steps or a pattern that has a set order and structure.

M

masonry The assembly of units, such as bricks and blocks, usually bonded by mortar.

mortar A mixture of cement, sand (fine aggregate) and water, with or without other additives.

N

non-residential construction The building of hotels, motels, shops, office areas, factories and so on.

O

oncosts Costs associated directly with the employment of workers, such as allowing for sick days, holiday pay, taxes and superannuation.

overhead An amount that is usually a percentage of costs on each job to allow for administration costs such as a phone, vehicle and insurances.

P

parallax error Where an object or point of measurement appears to shift or change position as a result of a change in position of the observer.

plumb Any vertical line or surface that if extended would line up with the centre of the Earth.

pneumatic tool A tool that is powered by means of compressed air via air hoses connected to an air compressor.

power tool A tool or machine that uses a power source other than human force to power or drive the device; usual power sources include 240-volt electricity, electric batteries, compressed air or gas, or explosive gas or gunpowder cartridges.

pre-cast concrete tilt up panels Concrete items that are manufactured in a cast or mould and can be constructed on site or off site. These are typically erected on site using a crane to form the walls of a structure.

prefabricated house frames Usually domestic house frames that are made in a factory away from the building site and transported to site when required.

profit margin The amount of money left over from a project once all the costs have been paid for the project, including oncosts, overheads, material costs and wages.

R

rafters An inclined roof framing, extending from the eave to the ridge, and carrying the outer roof coverings.

residential construction The building of separate private residences for people to live in; does not include common circulation areas such as blocks.

resin A natural or synthetic compound or a mixture of compounds.

S

safe work method statement (SWMS) A document that describes how work is to be carried out, identifies the work activities assessed as having risks, and describes the control measures that will be applied to those work activities. It also includes a description of the equipment used in the work, the standards or codes to be complied with, and the qualifications of the personnel doing the work.

safe working load (SWL) The maximum load that can be safely handled by a lift, hoist, crane or other lifting device.

skin of brickwork A wall of brickwork that is a layer of bricks; two brick walls adjacent to each other is considered to be two skins of brickwork.

solvent Water or organic liquid that is usually volatile and is used to dissolve or disperse other material.

spalding A condition in which concrete becomes pitted, flaked or broken up caused by rusted steel reinforcement or repeated freezing and thawing.

specifications The precise construction requirements for a proposed structure.

spoil Excavated material that is unsuitable for, or surplus to, the requirements of the works.

stormwater Water that is collected or gathers from rainfall.

structural integrity The ability of a material to maintain its original structural qualities.

structurally sound The ability of a member to maintain its structural function.

stud A vertical member in a frame or partition of a building to which internal lining and/or external cladding material is affixed.

substrate Any material used as a base to which other components can be adhered.

T

traverse A survey consisting of a continuous series of connected straight survey lines, whose lengths and bearings are measured at each survey station or nominated point. A 'closed traverse' is when the lines form a complete circuit between two known points.

W

waterproof membrane A layer, usually of seamless polymeric material, placed beneath the tiling to prevent penetration of liquid water into the background.

wellbeing A state of good health and comfort.

work health and safety All of the factors and conditions that affect health and safety in the workplace, or could affect health and safety in the workplace. These affect employees (permanent and temporary), contractors, visitors and anyone else who is in the workplace.

workability The ease with which freshly mixed mortar can be handled, placed and finished.

INDEX

240-volt power supply – mains and generators 340–1
415-volt (three-phase) power supply 340

A

abrasive papers 329
ACC blocks 209
access and equity principles 6
accident reporting 97–8
accident report form 97, 98
   company form 99–100
accidents 88, 99, 134
   reportable 98–9
acid rain 211
acrylic paints 272
acute hazard 88
addition 60
adhesives 285
   cutting 286
   durability 286
   fire rating 286
   handling 286
   safety issues 381
   stacking 286
   storage 286
   uses 285–6
   wall tiles 368
adjacent side 63
adjustable shifting spanner 324, 325
administrative controls 87
advice about learning needs 12
aggregate
   coarse 269
   fine 269
   storage 270, 271
air breakers 342, 343
alignment bricks or blocks 213
all-terrain forklifts 286–7
alloy steel 275
aluminium planks 323
angle-head drill 335
annealed glass 282
anti-discrimination 6
apprenticeships 6
architects 7
area 58–9, 62
   formulas 60
artificial stones (oil stones) 321
AS 1216 Class labels for dangerous goods 93
AS 1288–2006 Glass in buildings – selection and installation 290
AS 1318 Use of colour for the marking of physical hazards 93
AS 1319 Safety signs for the occupational environment 93, 96
AS 1418 Cranes including hoists and winches 141

AS 1478.1 Chemical admixtures for concrete, mortar and grout – admixtures for concrete 290
AS 1684.2 Residential timber-framed construction – non-cyclonic areas 290
AS 1885.1 Measurement of occupational health and safety performance 98
AS 2294.1 Earth-moving machinery – protective structures – general 141
AS 2550 Cranes – safe use 141
AS 2626 Industrial safety belts and harnesses – Selection, use and maintenance 92
AS 2958.1 Earth-moving machinery – safety, Part 1: wheeled machines – brakes 141
AS 3700–2011 Masonry structures 207
AS 3958.1-2007 Ceramic tiles – guide to the installation of ceramic tiles 371, 377
AS 3972 General purpose and blended cements 216
AS 4687 Temporary fencing and hoardings 138
AS 4991–2004 Lifting devices 141
AS ISO 13007.1:2020 Ceramic tiles – grouts and adhesives Part 1: terms, definitions and specifications for adhesives 369–70
AS ISO 13007.2–2013 Ceramic tiles – grouts and adhesives Part 2: test methods for adhesives 382
AS ISO 13007.3–2013 Ceramic tiles – grouts and adhesives Part 3: terms, definitions and specifications for grouts 379
AS/NZS 1270 Acoustic–Hearing protectors 90
AS/NZS 1337.1 Eye protectors for industrial applications 89–90
AS/NZS 1715 Selection, use and maintenance of respiratory protective equipment 91
AS/NZS 1716 Respiratory protective devices 91
AS/NZS 1800 Occupational protective helmets – selection, care and use 89
AS/NZS 1801 Occupational protective helmets 89
AS/NZS 1873 Power-actuated (PA) hand-held fastening tools 141
AS/NZS 1891.1 Industrial fall-arrest systems and devices – harnesses and ancillary equipment 142
AS/NZS 1891.4 Industrial fall-arrest systems and devices – selection, use and maintenance 142
AS/NZS 2161.1 Occupational protective gloves – Selection, use and maintenance 91

AS/NZS 2210.1 Safety, protective and occupational footwear guide to selection, care and use 92
AS/NZS 4501.2 Occupational protective clothing – general requirements 92
AS/NZS 4667–2000 Quality requirements for cut-to-size and processed glass 290
AS/NZS 4668–2000 Glossary of terms used in the glass and glazing industry 290
AS/NZS 60745 Hand-held motor operated electric tools – safety – general requirements 141
asbestos 276, 279–82
   cutting 281
   durability 280
   fire rating 281
   state guidelines on removal and disposal 280
   uses 280
asbestos-containing material 135–6, 142, 280, 281
   handling 281
   stacking 281–2
   storage 282
asbestos-dust removal 346
asbestos waste 282
asking questions 31
assemble project components (basic construction project) 189–91
   assemble components to specification and quality, and check conformity to plans and specifications 191
   select assembly process 189–90
   set out, level and erect/install project in line, level and plumb 190–1
auger post-hole digger 330
Australian Building and Construction Commission (ABCC) 6
autoclaved aerated concrete (ACC) masonry 209
aviation snips 325, 326
awards 6

B

ball barrow 345
barricades and signage at appropriate locations 136–8
barrier creams 92
bars (reinforcement) 270
basic construction project
   assemble project components 189–91
   clean up 191–2
   get it right 195
   manufacture components 186–9
   pergola project (example) 188–9, 190–1
   review and prepare to undertake 182–6
battery gun 338
batts (insulation) 277

belt sanders 337
bench grinders 320, 322
best-practice principles 6
blankets (insulation) 277
blended cement (type GB) 216
block planes 320, 321
block sets 235
blocks 207, 208
    handling, stacking and storage 212–13
    see also concrete masonry/blocks
blown or poured-in insulation 277
blue metal 269
bobcat 212
bolsters 211, 212, 233–4
bolt cutters 274, 276, 326
breeze blocks 208–9
brick and block lifting tongs 241
brick barrows 213, 241, 345–6
brick bolsters 211, 212, 233–4
brick carriers 213
brick elevator 246
    routine maintenance 246
    safety 246
brick jointer trowel 237
brick jointers 238–9
brick saws 211, 212, 245–6
    blades 245
    foot pedal 245
    routine maintenance 246
    safety 246
    setting up 245
    tray 245
    water connection 245
brick tape measures 236
brickies trowel 237
bricklayers' (brickies) hammer 211, 212, 233
bricklaying and blocklaying materials 206–18
    clean up 218
    handle, store and stack materials 212–13
    jointing 217–18
    mortars 213–17
    preparing for work 206–12
bricklaying and blocklaying tools and equipment 231–47
    clean up 247
    hand tools 211–12, 232–9
    levelling tools 242–4
    planning and preparing for work 232
    power tools 244–6
    supplementary tools 239–42
bricklaying tools 237–9
bricks 207
    calculating number of 65, 66, 210–11
    classification 207, 208
    colours 207, 208
    common types determined by production process 207
    cutting 211–12
    fire rating 211
    handling, stacking and storing 212–13
    textures 207, 208
    thermal ratings 211
    types of 207–9
    see also concrete masonry/blocks; stone masonry

brickwork 217
    calculations 64, 65, 66, 211
    efflorescence 211
broad knife 279
brooms 351, 373
brushes see paint brushes
BS EN 1263-12002 Safety ncts safety requirements, test methods 142
bucket trowel 237, 238, 372
buckets 239, 371, 372
builders 7
Building and Construction General On-site Award 2010 6
building inspectors 7
building methods, improvements 5
bull floats 327
burning off rubbish 103
Bycol 214, 217

C

calcium silicate 207, 208
calculating costs of a project 165–70
calculations 58
    construction examples 63–6
    conversions 61–2
    percentages 61
    Pythagoras' theorem 62
    ratios 62
    trigonometry 62–3
    types of calculation 59–61
    types of measurement 58–9
callows 207
capability areas 11
carbon steel 275
carpentry and construction materials 257–89
    materials and components 259–86
    planning and preparation 258–9
    selecting appropriate materials 259
    transporting materials 286–8
carpentry tools and equipment 317–39
    cleaning up after use 351–2
    hand tools 319–31
    planning and preparation 318
    power and pneumatic tools 331–9
carrying (manual handling) 263, 264
casualty control 101
CCA (copper chromium arsenate) 135, 262
cement 215–16 269
    handling 271
    health implications of handling 216
    storage 269
    types of 216
cement mortar 214
cementitious grout (CG) 380
centre line length of trench 63, 64
centring 374–5
ceramic tiles 369
chalk line 235–6, 370
charge-out fee 166
chemical hazards 89
cherry pickers 346, 347
chisels 233, 320, 321
chronic hazard 88
circlip pliers 325
circular saws 332–3
    routine maintenance 332–3
    safety 333

circumference 59
civil sector 3
clamps 323
Class A fires and extinguishers 104
Class B fires and extinguishers 104
Class C fires and extinguishers 104
Class F fires and extinguishers 104, 105
Class F fires and extinguishers 104, 105
claw hammers 319, 320
clay masonry bricks 207
    colours 207, 208
    textured finishes 207, 208
clean up
    after installing wall tiles 380–1
    after laying bricks 218
    after using tools and equipment 351–2
    basic construction project 191–2
    bricklaying and blocklaying tools and equipment 247
    construction materials 288–9
    worksite 192, 351
clear-framed spectacles 90
clear wide-vision goggles 90
clients 7
clinkers 207
clothing, protective 92, 136, 185
club hammer 233
coarse aggregate (blue metal) 269, 271
codes of practice and guidelines (WHS) 82
cold chisels 234
coloured oxides 217
combination square 319
combined picture and word signs 93
commercial sector 3
commons (bricks) 207
communication 30, 31–4, 35, 37, 38
company accident report forms 99–100
company goals and objectives 6
compo mortar 214
compound mitre saws 333
compressed air nailers 338
compressed fibre-cement sheeting 376
compressors 341–2
    routine maintenance 342
concrete 268–9
    environmental issues 272
    fire rating 271
    reinforced 270
    reinforcement 270, 271–2
concrete components 268–72
    cement 269, 271
    coarse aggregate (blue metal) 269, 271
    fine aggregate (sand) 269, 271
    handling 271
    water 269
concrete edging tools 327
concrete masonry/blocks 208
    breeze blocks 208–9
    cutting 211–12
    handling, storing and stacking 212–13
    for retaining walls 208, 209, 221
concrete mixers 342–4
    routine maintenance 344
    safety 344
concrete quantities, calculating 64, 65
concrete screeds 37
concrete vibrators 344–5
    routine maintenance 345

INDEX **401**

concreting tools 326–7
construction/erection stage (construction process) 191
construction hazards 84, 88
construction induction cards 82, 83
construction industry, scope and nature 2–4
construction job roles, occupations and trade callings 4
construction managers 7
construction materials and components 259–86
construction phase, measuring instruments for 53–6
construction process 190
    pergola project (example) 190–1
    step 1 – set-out and levelling 190
    step 2 – excavation as required 190–1
    step 3 – construction/erection 191
construction project, basic 181–91, 195
construction specification system 160, 161
controlled measures 4
controlling risks 86–7
    developing specific control measures 87
    elimination 86
    how to develop and implement control options 87
    implementing controls 87
    substitution, isolation and engineering controls 86–7
conversions 61–2
Core Skills for Work Development Framework 11, 12
cored unit bricks 207
corner blocks 235
corner tools 279
cornices 278
cosine function (cos) 63
cost analysis 162
cost calculation for a project 165–70
coursing chart 211
cranes 87, 212, 263, 274, 284, 286
cross pein hammers 320
cross-laminated timber (CLT) 265
crosscut saws 321, 322
crowbars 330
customer variations 8
cutting
    adhesives 286
    asbestos 281
    engineered timber products 266–7
    glass 284
    mortar 215
    plasterboard sheeting 278–9
    steel 275, 278
    timber 263
cutting tools (masonry work) 211–12, 233–4

D

danger (hazard) signs 94, 95–6
dangerous occurrences 99
datum 374
dead loads 240
decorative walls 209
development needs, identifying 11–13
diagonal pattern 373
diamond-dust blade wet-cutting brick saw 211, 212
diamond pattern 373–4
diamond stones 321, 322
diesel generators 341
disc sanders 337
disposable dust masks 90–1
division 60–1
documentation 8, 38–9
    estimating and costing process 170
    see also workplace documentation
double open-end spanner 34
double-ended scutch hammer 233
double-ring end spanner 324, 325
draftspersons 7
drill types 335–6
drop saws 333
    unsafe use (get it right) 109
drum sanders 337
drums 239
dual purpose step/extension ladder 349
durability
    adhesives 286
    asbestos 280
    engineered timber products 266
    glass 284
    of masonry 211
    paints 274
    reinforced concrete 270
    sealants 274
    timber 261–2
dust masks 90–1
dust suppression 289
dustpan 373
duty of care requirements 83
    employees/workers 83
    PCBU/principal contractors 83
    PCBUs/employers 83
    subcontractors 83

E

ear muffs 90
ear plugs 90
    fitting instructions 137
earth leakage boxes 183
economic considerations of materials 259
edging tools 327
efflorescence 211
electric demolition hammers 342
electric drills 335–7
    routine maintenance 336
electric planers 333–4
    get it right 355
    routine maintenance 334
    safety 334
electrical equipment, safety and accident prevention tags 96, 332
electrical fires 103
electronic format 5
elevated work platforms 346–7
    routine maintenance and safety 347
elevations 158, 160
elimination of risk 86
emergency contacts 144
emergency coordinator 101
emergency information signs 94, 96
emergency plan 101, 144
emergency response procedures 100–7, 144
employability skills 2, 30, 52, 80, 132, 158, 182, 206, 232, 258, 318, 366
employees/workers
    duty of care 83
    roles and responsibilities 6
employers *see* PCBUs/employers
employment conditions 6
employment impacts of technology, work processes and environmental issues 4–6
engineered timber products 285
    common trade names 266
    cutting 266–7
    durability 266
    environmental issues 268
    fire ratings 266
    handling 267
    manufacture 265–6
    stacking 267
    storing 268
    uses 266
engineering controls 87
environmental and resource efficiency requirements 13–15
environmental hazards, identifying and reporting 14
environmental issues
    concrete 272
    in the construction industry 5–6, 184, 192
    engineered timber products 268
    insulating materials 277
    steel 276
    timber 264
environmental legislation and regulations
    breaches 15
    compliance with 184
    strategies to comply with 184
equal employment opportunity 6
equipment *see* plant and equipment; tools and equipment
estimation and costing
    calculate costs 165–70
    document details and verify where necessary 170
    Garage Project example 165–6, 167–70, 171
    gather information 158–64
    get it right 173
    materials, time and labour estimation 164–6
ethical standards of the company 6
evacuation diagram 101, 102
evacuation practice 144
evacuation process 101
excavation
    calculating volume of spoil to be excavated 63–4
    as required (construction process) 190–1
excavation tools 329–30
    light digging, shovelling, cleaning out and spreading tools 329
    trimming, detailing and finishing tools 329–30

explosive-powered tools (EPTs) 338–9
   routine maintenance 339
   safety 339
extension ladders 347
extension leads 331–2
eyes/face protection 89–90, 184, 284

F

F clamps 324
face (bricks) 207
face shields 90
fall protection harness 92
faux stone wall panels 210
feedback 40
fencing requirement checklist for construction site 138, 139
fibre-cement sheeting 376
fibreboard 265, 266
fibreglass reinforcement, storage 271–2
fibrous plaster 5
fillers 273
filling knives 328
fine aggregate (sand) 269, 271
finishing uses of timber 261
fire blankets 106
fire extinguishers
   and class of fire 104–5
   guide 105
   operating and using 105–6
fire plans 103
fire rating
   adhesives 286
   bricks and masonry 211
   concrete 271
   engineered timber products 266
   glass 284
   insulating materials 276
   mortar 215
   paints 274
   plasterboard sheeting 279
   sealants 274
   steel 275
   timber 262–3
fire safety equipment 102–6
   types and purpose 102–3
firefighting equipment 106
   extinguishers 104–6
firefighting equipment signs/fire signs 94, 96
fires
   classes of 104–5
   common reasons for fires on building sites 103
   in the event of a fire 104
   preventing and fighting 104
   three elements of 103, 104
first aid kits 101–2
   kit B contents 102–3
   kit B stored contents 102, 103
first-aid officers 97, 101
flange 234
flat finish 272
flat steel trowel 279
float glass 282
floats 326, 327
floor plans 158, 159
floor punches 320, 321

floor scrapers 242
floor tiles
   calculating quantities 65, 66
   differences from wall tiles 368
foil (insulation) 277
foot protection 92
footings 64, 191
forklifts 263, 267, 286–7
forks 330
formal methods of recording suggestions for improvement 40
formulas 60
415-volt (three-phase) power supply 340
four-fold rule 56, 323
friable asbestos 135, 280
friction and disharmony in teams 10–11
front-end loaders 212
fully vitrified tiles 368, 369
furrow (in the mortar) 237

G

G clamps 324
Gantt chart 31, 33
Garage Project example 165–6, 167–70, 171
gas nailers 338
gathering information (estimation and costing) 158–64
gauge heights 236
gauge rods 236–7, 374–5
general construction induction training 82–3
   construction induction cards 82, 83
general foreperson 8
general purpose blended cement (type GB) 269
general purpose limestone cement (type GL) 216
general purpose Portland cement (type GP) 216, 269
   recommendations for handling 216
generators 340–1
   routine maintenance 341
generic skills needed in the construction industry 12
get it right 17, 43, 69, 109, 147, 173, 195, 221, 291, 355, 383
glass 282
   cutting 284
   durability 284
   fire rating 284
   handling 284–5
   stacking 285
   storing 285
   types of 282
   uses 283
glass blocks 282–3
gloss finish 272
glossary 397–9
gloves 91, 212–13, 263, 271
glues 285
Glulam (glue-laminated) beams 265
goggles 90
granite tiles 368, 369
greenhouse gas emissions 14
grouting tiles 379–80
   procedure 380, 381

grouts
   classification and designation 379
   colouring 379
grubbing mattock 330
Gyprock™ sheets 53
gypsum (from plasterboard) reuse 288

H

hacking knives 328
hacksaws 274, 276
half-moon ring spanner 325
hammers 319–20
   for masonry work 211, 212, 233–4
hand planes 320, 321
hand sanders 279
hand saws 263, 321, 322
hand signals 38
hand tools 185, 186, 232
   abrasive papers 329
   bricklaying tools 237–9
   for bricks and masonry 211–12, 232–8
   for carpentry 319–31
   concreting tools 326–7
   excavation tools 329–30
   hoses 327
   maintenance 192, 351–2
   metalworking tools 324–6
   painting tools 272–3, 327–9
   preparing for work 232
   setting-out tools 235–6
   spirit levels 243, 330–1
   woodworking tools 319–24
handling
   adhesives 286
   asbestos-containing material 281
   bricks, blocks and masonry 212–13
   engineered timber products 267
   glass 284–5
   insulating materials 277
   mortar 215
   paints 274
   plasterboard sheeting 279
   steel 274–6
   timber 263–4
hard-surface trucks 287
hardboard 265, 266
hardwood trees 260
   structure 260
   types in use 260–1
hazard identification 84, 183
   assessment and reporting in the work area to designated personnel 132–3
   contribute to WHS, hazard, accident or incident reports 134
   report safety risks in the work area based on identified hazards 133
hazard report form 133, 134
hazard warning signs 94, 95
hazardous materials 135–6, 192
   asbestos-containing material 135–6, 142, 279–82
   found in existing structures 135
   identification in line with legislative and workplace requirements 135
   use appropriate signs and symbols to secure 135

hazards 88
  reporting 132–3, 142
  types of and potential harm 133
  on worksites, control risks and hazards effectively and immediately 135
Health and safety committee members 97
health and safety communication and reporting process 93–100
health and safety personnel 97, 100–1, 144
  roles 101
health and safety representatives 97
hearing protection 90
heartwood of timbers, natural durability classification 262
Hebel blocks 209
HEPA vacuum cleaners 346
herringbone pattern 374
hiab 212, 263, 264
hierarchy of control 86, 183
high-risk construction work 83–4
high-strength, low-alloy steel 275
Hilti gun 338
historical cost analyses 162
hoardings 138
hollow unit bricks 207
hopscotch design 374
horizontal drum mixers 343
horizontally cored unit bricks 207
hose reels 106
hoses 327
hydrostatic levels 243
hypotenuse 63

I

'I' beams 265
immersion vibrators 344
impact drills/drivers 335
implementing controls 87
improvements for future projects, suggestions for 40
incident and emergency response procedures 100–7
  fire safety equipment types and purpose 102–6
  first aid procedures 101–2, 103
  roles of designated health and safety personnel 100–1
incident reporting 97–8, 142
incidents 88, 99, 134
  reportable 98–9
inclined drum mixers 343
industrial sector 3
industrial vacuum cleaners 346
  routine maintenance and safety 346
industrial work platforms 346–7
industry structure 2–4
infrastructure construction 2, 3
injury management 100
innovation 12
insulated combination pliers 325
insulated diagonal cutters 325
insulating materials 276
  environmental issues 277
  fire rating 276
  handling 277
  storing 277
  types and uses 276, 277

ISO 6165 Earth-moving machinery – basic types – identification and terms and definitions 141
ISO 6746-1 Earth-moving machinery – definitions of dimensions and codes – Part 1: base machine 141
ISO 6746-2 Earth-moving machinery – definitions of dimensions and codes – Part 2: equipment and attachments 141
ISO 7133 Earth-moving machinery – tractor scrapers – terminology and commercial specifications 141
isolating the hazard from people 87

J

jack planes 320, 321
jackhammers and breakers 342, 343
  routine maintenance 342
jemmy bars 322
jewellers' snips 325, 326
jig saws 334
  routine maintenance 334
  safety 334
job roles 4
job safety analysis (JSA) 87, 135, 138, 192, 231
joint rakers 238
jointer planes 320, 321
jointing 217–18
  finishes 217, 218, 238

K

'Kango hammers' 342
keyhole saw 279

L

L-shaped area 58, 59
labour constant 165
ladders and trestles 347–9
  safe use 349
laminated glass 282
laminated veneer lumber (LVL) 265
land surveyors 8
language, literacy and numeracy skills 11
larry 241, 242
laser distance-measuring devices 57
laser levels 243–4
latex paints 272
lead-based paint 135
learning needs, identifying 12
length 58, 62
  conversions 61
level line 370, 374
levelling tools 242–4
licensing requirements 6–7
lifting (manual handling) 263, 264, 287, 288
lifting tongs 241
lightweight mobile scaffolding 347
lime 216
lime mortar 214
linear measurements 56
  measuring equipment 56–8
  parallax error 57–8
liquid volume 62
listening carefully 31
live loads 240

logical sequence of work 31
logs, cross-section 260
long tape measures 56–7
long-handled round-mouth shovel 239, 329, 330
long-handled square-mouth shovel 330
LOSP (light organic solvent-borne preservative) 262
lump hammers 211, 212, 233

M

maintenance of physical control measures 87
man lifts 347
mandatory (must do) signs 94
manual handling 263, 264, 267, 268, 274, 287–8
manufacture components for basic construction project 186–7
  check for accuracy, quality and suitability 188, 189
  select processes to manufacture components 187
manufacturers' installation instructions 35, 36, 377
manufacturers' specifications 160–2, 163, 377
marking gauges 322
mash hammer 233
masonry products
  handling, stacking and storage 212–13
  see also bricks; concrete masonry/blocks; stone masonry
masonry structures 231
mass 59, 62
material quantity requirements 186, 210–11, 214–15
material rates 162, 164
material resources, monitoring 13
material reuse *see* recycling/reuse
materials 4
  clean up 288–9
  construction materials and components 259–86
  estimation 165–6
  selecting appropriate 259
  selection in accordance with workplace procedures 138
  see also specific types e.g. glass
mattocks 330
measurement techniques and methods 52, 53
measurements
  linear measurements 56–8
  obtaining measurements 52–5
  types of 58–9
measuring equipment
  construction phase 53–6
  linear measurements 56–8
  pre-construction phase 53
metal planks 241, 323
metal toolboxes 240
metalworking tools 324–6
mild steel 274
mineral fibre (insulation) 277
mineral turpentine (turps) 272
misinterpreting information 34

mixers
    concrete mixers 342–4
    mortar mixers 244–5, 249
monitoring resource efficiency 13
mortar 213–17
    commonly used 214
    cutting 215
    fire rating 215
    handling 215
    plasticisers 217
    quantity requirements 214–15
    workability 216
mortar boards 240
mortar components 215
    cement 215–16
    lime 216
    oxides 216–17
    sand 216
mortar joint raker 238
mortar mixers 244–5
    getting it right 249
mosaic tiles 368
multiplication 61

N
nail guns 4, 31, 338
nail punches 320, 321
naked flames 103
National Classification system 161
National Construction Code 283, 284
National Standard for Workplace Injury and Disease Recording 98
natural stone masonry 210, 211
natural stones (oil stones) 321
near misses 99
needle-nose pliers 325
nitrogen-powered guns 338
non-friable asbestos 135, 280, 281
non-residential construction 2, 3
nonverbal communication 38
notched trowels 371, 379

O
obtaining measurements 52–5
    linear measurements 56–8
    select and apply method for task requirements 52–3
    select equipment, check for serviceability and rectify or report faults 53–6
occupations 4–5
oil paints (oil-based paints) 272
oil stones 321, 322
oncosts 166
online estimating programs 54
opposite side 63
orbital sanders 337
organisation policy, guidelines and requirements 6
organisational requirements 6
overheads 166–7
oxides 216–17

P
pad saws 321, 322
paint brushes 242, 273, 328
    cleaning and storage 273, 274

painting tools 272–3, 327–9
paints 272
    durability 274
    fire rating 274
    forms of 272
    handling 274
    storage 274
    uses 272
pallet trolleys 267, 287
    safety and maintenance 287
pan mixers 343–4
panel saws 321, 322
parallax error 57–8
particleboard 265, 266
PASS (fire extinguisher operation) 105–6
patterned glass 282
PCBU/principal contractors, duty of care 83
PCBUs/employers
    accident report forms 99–100
    duty of care 83
    emergency plan 101
    first aid facilities 101–2
    responsibility to provide PPE and training 88–9
    roles and responsibilities 6
    workers compensation insurance 100
pencils 371
percentages 61
performance levels, variations to 8–9
pergola project
    construction process 190–1
    plans and specifications 188–9
perimeter 59, 64
perpend 217, 218
personal protective equipment (PPE) 7, 87
    cleaning and maintenance 92
    for construction work 185
    equipment and clothing 89–92
    for handling glass 284–5
    identifying, wearing, correctly fitting, using and storing 136, 137
    purpose and use 88–9
    selecting and using for each part of the task 184–5
physical characteristics of materials 259
physical hazards 89
picks 330
picture signs 93
pin sets 235, 236
pincers 322–3
pinch bars 322
pinwheel pattern 374
pistol-grip drill 335
pistol-grip drill with side handle 335, 336
placement of safety signs 97
planks 240, 241, 323
planning work task activities 30–7
    clarify work task, WHS and equipment requirements with supervisor 31–4
    determine work task, WHS and equipment requirements for the work 31, 33
    read and/or interpret a work order 30–1, 32
    review planning of activities to establish the effectiveness of the process 39–40

steps to complete work task requirements 34–5
    steps to ensure efficient conduct of work 35
    task steps in conjunction with team members 35–7
plans and specifications 158–60
    checking conformity of work with 191
    pergola project 188–9
    review 182–3
plant and equipment 339–49
    safety 339
plasterboard hammers 320
plasterboard sheeting 277–8
    cutting and fixing 278–9
    fire rating 279
    handling 279
    recycling 288
    securing prior to wall tiling 377
    storage 279
    tools 279
    types and uses 278
    for wet areas 376–7
plasticisers 214, 217
platform-style step ladder 347
platform trolleys 287
pliers 325
plugging chisels 234
plumb 370, 374
plumb bobs 242–3
plumb line 370, 374
plywood 265, 266, 376
pneumatic drills 335, 336
pneumatic jackhammers and breakers 342, 343
pneumatic nailers 338
pneumatic tools 331, 336, 338, 342
podgers 324
poker vibrators 344–5
polychlorinated biphenyls (PCBs) 135
polystyrene (insulation) 277
pop rivet guns 325, 326
porcelain tiles 368, 369
portable barriers 138
portable electric sanders 337–8
    routine maintenance 337–8
    safety 338
portable fire extinguishers
    and classes of fire 105–6
    guide 105
    operating and use 105–6
portable routers/trimmers 335
    routine maintenance 335
    safety 335
portable site compressors 341
portable site generators 340–1
post-hole shovel 330
power saws 263
power screwdrivers 336
power supply 340–1
power tools 186, 331–9
    for bricks and masonry 244–6
    checking 185
    preparing for work 232
    safety checklist 349–50
powered nailers 4, 31, 338
    routine maintenance 339
    safety 339

pre-construction phase, measuring instruments for 53
prefabricated house frames 87
pre-mixed mortars 214, 215
preparation (for tasks) 35
primer 272
principal certifying authority 8
procedures
    for assessing first aid 101–2
    for reporting hazards, incidents and injuries 97–100, 142
    for responding to incidents and emergencies 100–7
    safe operating procedures 183–4
production rates 162
profile clamps and braces 236
profile pole 236
profit 167
profit margin 167
progressive massive fibrosis 216
prohibition (must not do) signs 94
project managers 8
project planning cycle of review 39
protective clothing 92, 136, 185
putty knives 327, 328
Pythagoras' theorem 62, 63

Q

quality assurance officers 8
quantity requirements 186
    mortar 214–15
    stone masonry 210–11
quantity surveyors 8
quick-action clamps 324
quick/pocket square 319

R

Ramset™ gun 338
range finders 57
ratios 62
reaction resin grout (RG) 380
reading instructions, plans, job sheets, etc. 31
reconstituted stone 210
recording and calculating costs 167
rectangular shape 58, 59
recycling/reuse 192, 264, 268, 276, 288
    financial benefits 288–9
reflective film 276
regular shape 58, 59
regulations (WHS) 81, 82
Regyp 288
reinforced concrete 270
    durability 270
    strength 270
reinforcement 270, 271
    storage 271–2
reinforcement steel 275
reportable accidents/incidents 98–9
reporting hazards, incidents and injuries, procedures for 97–100, 142
reporting processes (effective reporting) 8–9
requesting additional support 9
residential construction 2–3
    trades 4, 5

resource efficiency requirements 13–15
    identify and report issues 14
    suggestions to improve workplace practices 14
respirators 91, 216
responsibilities and duties (of employees and employers) 6
responsibility for own workload 7–9
restriction signs 94
retaining walls 208, 209
    get it right 221
retractable metal tape 56
review and prepare to undertake basic construction project 182–6
    determine quantity of material 186
    prepare all work to comply with project and regulatory requirements 183–4
    review plans and specifications 182–3
    select and use PPE for each part of the task 184–5
    select tools and equipment, check for serviceability and report any faults 185–6
review of control measures 87
review planning of activities to establish the effectiveness of the process 39–40
right-angle triangle 62, 63
ring and open-end spanner 324, 325
rip saws 321, 322
risk 88
risk assessment 85–6, 132–4, 183
    how the hazard may cause harm 86
    likelihood of harm occurring 86
    potential severity of harm 86
risk control 86–7
    following safe work practices and duty of care requirements 133–4
risk identification and assessment on construction sites 132–4
risk management principles 84
    how to assess risks 85–6
    how to control risks 86–7
    how to identify hazards 84
    how to review controls 87
risk-management process 88
risk matrix diagram 85, 88
river gravel 269
roller tube 329
rollers 273, 274, 328–9
rolling tray 329
roofing square 319
rotary laser levels 243–4
round-nose shovels 239, 329, 330
rulers (rules) 56, 323, 370

S

sabre saws 334
safe operating procedures 183–4
Safe Work Australia 81
    falls from ladders 347
    identifying hazards 84
    Model Code of Practice: how to manage work health and safety risk 132–3
    Model Code of Practice: managing electrical risks in the workplace 331–2

    Model Code of Practice: managing the risk of falls at workplaces 342
    Model Code of Practice: managing the work environment and facilities 82, 144
    risk management steps 84–7
    working with silica and silica-containing products 216
safe work method statement (SWMS) 31, 83, 87, 135, 138, 231
    in high-risk construction work 83–4
safe work practices
    applying 141–3
    carry out tasks in a manner that is safe for operators, other personnel and the general community 141
    in construction 7, 83–4
    for dealing with accidents, fire and other emergencies 144
    do not use tools and equipment in areas containing identified asbestos 142
    and duty of care requirements for controlling risks 133–4
    follow procedures and report hazards, incidents and injuries 97–100, 142
    follow worksite safety signs and symbols 142
    maintain worksite area to prevent incidents and accidents, and meet environmental requirements 142
    plan and prepare for 136–40
    site housekeeping inspection checklist 142
    use plant and equipment in accordance with specifications, regulations and Australian Standards 141–2
safety 336–7
safety and accident prevention tags for electrical equipment 96
safety data sheets (SDSs) 138, 140, 160, 163, 192, 216, 381
safety fencing 138, 183, 184
safety footwear 92, 184
safety helmets 89, 184
safety plans 183–4
safety risks, in the work area based on identified hazards 133
safety signs and symbols 93–7, 339
    and barricades 136–7
    categories by colour and shape 94–6
    following worksite requirements for 142
    placement 97
    safety and accident prevention tags for electrical equipment 96, 332
    to secure hazardous materials 135
safety trimming knife 323
sand 216, 269
    for concrete work 269, 271
sandstocks 207
sapwood 262
sash clamps 324
satin finish 272
saw stools 323–4
scaffolding 240–1, 347
scale rule 56
scheduling programs 35
scissor lifts 346

scrapers 373
scutch chisels 234
scutch hammers 211, 212, 233
sealants 273
    durability 274
    fire rating 274
sealer (paint) 272
sections 158
selecting appropriate materials 259
    economic considerations 259
    physical characteristics 259
sequencing of work tasks 31, 35
    and allocate time and resources to each step 38
set-out and levelling (construction process) 190
setting out wall tiles 373–5
    centring 374–5
    patterns 373–4
    synchronisation at wall junctions 375
    wet areas 375
setting-out tools 235–6
shave hooks 328
sheet reinforcement 270, 271
short-handled round-mouth shovel 330
short-handled square-mouth shovel 239–40, 330
shovels 239–40, 329–30
silica dust, health implications of 216
silicosis 216
sine function (sin) 63
single ladder 348
single-ended scutch hammer 233
single open-end spanner 34
site clean-up 192, 351
site housekeeping inspection checklist 142
site inspection 160, 258–9
site managers 8
skills development needs 11–13
    advice about learning needs 12
    core skills for work 11
    generic skills needed 12
    keeping up to date 12–13
    language, literacy and numeracy skills 11
    technical or discipline-specific skills 11–12
skin of brickwork, calculations 64, 65, 66
slate tiles 368, 369
sledgehammers 320
small tool 279
smokers 103
smoother planes 320, 321
software programs 5, 53, 54, 55
softwood trees 261
    structure 261
    types in use 261
solid unit bricks 207
solvents 52, 272
spacers 372
spades 330
spanners 324–5
special purpose cements 216
special purpose unit bricks 207
specifications 160, 161, 162
    manufacturers' specifications 160–2, 163
    see also plans and specifications
spirit levels 243, 330–1, 370, 371
spoil, calculating volume of 63–4

sponge pads 273
sponges 373, 380
spray guns 273, 274
spring clamps 324
spud bar 330
square-drive ratchet handle and socket 325
square finish (mortar) 238
square-mouth shovels 239–40, 329, 330
squares 319
squeegees 380
stacking
    adhesives 286
    asbestos-containing material 281–2
    bricks, blocks and masonry 213
    engineered timber products 267
    glass 285
    timber 25, 263–4
stainless steel 275
Stanley knife 323
static laser level 244
steel 274
    cutting 275, 278
    environmental issues 276
    fire rating 275
    handling 274–6
    storage 276
    types and uses 274, 275
steel fabric 270
steel floats 326, 327
steel products 275
steel reinforcement, storage 271
step ladders 347
stiff-bristle brooms 351
stone masonry 210
    cutting 211–12
    durability 211
    fire rating 211
    handling, storing and stacking 212–13
    size and quantity 210–11
    thermal ratings 211
stone wall configuration 210
storage
    adhesives 286
    aggregate 270
    asbestos-containing material 282
    bricks, blocks and masonry 213
    engineered timber products 268
    get it right 291
    glass 285
    insulating materials 277
    paints 274
    plasterboard sheeting 279
    reinforcement 271–2
    steel 276
    timber 264, 265
    tools and equipment 192
stormwater pollution 13
straight-blade tin snips 276
straight edge 372, 378
straw brooms 351
strength
    reinforced concrete 270
    timber 262
stretcher bond pattern 374
string lines 235
structural integrity 215
structural steel 275

structural uses of timber 261
subcontractors, duty of care 83
substituting the hazard with something safer 86
substrate (wall tiling) 375–7
subtraction 60
sun shades 89
supervision 87
supervisors 97
synthetic mineral fibres (SMFs) 135

T

tangent function (tan) 63
tape measures 56–7, 323, 370, 371
    brick tape measure 236
    get it right 69
teams 9–11
    formal methods of recording suggestions for improvement 40
    friction and disharmony in 10–11
    get it right 17
    individual roles 9
    initiate improvements 10
    plan task steps in conjunction with other members 35–7
    plan work activities and deadlines with group members and other affected workers 7–8
    providing assistance and encouragement within 9–10
    roles for a construction project 7–8
    site goals 9
technical or discipline-specific skills 11–12
technological advances in the construction industry 4–5
technology, use of 34
temporary electrical power 340
temporary fencing 138, 183, 184
tenon saws 321, 322
terracotta tiles 368, 369
thermal ratings, bricks and masonry materials 211
tile adhesive 368–70
    classification and designation 370
    quantities 368
tile cutters 372–3
tiling (floor tiles) 368
    calculating quantities 65, 66
tiling (wall tiles) 365–83
tilting drum mixers 343
timber 259–60
    cutting 263
    durability 261–2
    environmental issues 264
    fire rating 262–3
    handling 263–4
    identification 260–1
    preservative treatment 262
    stacking 263–4, 265
    storage 264, 265
    strength 262
    uses 261
timber mallets 319, 320
timber planks 323
timber toolboxes 240
time and labour, estimation 165, 166
tin snips 239, 274, 276, 325, 326
tool carriers 240

toolbox talks 9
tools and equipment 4
    abrasive papers 329
    bricklaying and blocklaying 211–12, 231–47
    carpentry 317–39
    check for serviceability and report any faults 185
    cleaning up after using 192, 351
    concreting tools 326–7
    cutting steel 276
    excavation tools 329–30
    hoses 327
    maintenance 192, 351–2
    metalworking tools 324–7
    painting tools 272–3, 327–9
    planning and preparing for work 232, 318
    plasterboard sheeting 279
    power and pneumatic tools 331–9, 349–50
    registers 192
    selection in accordance with workplace procedures 138
    setting-out tools 235–6
    spirit levels 243, 330–1
    storing, securing and reporting any faults 192
    usage 186
    using safely 349–51
    wall tiling 370–3
    woodworking tools 319–24
    working at peak efficiency 13
    see also hand tools; plant and equipment
toughened glass 282
trades 4
    residential construction 4, 5
trailer lifts 346
trailer-mounted site compressors 341–2
training, instruction and information (control measures) 87
training packages, by industry area 12
transporting materials 286–8
treated timber 135, 262
    environmental issues 264
tree age 260
trench mesh 270, 271
trenching shovel 330
trestles 349
trigonometric functions 62–3
trigonometry 62–3
    using (example) 63
trowels 237–8, 326, 368, 371, 372
trundle wheels 57
try planes 320, 321
try square 319
tuck pointers 237, 238
twin screedboard vibrators 345
240-volt power supply – mains and generators 340–1
two-wheeled barrow 345, 346

U

undercoat 272
urea-formaldehyde foam 276
utility knife 278, 279, 323

V

vacuum lifting apparatus 284
variations and difficulties affecting own performance and report these issues 8
vibrating table 345
vinyl paints 272
visual cues 31
volume 59, 62
    calculations 63–5
    formulas 60

W

wall tile installation 365–83
    clean up 380–1
    get it right 383
    grouting tiles 379–80
    installing tiles 378–9
    placement of tiles in preparation for laying 377–8
    preparation 366–73
    preparing tiles for installation 378
    setting out tiles 373–5
    substrate 375–7
wall tiles
    adhesive quantities 368
    adhesives 368–70
    aesthetic faults 377
    assessing quality 377
    calculating quantities 366–8
    differences from floor tiles 368
    selection of appropriate materials 368–70
    structural faults 377
wall tiling tools and equipment
    for clean-up 373
    for installation 371–3
    for preparation 370–1
wardens 97, 101
Warrington hammers 320
waste disposal/waste management plan 192, 289
water (for concrete mixing) 269
water-based paints 272
water level 243
waterproof membranes 375
weep holes 218
wet areas (wall tiling) 375
wet saw unsafe use (get it right) 147
wheelbarrows 241, 345–6
    routine maintenance 345–6
    safety 346
window schedule 160, 162
wood chisels 320, 321
wood fibreboards (insulation) 277
wood plastic composites 265
wood preservatives 262
wooden floats 326
woodworking tools 319–24
word-only signs 93
work conditions 6–7
work health and safety 4, 7, 31, 35
Work Health and Safety Act 2011 (WHS Act) 80, 81, 134
Work Health and Safety Acts and Regulations
    current state and territory 81
    duty of care requirements 83

work health and safety legislative requirements of construction work 80–1
    basic roles, responsibilities and rights of duty holders 83
    compliance with 183, 184
    current state and territory WHS Acts and Regulations 81
    development of state and territory WHS Acts 80–1
    general construction induction training 82–3
    reasons for development of WHS laws 80–1
    regulations, codes of practice and guidelines 82
    safe work practices in construction 7, 83–4
Work Health and Safety Regulations 2011 (Cth) 82
work orders 30–1, 32
    get it right 43
work procedures (control measures) 87
work processes, improvements in 5
work task activities, planning 30–7
    analyse task and identify work steps to ensure efficient conduct of work 35
    clarify work task, WHS and equipment requirements with supervisor 31, 34
    determine work task, WHS and equipment requirements 31
    plan steps to complete work task requirements 34–5
    plan task steps in conjunction with team members 35–7
    read and/or interpret work orders 30–1, 32
    sequencing of work 31–3, 35
work task organisation 37–40
    complete records of task planning and progress 38–9
    provide feedback and suggestions on improvements for future projects 40
    review planning of activities to establish the effectiveness of the process 39–40
    sequence work task steps and allocate time and resources to each step 38
work teams 7–8, 9–11, 35–7, 40
worker's compensation insurance 100
    eligibility for insurance 100
Worker's Compensation Insurance Accident form 99
workplace documentation 35, 39
workplace environmental practices, report breaches of 15
workplace practices, improvements to 14
wrecking bars 322
wrist guard PPE 284–5

Y

yard brooms 351